Mathematics and Democracy

Mathematics and Democracy

DESIGNING BETTER VOTING AND FAIR-DIVISION PROCEDURES

Steven J. Brams

PRINCETON UNIVERSITY PRESS · PRINCETON AND OXFORD

Copyright © 2008 by Princeton University Press
Published by Princeton University Press, 41 William Street,
Princeton, New Jersey 08540
In the United Kingdom: Princeton University Press,
3 Market Place, Woodstock, Oxfordshire OX20 1SY

ISBN-13: 978-0-691-13320-1
ISBN-13 (pbk.): 978-0-691-13321-8

Library of Congress Cataloging-in-Publication Data

Brams, Steven J.
Mathematics and democracy : designing better voting and fair-division procedures /
Steven J. Brams.
p. cm.
Includes bibliographical references and index.
ISBN 978-0-691-13320-1 (hardcover : alk. paper)—ISBN 978-0-691-13321-8 (pbk. : alk.
paper)   1. Voting—Mathematical models.   2. Elections—Mathematical models.
3. Public finance—Mathematical models.   I. Title.
JF1001.B73   2008
324.601'513—dc22   2007023584

This book has been composed in Times Roman
Printed on acid-free paper. ∞
press.princeton.edu

Printed in the United States of America

1   2   3   4   5   6   7   8   9   10

Political science is the science not only of what is,
but of what ought to be.

—Edward H. Carr, *The Twenty-Years' Crisis, 1919–1939*

Contents

Preface

In this book, I try to show how mathematics can be used to illuminate two essential features of democracy:

- how individual preferences can be aggregated to give a social choice or election outcome that reflects the interests of the electorate; and
- how public and private goods can be divided in a way that respects due process and the rule of law.

Whereas questions of aggregation are the focus of *social choice theory*, questions of division are the focus of *fair division*.

Democracy, as I use the term, will generally mean representative democracy, in which citizens vote for representatives, from a president on down. But I also analyze referendums, in which citizens vote directly on propositions, just as they did in assemblies in ancient Greece.

I analyze *procedures*, or rules of play, that produce outcomes. By making precise properties that one wishes a voting or fair-division procedure to satisfy, and by clarifying relationships among these properties, mathematical analysis can strengthen the intellectual foundations on which democratic institutions are built.

Because there may be no procedure or institution that satisfies all the properties one might desire, I examine trade-offs among the properties. In the case of some procedures, I also consider practical problems of implementation and discuss experience with those that have been tried out.

INSTITUTIONAL DESIGN AND ENGINEERING

The voting and fair-division procedures I analyze foster democratic choices by giving voters better ways of expressing themselves, by electing officials who are more likely to be responsive to the electorate, and by allocating goods to citizens that ensure their shares are equitable or preclude envy. In some cases I criticize current procedures, but most of the analysis is constructive—I suggest how these procedures may be improved.

Designing procedures that satisfy desirable properties, or showing the limits of doing so, is sometimes referred to as *institutional design* or *mechanism design.* I present empirical examples to illustrate this approach, but the bulk of the analysis is theoretical.

The product of such analysis is normative: the prescription of new procedures or institutions that are superior, in terms of specified democratic criteria, to ones that arose more haphazardly. Like engineering in the natural sciences, which translates theory (e.g., from physics) into practical design (e.g., a bridge), engineering in the social sciences translates theory into the design of political-economic-social institutions that better meet the criteria one deems important.

This book is divided into two parts.

PART 1. VOTING PROCEDURES

One cornerstone of democracy is honest and periodic elections, wherein there is meaningful competition among parties, interest groups, or individuals for political office. Several of the voting procedures that I analyze are relatively new and not well known, but they offer significant advantages over extant procedures. Common to many of them is approval balloting, whereby voters can approve of as many candidates or alternatives as they like without having to rank them.

Approval balloting may take different forms. Under approval voting, the candidate or alternative with the most votes wins. Under other methods of aggregating approval votes, different candidates or alternatives may win. These methods maximize different objective functions, or constrain outcomes in certain ways, in order to achieve certain ends, such as the proportional representation of different interests in the electorate.

Most social-choice analyses assume rational individuals, who select the most effective or efficient means to satisfy their goals, and examine the implications of these individual choices on collective choices. *Game theory* is an important tool in such analyses, especially in identifying outcomes that are stable or in equilibrium and institutions that support the equilibria one finds. These institutions do not necessarily dispel conflict but manage it so that political life continues, with politicians never being sure that they will continue in office but quite sure that the institutions will persist.

PART 2. FAIR-DIVISION PROCEDURES

As central as elections are to the performance of a democracy, a democracy would be a sham if the politicians elected were not restricted by due process and the rule of law. Ideally, democracies treat all citizens the same way—at least when governed by a constitution or other laws—particularly with respect to their civil rights and certain freedoms, such as freedom of association and freedom of religion.

The equal treatment of citizens depends in part on their receiving fair shares of things that must be divided among them, not just on the efficiency of outcomes. Accordingly, I analyze different procedures of fair division—applicable to both divisible and indivisible goods—and study their distributional consequences.

Fairness requires that one take into account the different preferences or claims of players who have a stake in an outcome. Step-by-step rules or algorithms to implement the fair division of goods, which may be homogeneous (like money) or heterogeneous (like land with different objects on it), are analyzed. Questions that relate to the fair division of people or groups include: What political parties are best suited to form a government? Which parties should get what cabinet ministries in the government?

USE OF MATHEMATICS AND SCOPE

While a good mathematical background will make reading the theoretical parts of this book easier, several chapters are accessible to those with little mathematical training. When a topic in a chapter goes beyond the level of the rest of the chapter or is a digression from its main theme, I discuss it in an appendix to the chapter.

Of course, some chapters are inherently more analytic or mathematical than others, so the reader may want to skip those that cause difficulty. In fact, I encourage selective reading of the chapters, most of which are relatively self-contained and can be read independently of others. A glossary at the back of the book provides a quick reference to the most important concepts I use.

I have certainly not covered all institutions in the public sphere. For example, there is now a large literature on redistricting, or the drawing of district boundaries after a census; on auctions, which governments employ to license or sell such things as oil leases and parts of the electronic spectrum; and on matching algorithms, which are used in the selection of schools by children and parents and hospital residencies by doctors. There is also a substantial qualitative literature on problems of implementing and evaluating democratic reforms.

THEORY INTO PRACTICE

I hope this book encourages readers to use mathematics in the service of democracy. But beyond showing that there are better procedures in principle, I recommend taking the additional step of trying to implement these procedures to determine whether their advantages in theory translate into advantages in practice. If this is not the case, then the theory will need to be revised, restarting the process of both discovery and testing until it reaches fruition in more robust democratic institutions.

ACKNOWLEDGMENTS

Most of the chapters in this book are based on previously published but widely scattered articles. I am grateful to the journals in which they appeared for granting permission to use them in somewhat modified form in this book.

The bulk of these articles are coauthored. I am very pleased to acknowledge the many contributions that my coauthors—Peter C. Fishburn, Michael A. Jones, Todd R. Kaplan, D. Marc Kilgour, Daniel L. King, Christian Klamler, Richard F. Potthoff, M. Remzi Sanver, Alan D. Taylor, Jeffrey L. Togman, and William S. Zwicker—made to them and indirectly to this book. For consistency I have used the first-person singular ("I") throughout this book, but "we" would have been appropriate for most of the chapters. Finally, I am grateful for the valuable suggestions of Javier M. Duarte, Leah Graboski, Jack H. Nagel, Mahendra Prasad, Warren D. Smith, James C. Zorich, and several anonymous referees, for the expert copy-editing of Alice Calaprice, for the fine production of the book by Natalie Baan, and for the strong support of my editor at Princeton University Press, Chuck Myers.

PART 1. *Voting Procedures*

1

Electing a Single Winner: Approval Voting in Practice

1.1. INTRODUCTION

It may come as a surprise to some that there is a science of elections, whose provenance can be traced back to the Marquis de Condorcet in eighteenth-century France, Charles Dodgson (Lewis Carroll) in nineteenth-century England, and Kenneth Arrow in twentieth-century America. Since Arrow published his seminal book, *Social Choice and Individual Values*, more than fifty years ago (Arrow, 1951, 1963)—for which in large part he received the Nobel Memorial Prize in Economics in 1972—there have been thousands of articles and hundreds of books published on everything from the mathematical properties of voting systems to empirical tests of the propensity of different systems to elect centrist candidates.[1]

The 2000 U.S. presidential election highlighted, among other things, the frailties of voting machines and the seeming arbitrariness of such venerable U.S. institutions as the Electoral College and the Supreme Court. Political commentary has focused on these aspects but given very little attention to alternative voting systems, about which the science of elections has much to say.

Several alternative systems for electing a single winner have been shown to be far superior to *plurality voting* (PV)—the most common voting system used in the United States as well as in many other places—in terms of a number of criteria. PV, which allows citizens to vote for only one candidate, suffers from a dismaying flaw. In any race with more than two candidates, PV may elect the candidate least acceptable to the majority of voters. This frequently happens in a three-way contest, when the majority splits its votes between two centrist candidates, enabling a candidate on the left or right to defeat both centrists. PV also

Note: This chapter is adapted from Brams and Fishburn (2005) with permission of Springer Science and Business Media; see also Brams (2002, 2006b) and Brams and Fishburn (1992a).

[1] For background on the formal theory underlying social choice and collective decision-making, see Austen-Smith and Banks (1999, 2005) and Arrow, Sen, and Suzumura (2002). Recent texts include Grilli di Cortono, Manzi, Pennisi, Ricca, and Simeone (1999), Hodge and Klima (2005), Gaertner (2006), and Nurmi (2006).

forces minor-party candidates into the role of spoilers, as was demonstrated in the 2000 presidential election with the candidacy of Ralph Nader. Nader received only 2.7 percent of the popular vote, but this percentage was decisive in an extremely close contest between the two major-party candidates.

Of the alternatives to PV, I recommend *approval voting* (AV), on both practical and theoretical grounds, in single-winner elections. Proposed independently by several analysts in the 1970s (Brams and Fishburn, 1983, 2007, ch. 1), AV is a voting procedure in which voters can vote for, or approve of, as many candidates as they wish in multicandidate elections—that is, elections with more than two candidates. Each approved candidate receives one vote, and the candidate with the most votes wins.

The candidate with the most votes need *not* win in an election. Merrill and Nagel (1987) make the useful distinction between a balloting method, which describes how voters can legally vote (e.g., for one candidate or for more than one), and a decision rule that determines a winner (e.g., the candidate with a plurality wins, or the candidate preferred to all others in pairwise comparisons wins). For convenience, I use the shorthand of AV to mean approval balloting with a plurality decision rule, but I consider other ways of aggregating approval votes later.

In the United States, the case for AV seems particularly strong in primary and nonpartisan elections, which often draw large fields of candidates. Here are some commonsensical arguments for AV that have been made:

1. *It gives voters more flexible options.* They can do exactly what they can under PV—vote for a single favorite—but if they have no strong preference for one candidate, they can express this fact by voting for all candidates they find acceptable. In addition, if a voter's most preferred candidate has little chance of winning, that voter can vote for both a first choice and a more viable candidate without worrying about wasting his or her vote on the less popular candidate.

2. *It helps elect the strongest candidate.* Today the candidate supported by the largest minority often wins, or at least makes the runoff if there is one. Under AV, by contrast, the candidate with the greatest overall support will generally win. In particular, *Condorcet winners*, who can defeat every other candidate in separate pairwise contests, almost always win under AV, whereas under PV they often lose because they split the vote with one or more other centrist candidates.

3. *It will reduce negative campaigning.* AV induces candidates to try to mirror the views of a majority of voters, not just cater to minorities whose voters could give them a slight edge in a crowded plurality contest. It is thus likely to cut down on negative campaigning, because candidates will have an incentive to try to broaden their appeals by reaching out for approval to voters who might have a different first choice. Lambasting such a choice would risk alienating this candidate's supporters and losing their approval.

4. *It will increase voter turnout.* By being better able to express their prefer-ences, voters are more likely to vote in the first place. Voters who think they might be wasting their votes, or who cannot decide which of several candidates best represents their views, will not have to despair about making a choice. By not being forced to make a single—perhaps arbitrary—choice, they will feel that the election system allows them to be more honest, which will make voting more meaningful and encourage greater participation in elections.

5. *It will give minority candidates their proper due.* Minority candidates will not suffer under AV: their supporters will not be torn away simply because there is another candidate who, though less appealing to them, is generally con-sidered a stronger contender. Because AV allows these supporters to vote for *both* candidates, they will not be tempted to desert the one who is weak in the polls, as under PV. Hence, minority candidates will receive their true level of support under AV, even if they cannot win. This will make election returns a better reflection of the overall acceptability of candidates, relatively undistorted by insincere or strategic voting, which is important information often denied to voters today.

6. *It is eminently practicable.* Unlike more complicated ranking systems, which suffer from a variety of theoretical as well as practical defects, AV is simple for voters to understand and use. Although more votes must be tallied under AV than under PV, AV can readily be implemented on existing voting machines. Because AV does not violate any state constitutions in the United States (or, for that matter, the constitutions of most countries in the world), it requires only an ordinary statute to enact.

Voting systems that involve ranking candidates may appear, at first blush, more appealing than AV. One, the Borda count, awards points to candidates according to their ranking. Another is the Hare system of *single transferable vote* (STV)—with variants called the "alternative vote" and "instant runoff"—in which candidates receiving the fewest first-choice votes are progressively eliminated. Their votes are transferred to second choices—and lower choices if necessary—until one candidate emerges with a majority of voters.

Compared with AV, these systems have serious drawbacks. The Borda count fosters "insincere voting" (for example, ranking a second choice at the bottom if that candidate is considered the strongest threat to one's top choice) and is also vulnerable to "irrelevant candidates" who cannot win but can affect the outcome. STV may eliminate a centrist candidate early and thereby elect one less acceptable to the majority. It also suffers from "nonmonotonicity," in which voters, by raising the ranking of a candidate, may actually cause that candidate to lose—just the opposite of what one would want to happen. I give examples of these drawbacks in the appendix to chapter 2.

As cherished a principle as "one person, one vote" is in single-winner elec-tions, democracies, I believe, can benefit more from the alternative principle of "one candidate, one vote," whereby voters make judgments about whether each candidate on the ballot is acceptable or not. The latter principle makes the tie-in of a vote not to the voter but rather to the candidates, which is arguably more egalitarian than artificially restricting voters to casting only one vote in multicandidate races. This principle also affords voters an opportunity to express their intensities of preference by approving of, for example, all candidates except one they might despise.

Although AV encourages sincere voting, it does not altogether eliminate strategic calculations. Because approval of a less-preferred candidate can hurt a more-preferred approved candidate, the voter is still faced with the decision of where to draw the line between acceptable and unacceptable candidates. A rational voter will vote for a second choice if his or her first choice appears to be a long shot—as indicated, for example, by polls—but the voter's calculus and its effects on outcomes is not yet well understood for either AV or other voting procedures.

While AV is a strikingly simple election reform for finding consensus choices in single-winner elections, in elections with more than one winner—such as for a council or a legislature—AV would not be desirable if the goal is to mirror a diversity of views, especially of minorities; for this purpose, other voting systems should be considered, as I will discuss in later chapters.

On the other hand, minorities may derive indirect benefit from AV in single-winner elections, because mainstream candidates, in order to win, will be forced to reach out to minority voters for the approval *they* (the mainstream candidates) need in order to win. Put another way, these candidates must seek the *consent* of minority voters to be the most approved, or consensus, choices. While promoting majoritarian candidates, therefore, AV induces them to be responsive to minority views.

1.2. BACKGROUND

In this chapter, I describe some uses of AV, which began in the thirteenth century. However, I concentrate on more recent adoptions of AV, beginning in 1987, by several scientific and engineering societies, including the

- Mathematical Association of America (MAA), with about 32,000 members
- American Mathematical Society (AMS), with about 30,000 members
- Institute for Operations Research and Management Sciences (INFORMS), with about 12,000 members
- American Statistical Association (ASA), with about 15,000 members
- Institute of Electrical and Electronics Engineers (IEEE), with about 377,000 members

Smaller societies that use AV include, among others, the Public Choice Society, the Society for Judgment and Decision Making, the Social Choice and Welfare Society, the International Joint Conference on Artificial Intelligence, the European Association for Logic, Language and Information, and the Game Theory Society (see chapter 5).

Additionally, the Econometric Society has used AV (with certain emendations) to elect fellows since 1980 (Gordon, 1981); likewise, since 1981 the selection of members of the National Academy of Sciences (1981) at the final stage of balloting has been based on AV. Coupled with many colleges and universities that now use AV—from the departmental level to the schoolwide level—it is no exaggeration to say that several hundred thousand individuals have had direct experience with AV.

Probably the best-known official elected by AV today is the secretary-general of the United Nations (Brams and Fishburn, 1983). AV has also been used in internal elections by the political parties in some states, such as Pennsylvania, where a presidential straw poll using AV was conducted by the Democratic State Committee in 1983 (Nagel, 1984).

Bills to implement AV have been introduced in several state legislatures (see section 1.2). In 1987, a bill to enact AV in certain statewide elections passed the Senate but not the House in North Dakota. In 1990, Oregon used AV in a statewide advisory referendum on school financing, which presented voters with five different options and allowed them to vote for as many as they wished (Wright, 1990).

In the late 1980s, AV was used in some competitive elections in countries in Eastern Europe and the Soviet Union, where it was effectively "disapproval voting," because voters were permitted to cross off names on ballots but not to vote for candidates (Shabad, 1987; Keller, 1987, 1988; White, 1989; Federal Election Commission, 1989). But this procedure is logically equivalent to AV. Candidates not crossed off are, in effect, approved of, although psychologically there is almost surely a difference between approving and disapproving of candidates.

With this information as background, I trace in section 1.3 my early involvement, and that of several associates, with AV. In section 1.4 I discuss how AV came to be adopted by the different societies.

In section 1.5 I report on empirical analyses of ballot data of some professional societies that adopted AV; they help to answer the question of when AV can make a difference in the outcome of an election. In section 1.6 I investigate the extent to which AV elects "lowest common denominators," which has concerned even supporters of AV. In section 1.7 I discuss whether voting is "ideological" under AV.

The confrontation between the theory underlying AV, which is rigorously developed in chapter 2, and practice offers some interesting lessons on "selling" new ideas. The rhetoric of AV supporters (I include myself), who have put forward the kinds of arguments outlined in section 1.1, has been opposed not only by those supporting extant systems like plurality voting (PV)—including incumbents elected under PV—but also by those with competing ideas, particularly proponents of other voting systems like the Borda count and the Hare system of single transferable vote.

I conclude that academics probably are not the best sales people for two reasons: (1) they lack the skills and resources, including time, to market their ideas, even when they are practicable; and (2) they squabble among themselves. Because few if any ideas in the social sciences are certifiably "right" under all circumstances, squabbles may well be grounded in serious intellectual differences. Often, however, they are not.

1.3. EARLY HISTORY

In 1976, I was attracted by the concept of "negative voting" (NV), proposed in a brief essay by Boehm (1976) that was passed on to me by the late Oskar Morgenstern. Under NV, voters can either vote for one candidate or against one candidate, but they cannot do both. Independently, Robert J. Weber had begun working on AV (he was apparently the first to coin the term "approval voting").

When Weber and I met in the summer of 1976 at a workshop at Cornell University under the direction of William F. Lucas, it quickly became apparent that NV and AV are equivalent when there are three candidates. Under both systems, a voter can vote for just one candidate. Under NV, a voter who votes against one candidate has the same effect as a voter who votes for the other two candidates under AV. And voting for all three candidates under AV has the same effect as abstaining under both systems.

When there are four candidates, however, AV enables a voter better to express his or her preferences. While voting against one candidate under NV has the same effect as voting for the other three candidates under AV, there is no equivalent under NV for voting for two of the four candidates. More generally, under NV a voter can do everything that he or she can do under AV, but not vice versa, so AV affords voters more opportunity to express themselves.

Weber and I wrote up our results separately, as did three other analysts who worked independently on AV in the 1970s (discussed in Brams and Fishburn, 1983, 2007; see also Weber, 1995). But the *idea* of AV did not spring forth, full-blown, only about thirty years ago; its origins go back many centuries. Indeed, AV was actually used, beginning in the thirteenth century, in both Venice (Lines, 1986) and papal elections (Colomer and McLean, 1998); it was also used in elections in nineteenth-century England (G. Cox, 1987), among other places.

In the summer of 1977, Peter C. Fishburn, then at Pennsylvania State University and later at Bell Telephone Laboratories, and I met at a conference on Hilton Head Island, South Carolina, under the direction of James S. Coleman.

We then began a long collaboration, which resulted in one book (Brams and Fishburn, 1983, 2007) and many articles on AV and other voting procedures (Brams and Fishburn, 2002).

Our first article (Brams and Fishburn, 1978) was a formal analysis of the properties of AV that included, as an illustration, its application to the 1968 U.S. presidential election, in which there were three significant candidates (Richard M. Nixon, Hubert H. Humphrey, and George Wallace). Our analysis of this election was based on empirical research of my former Yale student, D. Roderick Kiewiet (1979), who showed that Nixon's popular-vote and electoral-vote victory in 1968 would have been much more substantial under AV than it was under PV.[2]

Even at this early stage AV generated academic controversy (Tullock, 1979; Brams and Fishburn, 1979), which I will say more about later. Nevertheless, Fishburn and I became convinced that AV is a simple and practicable election reform that could ameliorate, if not solve, serious problems in multicandidate elections.

I began a "campaign" in 1979 to get AV adopted in public elections, beginning with New Hampshire's first-in-the-nation presidential primaries in February 1980, which had multiple candidates running in both the Democratic and Republican primaries. Although my efforts received both national coverage (e.g., in the *New York Times* and *Los Angeles Times*) and local coverage in several New Hampshire newspapers (e.g., the *Manchester Union-Leader* and *Concord Monitor*), I was not successful in getting an AV bill out of committee, despite being a native of New Hampshire ("prodigal son returns"), testifying before Senate and House committees in New Hampshire's General Court (legislature), and meeting with the governor. Later testimony I gave before legislative committees in other states (e.g., New York and Vermont) was similarly unavailing in effecting reform.

The pros and cons of AV compared with other voting systems have been debated over the last twenty-five years in numerous publications.[3] But this is not the subject of this chapter, except insofar as showing how the rhetoric has influenced the history of adoptions (and nonadoptions) of AV.[4]

[2] For other retrospective studies of multicandidate elections, such as the 1992 presidential election involving Bill Clinton, George Bush, and Ross Perot (Brams and Merrill, 1994), see the citations in Brams and Fishburn (2002).

[3] For a sampling of this debate, see the exchanges between Arrington and Brenner (1984) and Brams and Fishburn (1984a); Niemi (1984, 1985) and Brams and Fishburn (1985); Saari and Van Newenhizen (1988a, b) and Brams, Fishburn, and Merrill (1988a, b); Brams and Fishburn (2001) and Saari (2001a); and Brams and Herschbach (2001a, 2001b) and Richie, Bouricius, and Macklin (2001). Recent popular accounts of the controversy over voting systems by science writers include MacKenzie (2000), Guterman (2002), Klarreich (2002), and Begley (2003).

[4] Donald G. Saari has been a proponent of the Borda count, most recently in Saari (2001b). But I know of no recent adoptions of the Borda count, though it and a variant have been used in two small Pacific Island countries, beginning about thirty years ago (Reilly, 2002). Proponents

1.4. THE ADOPTION DECISIONS IN THE SOCIETIES

Elections are not a burning issue in most scientific societies, with participation rates often considerably below 50 percent of the membership and sometimes closer to about 10 percent. For the candidates, on the other hand, who are often luminaries in their disciplines, outcomes are usually more consequential and sometimes represent, especially if the office is the presidency, recognition of professional achievements over one's career.

It is not surprising, then, that candidates are willing to make subdued versions of what, in political life, would be called campaign statements. In the more rarefied atmosphere of an academic or professional society, these statements, which usually accompany a mailed or an electronic ballot, tend more to emphasize broad goals than specific programs, although candidates often pledge to undertake new initiatives. Most candidates, while listing their past offices and qualifications for the new office, generally do not seek to disparage the opposition.

Genteel as most of these campaigns are, candidates do, nonetheless, try to garner support by highlighting their qualifications and proposing new approaches or ideas that differentiate them from their opponents. When AV was first proposed as a reform in the four societies that adopted AV in the late 1980s, no candidates or factions, with one major exception, identified AV as a threat either to their candidacies or points of view.

Of course, after AV's use, there are winners and losers, and some losers, undoubtedly, see themselves as victims of this reform. In one society (The Institute of Management Sciences, or TIMS, before it merged with the Operations Research Society of America, or ORSA, to become INFORMS), this logic worked in reverse: the winner under PV, before AV was adopted, would almost certainly have lost under AV—and this became an argument made for the adoption of AV!

I hasten to add that this argument against PV was not a personal argument directed against the PV winner. Rather, the argument was that another candidate commanded broader support and thereby "deserved" to win.

Next I briefly recount the adoption decisions of the first four societies to use AV.

1. *Mathematical Association of America (MAA).* In 1985, the president of the MAA, Lynn Arthur Steen, who was familiar with work on AV, asked the Board of Governors of the MAA to consider adoption of AV in its biennial

---

of instant runoff voting (IRV), which is a voting system based on STV and supported by an organization, FairVote, recently succeeded in getting IRV enacted in elections in San Francisco and some other jurisdictions. As noted in Brams and Herschbach (2001b), IRV supporters have done little serious analysis to back up their claims, although other studies of STV (e.g., Dummett, 1984) have been more probing. On the other hand, FairVote does have human and monetary resources that few academics can match.

elections for president-elect and other national offices. After "heated but not acrimonious" debate (Steen, 1985), AV was approved by the board in 1985, passed by the membership in 1986, and used for the first time in the 1987 MAA elections.

Steen earlier had written an article in *Scientific American* (Gardner, 1980) on the mathematics of elections, in which he discussed AV. Before the MAA's consideration of AV, he asked me to look into the use of single transferable vote (STV) by the American Mathematical Society (AMS), the major research society of mathematicians.[5] I showed via two counterexamples that the "Instructions to Voters" accompanying the 1981 ballot used by the AMS to elect a nominating committee contained an erroneous statement about a property of STV (Brams, 1982a), which led to an exchange with Chandler Davis (1982), who had been a proponent of STV when it was adopted by the AMS several years earlier. The erroneous statement was deleted from future instructions, but AV was not adopted by the AMS until 1992.[6]

Both Steen's knowledge and his position as president of the MAA made him a crucial player in the MAA's adoption of AV. So, also, was Steen's successor as president of the MAA, Leonard Gillman, who was a strong advocate of AV and played an active role in its eventual implementation in the 1987 elections of the association. For example, he wrote a description of AV for mathematicians, which included results of his own analysis (Gillman, 1987).

2. *The Institute of Management Sciences (TIMS),* which is now part of IN-FORMS. The use of AV by TIMS in 1988 was preceded by an experiment in which members were sent a nonbinding AV ballot, along with the regular PV ballot, in the 1985 elections. Although the AV ballot did not count, 85 percent of the members who voted in these elections returned the AV ballot. This permitted Fishburn and Little (1988) to compare the results of voting under the two different systems.

On the basis of their empirical analysis, which will be discussed later, Fishburn and Little (1988) concluded that AV did a better job of electing Condorcet winners than did PV. Not only was the experiment "remarkably successful" (Little and Fishburn, 1986), but the results also convinced the council of TIMS to adopt AV in 1987, leading to its later adoption by INFORMS when it formed in 1995.

In fact, an argument for conducting the experiment in the first place was that management scientists should "practice what we preach" (Jarvis, 1984).

---

[5]The MAA is the more teaching oriented of the two major American mathematical societies at the college-university level.

[6]It was adopted in part because counting votes by hand under STV proved to be too onerous, and computerizing the counting was not feasible at the time. Even so, AV was adopted only for those offices of the AMS that did not require an amendment to the by-laws, which would have required considerable effort to enact; voting for other offices is still by PV (Daverman, 2002; and Fossum, 2002). Patently, pragmatic considerations played a key role in the AMS's choices.

Before deciding on its usage, TIMS decided to collect information it viewed as necessary to make an informed judgment about the applicability of the theoretical analysis of AV to its own elections.

Both the consideration and adoption of AV by TIMS were certainly helped by the fact that the president of TIMS in 1984–1985, John D. C. Little, was interested in AV and collaborated with Fishburn on the experiment and its analysis. Before undertaking the experiment, inquiries were made of the candidates to ask their permission to participate in it. Because of its research potential, all agreed, prefiguring AV's eventual adoption.

3. *American Statistical Association (ASA).* The former chair of the ASA's Committee on Elections, Richard F. Potthoff, had read about AV and brought it to the attention of his committee. This committee recommended its adoption first in internal ASA elections; the ASA Board of Directors approved this recommendation.

After AV's successful use in 1986 in three elections for council governors, the election of two editors to serve on the board, and the election of a board member to serve on the Executive Committee, the Committee on Elections recommended that AV be used in association-wide elections, which was approved by the board ("Amendment to ASA By-Laws," 1987) and ratified as an amendment in 1987. Unlike the other societies, the ASA has had no association-wide multicandidate elections since the adoption of AV, though some internal elections and single-winner section elections have had more than two candidates.

4. *Institute of Electrical and Electronics Engineers (IEEE).* The adoption of AV by the IEEE has a politically charged history (Brams and Nagel, 1991). Beginning in 1984, AV was considered, along with other voting systems, for possible use in multicandidate elections. But not until the 1986 elections— when a petition candidate, Irwin Feerst, ran against two candidates for president-elect who were nominated by the Board of Directors—did the issue of election reform take center stage. The reason is that Feerst, with 35 percent of the vote, defeated one of the two board-nominated candidates and came within 242 votes (of 52,405 cast) of defeating the other candidate. This result starkly illustrated to the board how vulnerable their nominees, who together might win a substantial majority in an election, are to a minority candidate if these nominees split the majority vote more or less evenly.

In 1987 the board reverted to nominating only one candidate for president-elect, breaking a tradition of nominating two candidates that it had begun in 1982. Feerst was instrumental in bringing the question of how many nominees the board must nominate to a vote of the entire membership in the 1987 election, in which he did not run and there were no other petition candidates. By a 57 percent majority, members supported a constitutional amendment requiring that the board nominate at least two candidates, but this fell short of the two-thirds majority needed to amend the IEEE's constitution.

Nevertheless, it was clear that there was strong member support for making IEEE elections more competitive, which renewed interest in AV when the board returned to nominating two candidates and allowed petition candidates to run as well. In 1987, I was invited by the then president of the IEEE, Henry L. Bachman, to attend an Executive Council meeting to discuss AV.

Unable to do so, I suggested that Jack H. Nagel of the University of Pennsylvania, who had done extensive research on AV, take my place. Nagel did; he also attended a later meeting of the full Board of Directors, which adopted AV in November 1987. (AV had previously been used in internal IEEE elections, sometimes in modified form.) With its adoption, the board voted to nominate at least two candidates for each office.

When the IEEE's adoption of AV was announced at a December 1987 IEEE press conference in New York City that Nagel and I attended, Feerst objected strenuously to its use, arguing that it was a deliberate move to undermine his candidacy and the interests of "working engineers," whom he claimed to represent. But when Feerst ran in 1988 for president-elect under AV, he came in fourth in a field of four candidates.

To recapitulate, the paths to adoption of AV in the different societies have been diverse. Only in the MAA did the full-scale use of AV begin before it was first tried out in an experiment (TIMS) or in internal elections (ASA and IEEE).

The presidents of the MAA, TIMS, and the IEEE played active roles in AV's adoption in their societies, and each received assistance from an advocate of AV. In the ASA, on the other hand, it was writings on AV that sparked initial interest, which turned into adoption without much controversy.

Controversy was the hallmark of the IEEE deliberations. While the IEEE's adoption of AV was in part a response to a perceived threat to its established leadership, it is important to realize that the IEEE did not view it as its only alternative.

In fact, several other voting systems had been considered before AV was selected. For example, a runoff election between the two top contenders, if neither received a majority in the initial balloting under PV, was also seriously considered, but it was viewed as too costly to have a second round of voting and also would have required a constitutional change. Ultimately, a majority of board members concluded that AV better fit the needs of the organization than any other voting system, and that is why it was adopted.[7]

This quick overview does not do justice to the serious debates that occurred over the merits of AV, particularly in the MAA and the IEEE. Indeed, although there has been dissent over AV's use in some societies (Kiely, 1991), no society that adopted AV ever rescinded its decision, with one notable

---

[7] By no means do I suggest that AV is a panacea in all elections, especially those involving multiple winners. For such elections, as I will argue in later chapters, other voting systems seem better suited.

exception (the IEEE).[8] Looking at what AV has wrought in the societies that adopted it may offer some explanation of why it has been generally, but not universally, accepted.

1.5. DOES AV MAKE A DIFFERENCE?

Clearly, a new voting procedure makes a difference if it leads to the selection of a different winner. The best evidence we have that AV would have elected a different winner is from the 1985 TIMS experiment, in which ballot data for both the PV official elections and the AV nonbinding elections were compared (Fishburn and Little, 1988).

In one of the three 1985 elections, the official PV and actual AV ballot totals are shown in Table 1.1 for candidates A, B, and C. Also shown are the AV totals extrapolated from the 85 percent sample of members who returned their AV nonbinding ballots, which is a very high figure. The extrapolation is a straightforward one: approval votes are added to the actual AV totals for each candidate based on the propensity of the sample respondents who voted for one particular candidate on the PV ballot to vote for each of the other candidates on the AV ballot. This extrapolation is justified by the finding that there were no major differences in voting patterns on the official PV ballot between AV respondents and nonrespondents.

Observe that candidate C wins the official PV election by a bare eight votes (0.4 percent), but B would have won under AV by a substantial 170 votes (6.2 percent). By itself, the fact that C wins more plurality votes and B wins more approval votes does not single out one candidate as the manifestly preferred choice. But on the experimental ballot, voters were asked one piece of additional information: to rank the candidates from best to worst by marking next to their names 1 for their first choice, 2 for their second choice, and so on.

These data can be used to reconstruct who would defeat whom in hypothetical pairwise contests, which is not evident from the PV totals. For example, the fact that C edges out B in presumed first choices, based on the PV totals, does not mean that C would hold his or her lead when the preferences of the 166 A voters are taken into account. In fact, the experimental ballots of these 166 voters show that

[8] According to the IEEE executive director, Daniel J. Senese, AV was abandoned in 2002 because "few of our members were using it and it was felt that it was no longer needed." I responded in an e-mail exchange (June 2, 2002) that because "candidates now can get on the ballot with 'relative ease' [according to former IEEE president Henry L. Bachman in the same e-mail exchange] . . . the problem of multiple candidates [in the late 1980s] might actually be exacerbated . . . and come back to haunt you [IEEE] some day." There may be other reasons for the abandonment of AV, but I was not privy to such information.

TABLE 1.1
PV and AV Vote Totals in 1985 TIMS Election

Candidates	Official PV	Actual AV	Extrapolated AV
A	166	417	486
B	827	1,038	1,224
C	835	908	1,054
Total	1,828	2,363	2,764
No. of Voters	1,828	1,567	1,828

1. 70 provided rankings in the order ABC.
2. 66 provided rankings in the order ACB.
3. 3 provided no rankings but approved both A and B.
4. 27 made no distinction between B and C by rankings or approval.

In the B-versus-C comparison, it is reasonable to credit (1) and (3) to B (73 votes), (2) to C (66 votes), and (4) to neither candidate. When added to the PV totals, these credits give C (901 votes) exactly one more vote than B (900 votes). However, assuming the 27 voters in (4) split their votes between B and C in the pattern of the 139 voters (70 + 66 + 3) who ranked A first and also expressed a preference between B and C, B would pick up an additional vote (rounded to the nearest vote), resulting in a 914–914 tie.

This extrapolation indicates that there is not a single *Condorcet winner*, who can defeat all other candidates in pairwise contests. The usual reason one does not exist is that there is a *Condorcet paradox*, whereby majorities cycle. To illustrate, if there are three voters with preferences for candidates {X, Y, Z} of (1) XYZ, (2) YZX, and (3) ZXY, two out of three voters (voters 1 and 3) prefer X to Y, two out of three (voters 1 and 2) prefer Y to Z, and two out of three (voters 2 and 3) prefer Z to Y, showing that there is no candidate that is preferred to all others. Instead, each candidate can be defeated by one other, so majorities cycle.

In this election, however, it is a projected tie that precludes one candidate from defeating the others in pairwise contests. That there is no cycle, and that A in fact would lose to both B and C, is shown by ranking data in Fishburn and Little (1988).

While surprising, the lack of a single Condorcet winner should not obscure the fact that 170 more voters approved of B rather than C in the extrapolated AV returns, albeit C won the PV contest by eight votes. The reason for this discrepancy between the AV and PV results is that whereas C has slightly more *stalwart* supporters (i.e., those who vote only for one candidate) than B, supporters of the third candidate, A, somewhat more approved of B than C (44 percent to 40 percent). But B's big boost comes from the fact that substantially more of C's supporters approve of B than B's do of C, so B would have won handily under AV.

Is this desirable? In the absence of a Condorcet winner, Fishburn and Little (1988, pp. 559–560) concluded that "approval voting picks a clear winner on the basis of second choices. These show that B has a broader acceptance in the electorate than C. Therefore, the approval process, by eliciting more information from the voters, leads to the election of the candidate with the widest support." Although it is theoretically possible in close elections that the Condorcet winner will not be the most approved candidate, it rarely occurs.[9] However, the legitimacy of the AV winner may be questioned on other grounds.

1.6. DOES AV ELECT THE LOWEST COMMON DENOMINATOR?

One fear that has been expressed about the use of AV is that while it may help elect candidates more broadly representative than PV, these candidates could turn out to be rather bland and uninspiring. They may win simply because they offend the fewest voters, not because they excite the passions of many (Brams and Fishburn, 1988).

It is difficult to say whether, in principle, a compromise candidate is a better or worse social choice than a more extreme candidate who is the darling of some voters but the bane of others. In practice, fortunately, this dichotomous

---

[9] The 1999 election for president of the Social Choice and Welfare Society, which was decided by two approval votes among seventy-six cast, is one exception. The second-place AV candidate in this election would have defeated the AV winner by four votes in a head-to-head contest, based on candidate rankings. Brams and Fishburn (2001) deem this "nail-biting" election essentially a toss-up, whereas Saari (2001a) argues that most positional methods would have chosen the Condorcet winner (including the Borda count, wherein the Condorcet winner would have defeated the AV winner 60–59); see Laslier (2003) for more details on voting patterns in this election. A recent second exception, the 2006 election for president of the Public Choice Society, which was decided by one vote (Brams, Hansen, and Orrison, 2006), suggests that different Condorcet and AV winners are most likely in elections that are virtual dead heats. Regenwetter and Grofman (1998), using a random-utility model to reconstruct voter preferences in several elections—including some discussed here—show that winners under different voting systems almost always coincide. Laslier (2006) and Laslier and Vander Straeten (2003) analyzed data from a field experiment with AV in the 2002 French presidential election, which involved over 5,000 voters in two French towns, and conclude that AV was easily understood, readily accepted, and provided a more complete picture of the "political space." Baron, Altman, and Kroll (2005) show in a laboratory experiment that AV reduces parochialism, or bias toward one's own group. Earlier theoretical analyses as well as computer simulations (Brams and Fishburn, 1983; Lijphart and Grofman, 1984; Nurmi, 1987; Merrill, 1988) demonstrate that AV generally elects a Condorcet winner if there is one. If there is not one, as in the 1985 TIMS election experiment, then proponents of AV argue that AV provides a compelling way to resolve either a Condorcet paradox or a tie.

choice seems to arise rarely, as the data from the AV elections of the four societies demonstrate. Specifically, the winners under AV were candidates who were generally popular among *all* voters, however many candidates they voted for in the different elections. Thus, a divergence between forceful minority candidates, approved of by few, and "wishy-washy" majority candidates, approved of by many, is probably an infrequent event. But if elections are polarized, moderate voters under AV are likely to be decisive in swinging the outcome toward a less extreme candidate.

There are examples of elections in which the winner was not strong among all classes of voters. Consider the 1987 MAA election shown in Table 1.2 (Brams, 1988), wherein the votes received by the five candidates in this election are broken down by the votes each of the candidates received from voters who cast exactly one vote (1-voters), voters who cast exactly two votes (2-voters), and so on. Excluded from these totals are nine voters who voted for all the candidates, whose undifferentiated support obviously has no effect on the outcome.

In this election, 3,081 of the 3,924 voters (79 percent) were 1-voters, while the remaining 843 voters cast 1,956 votes, or an average of 2.3 votes each. Thus, the multiple voters cast 39 percent of the votes, though they constituted only 21 percent of the electorate.

Did the multiple voters make a difference? It would appear not, because the winner (A) received 28 percent more votes from 1-voters than the 1-voters' runner-up (D) did, just edged out B among 2-voters, but lost to several candidates among 3-voters and among 4-voters. A's victory, then, is largely attributable to the substantial margin received from 1-voters, not from the presumably more lukewarm support that A received from multiple voters.

Define a candidate who wins among all classes of voters—those who cast few votes (narrow voters) and those who cast many votes (wide voters)—as *AV-superior.* In the MAA election, assume narrow voters are those who cast one or two votes, and wide voters are those who cast three or four votes.

TABLE 1.2
AV Vote Totals in 1987 MAA Election

Candidates	1-Voters	2-Voters	3-Voters	4-Voters	Total
A	848	276	122	21	1,267
B	618	275	127	32	1,052
C	652	264	134	34	1,084
D	660	273	118	31	1,082
E	303	132	87	30	552
Total	3,081	1,220	588	148	5,037
No. of Voters	3,081	610	196	37	3,924

It turns out that the winner, candidate A, is not AV-superior, because he or she wins among narrow but not among wide voters. Does this vitiate A's winning status? In winning so decisively among 1-voters, whose preference intensities would seem to be greatest, it would be hard to argue that A is any kind of lowest common denominator. It should be noted, however, that some of the thirty-seven voters who voted for four of the five candidates probably also had intense preferences—but against the one candidate they chose to leave off their approved lists.

In twelve of the sixteen multicandidate AV elections analyzed in the four societies, the winners were AV-superior. In the four elections in which there was not an AV-superior winner, the pattern is similar to that in the 1987 MAA election shown in Table 1.2: the winner won by virtue of receiving greater support among narrow voters than among wide voters. These AV-nonsuperior winners, therefore, do not fit the mold of lowest common denominators—the choice of many wide voters but few narrow voters—but rather the opposite, which reinforces, not undermines, their legitimacy as winners.

The fact that the winners in three-quarters of the elections were AV-superior is perhaps not surprising, because one would expect such candidates would do better than losers across different types of voters. A little reflection, however, shows that this need not be the case. Paradoxically, a candidate may lose among every possible class of voters—that is, be *AV-inferior*—and still be the AV winner. For example, A might be the victor over C among narrow voters, and B might be the victor over C among wide voters. But C could emerge as the AV winner if A did badly among wide voters, B did badly among narrow voters, but C was a close second among both types.

No winners in the sixteen elections were AV-inferior. As already noted, even the support of the four AV-nonsuperior winners appeared to be more intense and heartfelt (i.e., from narrow voters) than that of the losers, so AV does not appear to elect lowest common denominators.

1.7. IS VOTING IDEOLOGICAL?

Consider again the 1987 MAA election. As can be calculated from table 1.2, 2-voters gave the candidates 22–26 percent of all their votes, 3-voters 10–16 percent, and 4-voters 2–5 percent. Venn diagrams (not shown here) indicate the shared support among the ten subsets of two candidates, the ten subsets of three candidates, the five subsets of four candidates, and the set of all five candidates. Examination of the *sources* of this support, as shown in the Venn diagrams, does not reveal any particular pairs, triples, or quadruples that received unusually great support, indicating that there was no obvious coalitional voting for certain subsets.

On the contrary, multiple votes are spread about as one would expect according to the null hypothesis that votes are distributed in proportion to the candidates' totals. In the case of A, for example, there were eighty-two shared votes with just B, ninety-one with just C, eighty with just D, and twenty-three with just E, which is roughly in accord with the candidates' overall totals. Indeed, every one of the thirty-two subsets in this election—including the 2.6 percent who abstained—got at least three votes.

The story is very different for the 1988 IEEE election shown in Table 1.3 (Brams and Nagel, 1991), wherein the approval vote totals are shown for all sixteen subsets of the four candidates in this race. Consider first the 3-voters, and note that nearly everyone in this category voted for ABD—5,605 voters, to be precise. By contrast, only 148, 143, and 89 voters, respectively, supported the other 3-subsets of ABC, ACD, and BCD that contain C.

Evidently, the numerous supporters of ABD voted against C by voting for everybody except C. This essentially negative kind of voting against C can also be seen in voting for the six 2-subsets. The three 2-subsets that do not include C (AB, AD, and BD) had an average of 4,027 voters each, whereas the three that included C (AC, BC, and CD) had an average of only 897 voters each.

In addition to the predominant clustering of support around A, B, and D, there are some subtle differences in the sharing of support. For each pair of candidates, Brams and Nagel (1991) computed an index of shared support by taking the ratio of ballots approving of both candidates in each pair by 2-voters and 3-voters to total ballots, excluding abstentions and votes for all four candidates. By this measure, A and D have the most affinity, with 22.9 percent shared support. They are followed by A and B, with 17.2 percent; and then by

TABLE 1.3
Numbers of Voters Who Voted for 16 Different Subsets in 1988 IEEE Election and AV Total

Subsets					
		None = 1,100			
	A = 10,738	B = 6,561	C = 7,626	D = 8,521	
AB = 3,578	AC = 659	AD = 6,679	BC = 1,425	BD = 1,824	CD = 608
	ABC = 148	ABD = 5,605	ACD = 143	BCD = 89	
		All = 523			
Totals					
	A = 28,073	B = 19,753	C = 11,221	D = 23,992	

B and D, with 13.9 percent. Although A, B, and D share much less support with C, B at 3.1 percent shares slightly more with C than do A (1.8 percent) and D (1.5 percent).

From these results, one might infer an underlying dimension on which D and C occupy opposite extremes, whereas A and B are located at intermediate positions. A is somewhat closer than B to D, but both B and A are much closer to D than to C, as shown in the following hypothetical continuum:

This representation corresponds to certain facts about the candidates. D and A were both board nominees, whereas C was a vociferous critic of IEEE officers, board, and staff. B, though like C a petition candidate, was in other ways close to the IEEE establishment, having previously served on the board. As for the slight distinction between D and A, judging from the candidates' biographies and statements it may reflect D's emphasis on technical research, which perhaps made D seem most distant from C, who sought to champion the working engineer.

Of the 54,204 ballots analyzed in this election, only 3,323 (6 percent) are "inconsistent" with the assumption that voters' preferences are based on the foregoing DABC ordering of candidates. *Inconsistent ballots* include approval of two nonadjacent candidates without including the adjacent candidate(s) between them, notably DC (608), AC (659), DAC (143), and DBC (89). Accounting for more than half the inconsistencies is the relatively minor inconsistency—in terms of perceived differences—represented by the pattern DB (1,824). Of the multiple voters, 17,435 (84 percent) cast ballots consistent with the hypothetical ordering.

Thus, candidates with obvious affinities tended disproportionately to share approval from multiple voters. In this sense voting was ideological: it reflected a pattern consistent with an underlying ordering of the candidates. Only in this election, however, was such a pattern found; far more typically, voting in the societies is nonideological, which is consistent with the null hypothesis alluded to earlier. But if AV is used in public elections, their more political character could well lead to the kind of ideological cleavages observed in the IEEE election—and the controversy that ensued.

It is important to note, however, that nonideological voting may mirror regularities not evident in the AV data themselves. As a case in point, the winner in the 1987 MAA election (Table 1.2) was a woman, and this pattern was repeated in the next MAA election in 1989. I have not analyzed data from the latter election, but the 1987 winner's victory, as shown earlier, cannot be impeached on grounds that she won mostly because of lukewarm support from wide voters. Nonetheless, because the female winners in 1987 and 1989 were the only women in each of the two races, it may be the case that they were helped by their uniqueness: by some they were perceived as the single best choice; by others they were seen as broadly acceptable.

1.8. SUMMARY AND CONCLUSIONS

AV has proved to be a practical and viable election reform in the four scientific and engineering societies that used it for the first time in 1987 and 1988. While AV supporters played a role in its adoption in three of the four societies (TIMS, MAA, and IEEE), none of its proponents was even aware of its consideration in the fourth society (ASA) until its adoption was imminent.

In all these societies, AV's adoption rested principally on the arguments—summarized in section 1.1—that it is preferable to PV in multicandidate races. In the IEEE, a petition candidate's near-win with vocal but only minority support certainly gave urgency to these arguments, accelerating AV's adoption after the board's attempt to limit the number of board-nominated candidates to one person met with the membership's disapprobation. Only in the case of the AMS's 1992 adoption of AV did practical considerations give it an edge over STV, and then only in some elections that were relatively easy to change.

The empirical analyses of election returns from the different societies indicate that AV may make a difference. So far it seems not to have elected candidates who can be characterized as lowest common denominators but instead candidates who either enjoyed support among all classes of voters, or who did particularly well among narrow voters whose support I presume to be more intense. Although voting seems generally nonideological in most society elections, a clear ordering of positions was identified in the IEEE election, and voting tended to be only for adjacent candidates in this ordering.

Condorcet winners almost always win under AV, with the only known exceptions being the 1999 Social Choice and Welfare election and the 2006 Public Choice Society election, which were near ties. If there is no single Condorcet winner, as was illustrated in the 1985 TIMS election experiment, then AV provides a way of determining which candidate receives the most support from all voters, not just those who rank this person first.

Not all societies that have been approached about adopting AV, including two that I belong to—the American Political Science Association (APSA) and the International Studies Association (ISA)—have been amenable to election reform, much less the adoption of AV. Significantly, these societies are dominated, or heavily populated by, academic political scientists; none holds competitive elections unless a petition candidate challenges the official slate (this has never happened in the ISA; in the APSA, the last challenge to a presidential candidate occurred almost forty years ago).

Among the lessons I draw from my experience is that the adoption of AV, and probably any election reform, requires key support from within an organization.

I never received this kind of support from politicians or political parties in my attempts to get AV adopted in public elections. By contrast, the society adoptions would not have occurred without influential members of each society favoring reform, sometimes for practical or, in the case of the IEEE, political reasons. Of course, they also needed to make their cases with arguments based on democratic principles; I like to believe that both the rhetoric of AV supporters as well as their analyses helped in this regard.

2

Electing a Single Winner:
Approval Voting in Theory

2.1. INTRODUCTION

In single-winner elections that attract multiple candidates, a voter may consider more than one candidate acceptable if there is no obviously "best" candidate. Even if there is, the voter may wish to approve of a more viable second choice if his or her first choice has little chance of winning.

To make "acceptability" more precise, in this chapter I extend the usual social-choice framework to include information not only on how voters rank candidates but also on where they draw the line between those they consider acceptable and those they consider unacceptable. This new information, of course, is precisely that which is elicited by approval voting (AV). AV gives voters the opportunity to be sovereign by expressing their approval for any set of candidates. In doing so, AV better enables voters both to elect, and to prevent the election of, candidates, as I will show in this chapter.

The set of possible outcomes under AV may be quite large. For example, Condorcet winners, who can defeat all other candidates in pairwise contests, are always AV outcomes, as are winners under several other voting systems (to be described later).

Although AV may lead to a plethora of outcomes, they are not haphazard choices that can easily be upset when voters are manipulative. The candidate elected by AV will often be a candidate at a *Nash equilibrium*, from which voters with the same preferences will have no incentive to depart. Moreover, AV can always elect a unique Condorcet winner (if one exists) as a *strong* Nash equilibrium, which yields outcomes that are invulnerable to departures by any set of voters. In allowing for other Nash-equilibrium outcomes, including even *Condorcet losers* who can be defeated by all other candidates in pairwise comparisons, AV does not rule out outcomes that may, on occasion, be more acceptable to voters.

Note: This chapter is adapted from Brams and Sanver (2006) with permission.

Saari and Van Newenhizen (1988a) and Saari (1994, 2001), among others, have argued that it is a vice that AV can lead to a multiplicity of outcomes.[1] I argue, on the contrary, that voters' judgments about candidate acceptability *should* take precedence over standard social-choice criteria such as electing a Condorcet winner, which may clash with these judgments.[2]

Because the notion of acceptability, based on the consent of voters to "accept" a candidate, is absent from preference-based systems, these systems do not distinguish voters who judge only a first choice acceptable from voters who judge all except a last choice acceptable. Indeed, *where* voters draw the line between approved and nonapproved candidates often reflects where the biggest drop-off is in the intensity of their preferences. In my view, this is critical information in the selection of a social choice, so I expand the standard social-choice framework to incorporate it.

In section 2.2 I define preferences and describe admissible and sincere strategies under AV. In section 2.3 I characterize AV outcomes and describe the "critical strategy profile"—a key concept in this chapter—that produces them. I then compare AV outcomes with those given by other voting systems. Among other things, I show that no "fixed rule," in which voters vote for a predetermined number of candidates, always elects a unique Condorcet winner, suggesting the need for a more flexible system.

The stability of outcomes under AV is analyzed in section 2.4. Besides the Nash-equilibrium results alluded to above, I show that Condorcet voting systems, which guarantee the election of Condorcet winners when voters sincerely rank candidates, may not elect Condorcet winners in equilibrium.

In section 2.5 I conclude on a normative note that voters, because they are sovereign, *should* be able to express their approval of any set of candidates. Likewise, a voting system *should* allow for the possibility of multiple acceptable outcomes, especially in close elections. That AV more than other voting systems is responsive in this way I consider a virtue. That it singles out as a strong Nash-equilibrium outcome a unique Condorcet winner may or may not be desirable.

I discuss these and other questions related to the nature of acceptable outcomes. Indeed, I suggest that "acceptability" replace the usual social-choice criteria for assessing the satisfactoriness of election outcomes.

[1] The critique of AV by Saari and Van Newenhizen (1988a) provoked an exchange between Brams, Fishburn, and Merrill (1988a, 1988b) and Saari and Van Newenhizen (1988b) over whether the plethora of AV outcomes more reflected AV's "indeterminacy" (Saari and Van Newenhizen) or its "responsiveness" (Brams, Fishburn, and Merrill).

[2] In fact, it is an old story that standard social-choice criteria may clash among one another. For example, even when there is a Condorcet winner who can defeat every other candidate in pairwise contests, there may be a different Borda-count winner (to be defined later) who, on the average, is ranked higher than a Condorcet winner. See Nurmi (1999, 2002) and Brams and Fishburn (2002) for other examples that have produced many of the so-called paradoxes of social-choice theory.

2.2. PREFERENCES AND STRATEGIES UNDER AV

Consider a set of voters choosing among a set of candidates. Individual candidates are denoted by small letters a, b, c. . . . A voter's *strict preference relation* over candidates is denoted by P, so aPb means that a voter strictly prefers a to b, which is denoted by the following left-to-right ranking (separated by a space): $a\ b$. In the subsequent analysis, I assume that all voters have strict preferences, so they are not indifferent among two or more candidates.[3]

In addition, every voter is assumed to have a *connected* preference: For any a and b, either $a\ b$ or $b\ a$ holds. Moreover, P is *transitive*, so $a\ c$ whenever $a\ b$ and $b\ c$. The list of preferences of *all* voters is called a *preference profile* **P**.

An *AV strategy S* is a subset of candidates. Choosing a strategy under AV means voting for all candidates in the subset and no candidates outside it. The list of strategies of all voters is called a *strategy profile* **S**.

The number of votes that candidate i receives at **S** is the number of voters who include i in the strategy S that each selects. For any **S**, there will be a set of candidates ("winners") who receive the greatest number of votes.

Voters are assumed to use admissible and sincere strategies. An AV strategy S of a focal voter is *admissible* if it is not dominated in a game-theoretic sense—that is, if there is no other strategy that gives outcomes at least as good as, and sometimes better than, S for all strategy profiles **S** of voters other than the focal voter. Brams and Fishburn (1978, 1983) show that admissible strategies under AV involve always voting for a most-preferred candidate and never voting for a least-preferred candidate.

An AV strategy is *sincere* if, given the lowest-ranked candidate that a voter approves of, he or she also approves of all candidates ranked higher. Thus, if S is sincere, there are no "holes" in a voter's approval set: everybody ranked above the lowest-ranked candidate whom a voter approves of is also approved; and everybody ranked below is not approved.[4]

---

[3] This restriction simplifies the analysis; its relaxation to allow for voter indifference among candidates has no significant effect on the findings.

[4] Excluding the strategies of voting for nobody and voting for everybody, sincere strategies are always admissible; insincere strategies may also be admissible, but if and only if there are four or more candidates. For example, if there are exactly four candidates, it may be admissible for a voter to approve of his or her first and third choices without also approving of a second choice (see Brams and Fishburn, 1983, pp. 25–26, for an example). However, the circumstances under which this happens are sufficiently rare and nonintuitive that I henceforth suppose that voters choose only sincere strategies. Relaxing this assumption to include admissible insincere strategies complicates the analysis but does not significantly alter the main findings. For refinements of sincerity that may be based on strategic considerations, see Merrill and Nagel (1987); for a different notion of sincerity and its relationship to stability, see Yunfeng, Chaoyuan, and Ting (1996).

A strategy profile **S** is said to be *sincere* if and only if the strategy *S* that every voter chooses is sincere, based on each voter's preference *P*. Sincere strategies are always admissible if "vote for everybody" is excluded, which I henceforth do. While the sincerity assumption is a simplification, it does not significantly alter the main findings.

As I will illustrate shortly, voters may have multiple sincere strategies, which is precisely what creates the indeterminacy that some analysts find problematic but which I find helpful. In expanding the opportunities for voters to express themselves—that is, exercise their sovereignty—AV abets the discovery of desirable social choices.

As an illustration of these concepts, assume that there are seven voters, who can be grouped into three different types and who vote for the set of four candidates $\{a, b, c, d\}$:

Example 2.1
1. 3 voters: $a\ b\ c\ d$
2. 2 voters: $b\ c\ a\ d$
3. 2 voters: $d\ b\ c\ a$

The three types define the preference profile **P** of all seven voters. For simplicity, I assume in this and later examples that all voters of each type choose the same strategy *S*.

Voters of type 1 have three sincere strategies: $\{a\}$, $\{a, b\}$, and $\{a, b, c\}$, which for convenience I write as *a*, *ab*, and *abc*. A typical sincere strategy profile of the 7 voters is $\mathbf{S} = (a, a, a, bc, bc, dbc, dbc)$, whereby the 3 voters of type 1 approve of only their top candidate, the 2 voters of type 2 approve of their top two candidates, and the 2 voters of type 3 approve of all candidates except their lowest ranked. The number of votes of each candidate at **S** is 4 votes for *b* and *c*, 3 votes for *a*, and 2 votes for *d*. Hence, AV selects candidates $\{b, c\}$ as the (tied) winners at **S**.

2.3. ELECTION OUTCOMES UNDER AV AND OTHER VOTING SYSTEMS

Given a preference profile **P**, consider the set of all candidates who *can* be chosen by AV when voters use sincere strategies. Call this set *AV outcomes*. Clearly, a candidate ranked last by all voters cannot be in this set, because it is inadmissible for any voter to vote for this candidate.

Define an *AV critical strategy profile* for candidate *i* at preference profile **P** as follows: Every voter who ranks *i* as his or her worst candidate votes only for the candidate that he or she ranks top. The remaining voters vote for *i and* all candidates they prefer to *i*.

Let $C_i(\mathbf{P})$ be the AV critical strategy profile of candidate i, which will play a crucial role in what follows. In Example 2.1, the critical strategy profile for candidate a is $C_a(\mathbf{P}) = (a, a, a, bca, bca, d, d)$, giving a 5 votes compared to 2 votes each for b, c, and d. It can easily be seen that $C_i(\mathbf{P})$ is sincere.

The next four lemmata provide a theoretical foundation for several of the subsequent propositions. They (i) show that under AV, candidate i cannot do better than at $C_i(\mathbf{P})$; (ii) characterize AV outcomes; (iii) characterize outcomes that can never be chosen under AV; and (iv) characterize outcomes that must be chosen under AV. The proofs of these lemmata demonstrate the correctness of these results.

Lemma 2.1. *Assume all voters choose sincere strategies. The AV critical strategy profile for candidate i, $C_i(P)$, maximizes the difference between the number of votes that i receives and the number of votes that every other candidate j receives.*

Proof. Clearly, no other sincere strategy profile yields candidate i more votes than its AV critical strategy profile $C_i(\mathbf{P})$. Now consider the number of votes received by any other candidate j at $C_i(\mathbf{P})$. Candidate j will receive no fewer and sometimes more votes if there are the following departures from $C_i(\mathbf{P})$:

1. A voter who ranked candidate i last, and therefore did not vote for him or her, votes for one or more candidates ranked below his or her top-ranked choice (possibly including candidate j); or
2. A voter who did not rank candidate i last or next-to-last votes for one or more candidates ranked below i (possibly including candidate j) unless i is ranked next-to-last.

In either case, candidate j never gets fewer, and may get more, votes when there are these departures from candidate i's critical strategy profile $C_i(\mathbf{P})$. Besides 1 and 2, the only other possible departure would be one in which a voter does not vote for candidate i, who is not ranked last, but only candidates above i. But this gives candidate i fewer votes than $C_i(\mathbf{P})$ and j possibly as many. ∎

Using Lemma 2.1, a simple way to determine whether any candidate i is an AV outcome follows from Lemma 2.2:

Lemma 2.2. *Candidate i is an AV outcome if and only if i is chosen at his or her critical strategy profile $C_i(P)$.*

Proof. The "if" part is a direct consequence of the fact that $C_i(\mathbf{P})$ is sincere. To show the "only if" part, suppose candidate i is not chosen by AV at $C_i(\mathbf{P})$. By Lemma 2.1, $C_i(\mathbf{P})$ maximizes the difference between the number of votes that i receives and the number of votes that any other candidate j receives, so there is no other sincere strategy profile at which i can be chosen by AV. ∎

Using Lemma 2.2, a characterization of candidates that *cannot* be AV outcomes is given by Lemma 2.3:

Lemma 2.3. *Given any preference profile **P** and any candidate i, i cannot be an AV outcome if and only if there exists some other candidate j such that the number of voters who consider j as their best choice and i as their worst choice exceeds the number of voters who prefer i to j.*

Proof. Given any preference profile **P** and any two candidates i and j, voters can be partitioned into three (disjoint) classes:

1. Those who prefer i to j
2. Those who consider j as the best choice and i as the worst choice
3. Those who prefer j to i but do not fall into class 2

At critical strategy profile $C_i(\mathbf{P})$, the voters in class 2 will vote for i but not j; those in class 3 will vote for j but not i; and those in class 3 will vote for both i and j. Setting aside class 3, which gives each candidate the same number of votes, candidate i cannot be selected at $C_i(\mathbf{P})$ if and only if the number of voters in class 2 exceeds the number of voters in class 1. Hence, by Lemma 2.2 candidate i cannot be an AV outcome at **P**. ∎

In effect, Lemma 2.3 extends Lemma 2.2 by saying precisely when candidate i will be defeated by candidate j and cannot, therefore, be an AV outcome.

Call a candidate *AV-dominant* if and only if, whatever sincere strategies voters choose, this candidate is the unique winner under AV. The final lemma "reverses" Lemma 2.3 by saying when candidate i must be, rather than can never be, an AV outcome.

Lemma 2.4. *Given any preference profile **P** and any candidate i, i is AV-dominant if and only if, given any other candidate j, the number of voters who consider i as their best choice and j as their worst choice exceeds the number of voters who prefer j to i.*

Proof. I begin with the "if" part. All voters who consider i as their best choice and j as their worst choice will vote for i and not for j under AV. Because this number exceeds the number of voters who prefer j to i—and would, in the worst situation for i, vote for j and not for i—i always receives more votes than j.

For the "only if" part, assume there exists some j such that the number of voters who prefer j to i equals or exceeds the number of voters who rank i as their best choice and j as their worst choice. If the voters who prefer j to i vote for j and not for i, and the voters who prefer i to j (without ranking i as their best and j as their worst choice) vote for both i and j, the number of votes that i receives will not be greater than the number of votes that j receives. Consequently, i will not be the unique winner under AV. ∎

AV can generate a multitude of outcomes. Consider again Example 2.1, in which I showed earlier that AV selects candidate a at $C_a(\mathbf{P})$. Similarly, AV selects candidates b and $\{b, c\}$, all with 7 votes, at critical strategy profiles $C_b(\mathbf{P}) = \{ab, ab, ab, b, b, db, db\}$ and $C_c(\mathbf{P}) = \{abc, abc, abc, bc, bc, dbc, dbc\}$. However, $C_d(\mathbf{P}) = \{a, a, a, b, b, d, d\}$, so candidate a (3 votes) rather than candidate d (2 votes) is chosen at candidate d's critical strategy profile.[5] In sum, the set of AV outcomes that are possible in Example 2.1 is $\{a, b, \{b, c\}\}$.

Although none of the three candidates is AV-dominant, candidate a would be if there were, for example, 2 $a\,b\,c$ voters, 2 $a\,c\,b$ voters, and 1 $b\,c\,a$ voter. Candidate a would always get 4 votes, whereas candidates b and c would, at best, get 3 votes each.

Call a candidate a *Pareto candidate* if there is no other candidate whom all voters rank higher. Example 2.1 illustrates three things about the tie-in of Pareto candidates and AV outcomes:

- a and b are Pareto candidates and AV outcomes.
- c is not a Pareto candidate but is a component of an AV outcome (it ties with b at $C_c(\mathbf{P})$).
- d is a Pareto candidate but not an AV outcome.

These observations are generalized by the following proposition:

Proposition 2.1. *The following are true about the relationship of Pareto candidates and AV outcomes:*

1. *At every preference profile $\mathbf{P}$, there exists a Pareto candidate that is an AV outcome or a component of an AV outcome.*
2. *Not every Pareto candidate is necessarily an AV outcome.*
3. *A non-Pareto candidate may be a component of an AV outcome but never a unique AV outcome.*

Proof. To show (1), take any preference profile $\mathbf{P}$. Assume that every voter votes only for his or her top choice. Then the one or more candidates chosen by AV, because they are top ranked by some voters, must be Pareto candidates. To show (2), it suffices to check the critical strategy profile $C_d(\mathbf{P})$ of Example 2.1, wherein candidate d is not an AV outcome but is a Pareto candidate because d is top ranked by the two type 3 voters.

---

[5] That d cannot be chosen also follows from Lemma 2.3: more voters (3) consider a as their best choice and d as their worst choice than prefer d to a (2).

In Example 2.1, I showed that c is not a Pareto candidate but is a component of an AV outcome. To show that a non-Pareto candidate can never be a unique AV outcome and prove (3), consider any **P** at which there exists a non-Pareto candidate i who is a component of an AV outcome. Take any sincere strategy profile **S** where this outcome is selected. Because i is not a Pareto candidate, there exists some other candidate j whom every voter prefers to i. Hence, every voter who voted for i at **S** must have voted for j as well, which implies that i and j tie for the most votes. Indeed, *all* candidates j who Pareto-dominate i will be components of an AV outcome at **S**. Because at least one of the candidates j that Pareto-dominate i must be ranked higher by one or more voters than all other candidates j, AV picks a Pareto candidate who ties candidate i. ∎

In Example 2.1, candidate b is the *Condorcet winner*, who can defeat all other candidates in pairwise contests, and candidate d is the *Condorcet loser*, who is defeated by all other candidates in pairwise contests. Not surprisingly, b is an AV outcome but d is not. However, consider the following 7-voter, 3-candidate example:

Example 2.2
1. 3 voters: $a\,b\,c$
2. 2 voters: $b\,c\,a$
3. 2 voters: $c\,b\,a$

Notice that the two type 2 and the two type 3 voters prefer candidates b and c to candidate a, so a is the Condorcet loser. But because the critical strategy profile of candidate a is $C_a(\mathbf{P}) = (a, a, a, b, b, c, c)$, a is an AV outcome—as are also candidates b and c, rendering all three candidates in this example AV outcomes.

The next proposition summarizes the Condorcet properties of AV outcomes:

Proposition 2.2. *Condorcet winners are always AV outcomes, whereas Condorcet losers may or may not be AV outcomes.*

Proof. If candidate i is a Condorcet winner, a majority of voters prefer i to every other candidate j. This implies that fewer voters rank j as their best choice and i as their worst choice, which by Lemma 2.3 implies that candidate i is an AV outcome. That a Condorcet loser may not be an AV outcome is shown by candidate d in Example 2.1, whereas candidate a in Example 2.2 shows that a Condorcet loser may be an AV outcome. ∎

Define a *fixed rule* as a voting system in which voters vote for a predetermined number of candidates, and the candidate(s) with the most votes is (are) elected. Under "limited voting," this predetermined number, which may be only one, is always less than the number of candidates to be elected.

Proposition 2.3. *There exists no fixed rule that always elects a unique Condorcet winner.*

Proof. Consider the following 5-voter, 4-candidate example:

Example 2.3
1. 2 voters: *a d b c*
2. 2 voters: *b d a c*
3. 1 voter: *c a b d*

Vote-for-1 elects {*a*, *b*}, vote-for-2 elects *d*, and vote-for-3 elects {*a*, *b*}. Thus, none of the fixed rules elects the unique Condorcet winner, candidate *a*. ∎

By contrast, several sincere strategies, including $C_a(\mathbf{P}) = (a, a, bda, bda, ca)$— in which different voter types vote for different numbers of candidates—elect *a*. Clearly, the flexibility of AV may be needed to elect a unique Condorcet winner, although AV may elect other candidates as well.

I next turn to scoring rules and analyze the relationship between the winner they select and AV outcomes. The best-known scoring rule is the Borda count (BC): Given that there are *n* candidates, BC awards $n-1$ points to each voter's first choice, $n-2$ points to each voter's second choice, . . . , and 0 points to his or her worst choice.

In Example 2.1, the BC winner is candidate *b*, who receives from the three types of voters a Borda score of $3(2) + 2(3) + 2(2) = 16$ points. In Example 2.2, the BC winner is also candidate *b*, who receives from the three types of voters a Borda score of $3(1) + 2(2) + 2(1) = 9$ points. In these examples, the BC winners coincide with the Condorcet winners, making them AV outcomes (Proposition 2.2), but this need not be the case, as I will illustrate shortly.

There are other *scoring rules* besides BC, so I begin with a definition. Given *m* candidates, fix a nonincreasing vector $(s_1, \ldots, s_m)$ of real numbers ("scores") such that $s_i \geq s_{i+1}$ for all $i \in \{1, \ldots, m-1\}$ and $s_1 > s_m$. Each voter's *k*th best candidate receives score s_k. A candidate's score is the sum of the scores that he or she receives from all voters.

For a preference profile **P**, a scoring rule selects the candidate or candidates that receive the highest score. A scoring rule is said to be *strict* if it is defined by a decreasing vector of scores, $s_i > s_{i+1}$, for all $i \in \{1, \ldots, m-1\}$.

Proposition 2.4 shows that all scoring-rule winners—whether they are Condorcet winners or not—are AV outcomes, but candidates that are selected by no scoring rule may also be AV outcomes.

Proposition 2.4. *At all preference profiles **P**, a candidate chosen by any scoring rule is an AV outcome. There exist preference profiles **P** at which a candidate is not chosen by any scoring rule but is, nevertheless, an AV outcome.*

Proof. To prove the first statement, take any preference profile **P** and any candidate i chosen by a scoring rule at **P**. Let $(s_1, \ldots, s_m)$ be the scoring-rule vector that results in the election of candidate i at **P**. By a normalization of the scores, we can without loss of generality assume that $s_1 = 1$ and $s_m = 0$.

Note that AV can be seen as a variant of a nonstrict scoring rule, whereby every voter gives a score of 1 to the candidates in his or her strategy set S (approved candidates) and a score of 0 to those not in this set. AV chooses the candidate or candidates with the highest score.[6]

Let $r_k(x)$ denote the number of voters who consider candidate x to be the kth best candidate at **P**. Because candidate i is picked by the scoring rule $(s_1, \ldots, s_m)$, it must be true that

$$s_1[r_1(i)] + s_2[r_2(i)] + \cdots + s_m[r_m(i)] \geq s_1[r_1(j)] + s_2[r_2(j)] + \cdots + s_m[r_m(j)] \quad (2.1)$$

for every other candidate j.

To show that the scoring-rule winner, candidate i, is an AV outcome, consider i's critical strategy profile $C_i(\mathbf{P})$. There are two cases:

Case 1. Voters rank candidate i last. Under a scoring rule, these voters give a score of 0 to candidate i, a score of 1 to their top choices, and scores between 0 and 1 to the remaining candidates. Under AV, these voters give a score of 0 to candidate i, a score of 1 to their top choices, and scores of 0 to the remaining candidates at $C_i(\mathbf{P})$.

Thus, candidate i does the same under the scoring rule as under AV (left side of inequality (2.1)), whereas all other candidates j do at least as well under the scoring rule as under AV (right side of inequality (2.1)). This makes the sum on the right side for the scoring rule at least as large as, and generally larger than, the sum of votes under AV, whereas the left side remains the same as under AV. Consequently, if inequality (2.1) is satisfied under the scoring rule, it is satisfied under AV at $C_i(\mathbf{P})$.

Case 2. Voters do not rank candidate i last. Under a scoring rule, these voters give candidate i a score of s_k if they rank him or her kth best. Under AV, these voters give a score of 1 to candidate i at $C_i(\mathbf{P})$. Thus, every s_k on the left side of inequality (2.1) is 1 for candidate i under AV, which makes the sum on the left side at least as large as, and generally larger than, the sum under a scoring rule. By comparison, the sum on the right side for all other candidates j under AV is less than or equal to the sum on the left side, with equality if and only if candidate j is preferred to candidate i by all voters.

---

[6] Of course, AV is not a scoring rule in the classical sense, whereby voters give scores to candidates according to the same predetermined vector. The restrictions on the vector that sincere strategies impose is that (i) the first component (score of the top candidate) be 1, (ii) the mth component (score of the bottom candidate) be 0, (iii) all components representing candidates at or above the lowest candidate a voter approves of are 1, and (iv) all components below the component representing this candidate are 0.

Consequently, if inequality (2.1) is satisfied under the scoring rule, it is satisfied under AV at $C_i(\mathbf{P})$.

Thus, in both cases 1 and 2, the satisfaction of inequality (2.1) under a scoring rule implies its satisfaction under AV at candidate i's critical strategy profile, $C_i(\mathbf{P})$. Hence, a candidate chosen under any scoring rule is also an AV outcome.

To prove the second statement, consider the following 7-voter, 3-candidate example (Fishburn and Brams, 1983, p. 211):

Example 2.4

   1. 3 voters: $a\,b\,c$
   2. 2 voters: $b\,c\,a$
   3. 1 voter: $b\,a\,c$
   4. 1 voter: $c\,a\,b$

Because candidate b receives at least as many first choices as a and c, and more first and second choices than either, every scoring rule will select b as the winner. But a is the Condorcet winner and, hence, an AV outcome by Proposition 2.2.[7] ∎

Note that candidate b in Example 2.4 is not AV-dominant: The number of voters who consider b as their best choice and a as their worst choice (2 voters), or b as their best choice and c as their worst choice (1 voter), does *not* exceed the number of voters who prefer a to b (4 voters) or c to b (1 voter) (see Lemma 2.4). Put another way, the critical strategy profile of candidate a, $C_a(\mathbf{P}) = (a, a, a, b, b, ba, ca)$, renders a the unique AV winner (5 votes), foreclosing the AV-dominance of candidate b (3 votes). Likewise, Condorcet winner a is also not AV-dominant, because the critical strategy profile of candidate b, $C_b(\mathbf{P}) = (ab, ab, ab, b, b, b, c)$, results in b's election (6 votes), foreclosing the AV-dominance of candidate a (3 votes).

Call a candidate *S-dominant* if he or she is the unique winner under every scoring rule, as candidate b is in Example 2.4.

Proposition 2.5. *At all preference profiles $\mathbf{P}$, an AV-dominant candidate is S-dominant, but not every S-dominant candidate is AV-dominant.*

Proof. Example 2.4 showed that not every S-dominant candidate is AV-dominant. To prove the first part of the proposition, note that a necessary and sufficient condition for candidate i to be S-dominant is that he or she receives at least as many first choices as every other candidate j, at least as many first and second choices as every other candidate $j, \ldots,$ and *more* first, second, $\ldots,$ and next-to-last choices as every other candidate j; otherwise, candidate i would not be assured of receiving more points than candidate j. But this condition, while

---

[7] Example 2.4 provides an illustration in which BC, in particular, fails to elect the Condorcet winner.

necessary, is not sufficient for a candidate to be AV-dominant (Lemma 2.4 gives a necessary and sufficient condition). ∎

In effect, Proposition 2.5 says that being AV-dominant is more demanding than being S-dominant. Whereas S-dominance counts choices at each distinct level (first, second, . . . , next-to-last) and requires that an S-dominant candidate never be behind at any level, and ahead at the next-to-last level, AV-dominance counts approval at different levels simultaneously (e.g., in the case of $C_a(\mathbf{P})$ in Example 2.4, the first level for some voters and the second level for other voters).

I next show the outcomes of two social-choice rules that are not scoring rules, the Hare system of single transferable vote (STV) and the majoritarian compromise (MC), are always AV outcomes (at their critical strategy profiles). But the converse is not true—AV outcomes need not be STV or MC outcomes.[8] Before proving this result, I illustrate STV and MC with a 9-voter, 3-candidate example:[9]

Example 2.5
  1. 4 voters: $a\ c\ b$
  2. 2 voters: $b\ c\ a$
  3. 3 voters: $c\ b\ a$

Under STV, the candidates with the fewest first-choice—and successively lower-choice—votes are eliminated; their votes are transferred to second-choice and lower-choice candidates in their preference rankings until one candidate receives a majority of votes. To illustrate in Example 2.5, because candidate b receives the fewest first-choice votes (2)—compared with 3 first-choice votes for candidate c and 4 first-choice votes for candidate a—b is eliminated and his or her 2 votes go to the second choice of the two type 2 voters, candidate c. In the runoff between candidates a and c, candidate c, now with votes from the type 2 voters, defeats candidate a 5 votes to 4, so c is the STV winner.

Under MC, first-choice, then second-choice, and then lower-choice votes are counted until at least one candidate receives a majority of votes; if more than one candidate receives a majority, the candidate with the most votes is elected.

---

[8] Ideally, of course, it would be desirable to prove this result for all voting systems, but we know of no general definition of a voting system that encompasses all those that have been used or proposed, in contrast to scoring systems and, as I will show later, Condorcet systems (Brams and Fishburn, 2002).

[9] These two voting systems, among others, are discussed in Brams and Fishburn (2002). MC, which is less well known than STV, was proposed independently as a voting procedure (Hurwicz and Sertel, 1999; Sertel and Yilmaz, 1999; Sertel and Sanver, 1999; Slinko, 2002) and as a bargaining procedure under the rubric of "fallback bargaining" (Brams and Kilgour, 2001b). As a voting procedure, the threshold for winning is assumed to be simple majority, whereas as a bargaining procedure the threshold is assumed to be unanimity, but qualified majorities are also possible under either interpretation.

Because no candidate in Example 2.5 receives a majority of votes when only first choices are counted, second choices are next counted and added to the first choices. Candidate c now receives the support of all 9 voters, whereas a and b receive 4 and 5 votes, respectively, so c is the MC winner.

Proposition 2.6. *At all preference profiles **P**, a candidate chosen by STV or MC is an AV outcome. There exist preference profiles **P** at which a candidate chosen by AV is neither an STV nor an MC outcome.*

Proof. To show that every STV outcome is an AV outcome, suppose candidate i is not an AV outcome at preference profile **P**. By Lemma 2.3, there exists a candidate j such that the number of voters who rank j as their best candidate and i as their worst candidate exceeds the number of voters who prefer i to j. A fortiori, the number of voters who consider j as their best candidate exceeds those who consider i as their best candidate.

This result holds for any subset of candidates that includes both i and j. Hence, STV will never eliminate j in the presence of i, showing that i cannot be an STV winner.

Neither can i be an MC winner, because j will receive more first-place votes than i. If this number is not a majority, the descent to second and still lower choices continues until at least one candidate receives a majority. Between i and j, the first candidate to receive a majority will be j, because j receives more votes from voters who rank him or her first than there are voters who prefer i to j. Thus, j will always stay ahead of i as the descent to lower and lower choices continues until j receives a majority.

To show that AV outcomes need not be STV or MC outcomes, consider Example 2.4, in which the Condorcet winner, candidate c, is chosen under both STV and MC. Besides c, AV may also choose candidate a or candidate b: a is an AV outcome at critical strategy profile $C_a(\mathbf{P}) = (a, a, a, a, b, b, c, c, c)$; and b is an AV outcome at critical strategy profile $C_b(\mathbf{P}) = (a, a, a, a, b, b, cb, cb, cb)$. ■

So far I have shown that AV yields at least as many, and generally more, (Pareto) outcomes than any scoring rule and two nonscoring voting systems. To be sure, one might question whether the three possible AV outcomes in Example 2.5 have an equal claim to being *the* social choice. Isn't candidate c, the Condorcet winner, BC winner, STV winner, and MC winner—and ranked last by no voters—the best overall choice? By comparison, candidate b is only a middling choice; and candidate a, who is the plurality voting (PV) winner, is the Condorcet loser.[10]

---

[10] Note that PV is a degenerate scoring rule, under which a voter's top candidate receives 1 point and all other candidates receive 0 points. By Proposition 2.4, sincere outcomes under PV are always AV outcomes, but not vice versa. As a case in point, candidate a is the sincere PV outcome in Example 2.5, whereas candidates b and c are also sincere AV outcomes.

Just as AV allows for a multiplicity of outcomes, it also enables voters to prevent them.

Proposition 2.7. *At every preference profile **P** at which there is not an AV-dominant candidate, AV can prevent the election of every candidate, whereas scoring rules, STV, and MC cannot prevent the election of all of them.*

Proof. In the absence of an AV-dominant candidate, there is no candidate who can be assured of winning, which implies that every candidate can be prevented from winning. To show that scoring rules, STV, and MC cannot prevent the election of all candidates when AV can, consider the following 3-voter, 3-candidate example:

Example 2.6
1. 1 voter: *a b c*
2. 1 voter: *b a c*
3. 1 voter: *c b a*

It is easy to see that there is no candidate that is AV-dominant in Example 2.6, based on Lemma 2.4. But to make perspicuous how AV can prevent the election of every candidate in Example 2.6—and why the other systems cannot—let "|" indicate each voter's dividing line between the candidate(s) he or she considers acceptable and those he or she considers unacceptable. If the three voters draw their lines as follows,

$$a \mid b\,c \qquad b\,a \mid c \qquad c \mid b\,a,$$

b and *c* will *not* be chosen (*a* will be). If the voters draw their lines as follows,

$$a \mid b\,c \qquad b \mid a\,c \qquad c\,b \mid a,$$

a and *c* will *not* be chosen (*b* will be). Thus, AV can prevent the election of every one of the three candidates, because none is AV-dominant.

By contrast, the Condorcet winner, *b*, wins under every scoring system, including BC, and also under MC. Under STV, either *a* or *b* may win, depending on which of the three candidates is eliminated first. Hence, only *c* is prevented from winning under these other systems, showing that AV is unique in being able to prevent the election of each of the three candidates. ∎

I have shown that AV allows for outcomes that BC, MC, and STV do not (e.g., *c* in Example 2.6 when there is a three-way tie). At the same time, it may preclude outcomes (e.g., *b* in Example 2.6) that other systems cannot prohibit. In effect, voters can fine-tune their strategies under AV, making outcomes responsive to information that transcends their preferences.

It is useful to know not only what outcomes can and cannot occur under AV but also what outcomes are likely to persist because of their stability. Although non-Pareto candidates cannot win a clear-cut victory under AV (Proposition 2.1), might it be possible for Condorcet losers to be AV outcomes *and* stable? I address this and related questions after defining two types of stability in the next section.

2.4. STABILITY OF ELECTION OUTCOMES

As earlier, assume that voters choose sincere strategies under AV. Now, however, suppose that they may not draw the line between acceptable and unacceptable candidates as they would if they were truthful. Instead, they may vote strategically in order to try to obtain a preferred outcome.

To determine what is "preferred," preference needs to be extended to sets. If a voter's preference is *a b*, I assume he or she prefers *a* to $\{a, b\}$, and $\{a, b\}$ to *b*. In assessing the stability of outcomes, this assumption seems very reasonable.

It is useful to distinguish two kinds of stability, the first of which is the following: Given a preference profile **P**, an AV outcome is *stable* if there exists a strategy profile **S** such that no voters of a single type have an incentive to switch their strategy to another sincere strategy in order to induce a preferred outcome.[11] In analyzing the stability of AV outcomes that do not involve ties,[12] we need confine our attention only to those outcomes stable at $C_i(\mathbf{P})$ because of the following proposition:

Proposition 2.8. *A nontied AV outcome i is stable if and only if it is stable at its critical strategy profile, $C_i(\mathbf{P})$.*

Proof. The "if" part follows from the existence of a strategy profile, $C_i(\mathbf{P})$, at which outcome *i* is stable. To show the "only if" part, assume candidate *i* is unstable at $C_i(\mathbf{P})$. At any other strategy profile **S′**, candidate *i* receives no more approval votes and generally fewer than at $C_i(\mathbf{P})$ by Lemma 2.1. Hence, those voters who switch to different sincere strategies to induce the election of a preferred candidate at **S** can also do so at **S′**. ∎

---

[11] Treating voters of one type, all of whose members have the same preference, as single (weighted) voters provides the most stringent test of stability. This is because any outcome that can be destabilized by the switch of individual voters (of one type) can be destabilized by the switch of all voters of that type, but the converse is not true: outcomes may be stable when some but not all voters of one type switch. This definition of stability precludes outcomes of the latter kind from being stable.

[12] To illustrate how ties may complicate matters, assume three voters have preferences *a b c*, *b c a*, and *c a b* and vote only for their first choices, which is not a critical strategy for any of them. Then the resulting tied outcome, $\{a, b, c\}$, will be stable if no voter prefers just its second choice to the tie. As this example illustrates, the stability of tied outcomes may depend on comparisons between singleton and nonsingleton subsets; to avoid this comparison, I assume nontied AV outcomes in several of the subsequent propositions.

The strategies of voters associated with a stable AV outcome at $C_i(\mathbf{P})$ define a Nash equilibrium of a voting game in which the voters have complete information about each other's preferences and make simultaneous choices.[13]

Neither candidate a nor candidate b is a stable AV outcome in Example 2.5. At critical strategy profile $C_a(\mathbf{P}) = (a, a, a, a, b, b, c, c, c)$ that renders candidate a an AV outcome, if the two type 2 voters switch to strategy bc, candidate c, whom the type 2 voters prefer to candidate a, wins. At critical strategy profile $C_b(\mathbf{P}) = (a, a, a, a, b, b, cb, cb, cb)$ that renders candidate b an AV outcome, the four type 1 voters have an incentive to switch to strategy ac to induce the selection of candidate c, whom they prefer to candidate b.

Although AV outcomes a and b in Example 2.5 are not stable at their critical strategy profiles, AV outcome c most definitely is stable at its critical strategy profile, $C_c(\mathbf{P}) = (ac, ac, ac, ac, bc, bc, c, c, c)$. No switch on the part of the four type 1 voters to a, of the two type 2 voters to b, or of the three type 3 voters to cb can lead to a preferred outcome for any of these types—or, indeed, change the outcome at all (because candidate c is the unanimous choice of all voters at c's critical strategy profile).

Not only can no single switch by any of the three types induce a preferred outcome for the switchers at $C_c(\mathbf{P})$, but no *coordinated* switches by two or more types can induce a preferred outcome. Thus, for example, if the ac-voters switched from ac to a, and the bc-voters switched from bc to b, they together could induce AV outcome a, which the four type 1 voters would clearly prefer to outcome c. But a is the worst choice of the two type 2 voters, so they would have no incentive to coordinate with the type 1 voters to induce this outcome.

That AV outcome c is, at the critical strategy profile of candidate c, invulnerable to coordinated switches leads to our second type of stability: given a preference profile $\mathbf{P}$, an outcome is *strongly stable* if there exists a strategy profile $\mathbf{S}$ such that no types of voters, coordinating their actions, can form a coalition K, all of whose members would have an incentive to switch their AV strategies to other sincere strategies in order to induce a preferred outcome.

I assume that the coordinating players in K are allowed to communicate to try to find a set of strategies to induce a preferred outcome *for all of them*. These strategies define a *strong* Nash equilibrium of a voting game in which voters have complete information about each other's preferences and make simultaneous choices.

Proposition 2.9. *A nontied AV outcome i is strongly stable if and only if it is strongly stable at its critical strategy profile, $C_i(\mathbf{P})$.*

---

[13] For an analysis of Nash equilibria in voting games under different rules and information conditions from those given here, see Myerson (2002) and references cited therein. Possible outcomes that different refinements of Nash equilibria may produce are described in De Sinopoli, Dutta, and Laslier (2006).

Proof. Analogous to that of Proposition 2.8. ∎

An AV stable outcome need not be strongly stable. To illustrate this weaker form of stability, consider AV outcome a in Example 2.1 and its critical strategy profile, $C_a(\mathbf{P}) = (a, a, a, bca, bca, d, d)$. The two type 2 voters cannot upset this outcome by switching from bca to bc or b, nor can the two type 3 voters upset it by switching from d to db or dbc. However, if these two types of voters cooperate and form a coalition K, with the two type 2 voters choosing strategy b and the two type 3 voters choosing strategy db, they can induce the selection of Condorcet winner b, whom both types prefer to candidate a. At critical strategy profile $C_a(\mathbf{P})$, therefore, AV outcome a is stable but not strongly stable, whereas AV outcome b is strongly stable at its critical strategy profile, $C_b(\mathbf{P}) = (ab, ab, ab, b, b, db, db)$.

If an AV outcome is neither strongly stable nor stable, it is *unstable*.

Proposition 2.10. *There may be no strongly stable or stable nontied AV outcomes—that is, every nontied AV outcome may be unstable.*

Proof. Consider the following 3-voter, 3-candidate example:[14]

Example 2.7
  1. 1 voter: *a b c*
  2. 1 voter: *b c a*
  3. 1 voter: *c a b*

The critical strategy profile that elects candidate a is $C_a(\mathbf{P}) = (a, b, ca)$. If voter 2 switches to bc, he or she can induce preferred outcome $\{a, c\}$. In a similar manner, it is possible to show that neither candidate b nor candidate c is a stable AV outcome. ∎

I next characterize strongly stable outcomes.

Proposition 2.11. *A nontied AV outcome is strongly stable if and only if it is a unique Condorcet winner.*

Proof. To prove the "if" part, suppose candidate i is a unique Condorcet winner at $\mathbf{P}$. We will show that i is a nontied AV outcome that is strongly stable at its critical strategy profile, $C_i(\mathbf{P})$. Clearly, i is a nontied AV outcome at $C_i(\mathbf{P})$ by Proposition 2.2. To show its strong stability, suppose there exists a coalition of voters K, comprising one or more types, that prefers some other candidate j to candidate i and coordinates to induce the selection of j. Because candidate i is a unique Condorcet winner, however, the cardinality of K is strictly less than the cardinality of coalition L, whose members prefer i to j. The members of L vote for i but not for j at $C_i(\mathbf{P})$. Hence, whatever sincere strategy switch

---

[14] Example 2.7 is the standard example of the Condorcet paradox, or cyclical majorities, in which there is no Condorcet winner.

the members of K consider at candidate i's critical strategy profile to induce the election of candidate j, j will receive fewer votes than i, proving that i is a strongly stable AV outcome.

To prove the "only if" part, suppose that candidate i is not a unique Condorcet winner. Consequently, there exists a candidate j and a majority coalition of voters K, comprising one or more types, that prefers j to i. I now show that i is not a strongly stable AV outcome at its critical strategy profile, $C_i(\mathbf{P})$, which by Proposition 2.8 shows that i is not a strongly stable AV outcome. Suppose AV does not elect i at $C_i(\mathbf{P})$. Then i is not an AV outcome and hence not a strongly stable one. Now suppose that AV elects i at $C_i(\mathbf{P})$. Because the members of K can change their strategies to elect j, whom they prefer to i, i is not a strongly stable AV outcome. ∎

That Proposition 2.11 does not apply to nonunique Condorcet winners is illustrated by the following example: There are two voters, 1 and 2, and two candidates, i and j; voter 1 ranks the candidates $i\,j$, and voter 2 ranks them $j\,i$. Then (i, j) is the critical strategy profile for both candidates. Obviously, neither one nor both voters can induce an outcome each prefers to the tied outcome $\{i, j\}$, so this outcome is strongly stable without there being a unique Condorcet winner.[15]

I next show that Condorcet losers as well as winners may be stable AV outcomes.

Proposition 2.12. *A unique Condorcet loser may be a stable AV outcome, even when there is a different outcome that is a unique Condorcet winner (and therefore strongly stable).*

Proof. Consider the following 7-voter, 5-candidate example:

Example 2.8
1. 3 voters: $a\,b\,c\,d\,e$
2. 1 voter: $b\,c\,d\,e\,a$
3. 1 voter: $c\,d\,e\,b\,a$
4. 1 voter: $d\,e\,b\,c\,a$
5. 1 voter: $e\,b\,c\,d\,a$

Candidate a is the Condorcet loser, ranked last by 4 of the 7 voters. But at its critical strategy profile, $C_a(\mathbf{P}) = (a, a, a, b, c, d, e)$, candidate a is a stable AV outcome, because none of the four individual voters, by changing his or her strategy, can upset a, who will continue to receive 3 votes.

Consider the critical strategy profile of candidate b, $C_b(\mathbf{P}) = (ab, ab, ab, b, cdeb, deb, eb)$, who receives 7 votes, compared with 3 votes each for a and e,

---

[15] Under a somewhat weaker definition of strong stability, the equivalence of strong Nash-equilibrium outcomes and Condorcet winners is shown for a large class of voting rules in Sertel and Sanver (2004).

2 votes for d, and 1 vote for c. Again, no single type of voter can upset this outcome, nor can any coalition, because candidate b is the unique Condorcet winner, making him or her strongly stable. ■

Whether a Condorcet loser, like candidate a in Example 2.8, "deserves" to be an AV winner—and a stable one at that—depends on whether voters have sufficient incentive to unite in support of a candidate like Condorcet winner b, who is the first choice of only one voter. If they do not rally around b, and the type 1 voters vote only for a, then a is arguably the more acceptable choice.

A *Condorcet voting system* is one that always elects a Condorcet winner, if one exists, when voters are sincere—that is, when they rank candidates according to their preferences. (Note that this use of "sincerity" is different from that of AV, which, as I noted earlier, allows for multiple sincere strategies.) A Condorcet winner, however, may not be elected as a Nash-equilibrium outcome under a Condorcet voting system, much less as a strong Nash-equilibrium outcome (as under AV).

Proposition 2.13. *No Condorcet voting system ensures the election of a unique Condorcet winner as a Nash-equilibrium outcome.*

Proof. Consider the following example, in which there is no Condorcet winner:

Example 2.9
1. 2 voters: $a\,d\,b\,c$
2. 2 voters: $b\,d\,c\,a$
3. 1 voter: $c\,a\,b\,d$

In the absence of a Condorcet winner, assume that different candidates may be chosen by a Condorcet voting system.

I first show that by changing the preference ranking of each of the three voter types, one at a time, in Example 2.9, different candidates can be rendered Condorcet winners. However, if a Condorcet voting system chooses a candidate or candidates preferred by this voter type in Example 2.9, then the Condorcet winner in the modified example is not a Nash-equilibrium outcome. Because it can be upset by some voter type changing its ranking to that in Example 2.9, this precludes the possibility of a Condorcet voting system's choosing a, b, c, or d in this example.[16]

To begin, assume that the preference ranking of the two type 2 voters in Example 2.9 is $b\,d\,a\,c$, but the other two types have the same preferences as shown in Example 2.9. Then candidate a is the Condorcet winner. If a Condorcet voting

---

[16] These choices do not preclude the possibility of the Condorcet voting system's choosing most subsets of candidates in Example 2.9, such as $\{b, c\}$, which may or may not be preferred to a Condorcet winner. Thus, Proposition 2.9 is applicable to social-choice functions, which are "resolute" social-choice rules that choose single candidates, and not to social-choice correspondences, which may choose nonsingleton subsets of candidates.

system would choose either candidate *b* or *d* in Example 2.9, then it would be in the interest of the type 2 voters to switch to *b d c a*, as given in Example 2.9, to obtain a preferred outcome.

Assume the preference ranking of the type 3 voter is *c d a b*. Then candidate *d* is the Condorcet winner. If a Condorcet voting system would choose candidate *c* in Example 2.9, then it would be in the interest of the type 3 voter to switch to *c a b d*, as given in Example 2.9, to obtain a preferred outcome.

Finally, assume the preference ranking of the two type 1 voters is *a c d b*. Then candidate *c* is the Condorcet winner. If a Condorcet voting system would choose candidate *c* in Example 2.9, then it would be in the interest of the type 1 voters to switch to *a d b c*, as given in Example 2.9, to obtain a preferred outcome.

In summary, I have shown that three of the four candidates in Example 2.9 can be rendered Condorcet winners by changing the preference ranking of one voter type. If this is the true ranking of these voter(s), it is always in their interest to misrepresent their preferences to those shown in Example 2.9, given that a Condorcet voting system chooses the preferred candidates postulated in each of the above cases. But to prevent this from happening in all the cases, one must preclude the possibility of a Condorcet voting system's choosing all four candidates in Example 2.9.

This undermines the Nash-equilibrium status of the unique Condorcet winners in the modifications of Example 2.9. Therefore, these Condorcet winners cannot be ensured of election as Nash-equilibrium outcomes. ∎

Proposition 2.13 casts doubt on the efficacy of Condorcet voting systems, such as those of Black or Copeland (Brams and Fishburn, 2002), to do what they purport to do *in equilibrium*. By contrast, AV always ensures that unique Condorcet winners can be elected as strong Nash-equilibrium outcomes.

2.5. SUMMARY AND CONCLUSIONS

AV outcomes subsume all outcomes that other voting systems I examined choose, including scoring rules like the Borda count (BC), single transferable vote (STV), and the majority compromise (MC). In addition, they include outcomes that none of these other systems selects, rendering AV outcomes more inclusive.

AV outcomes, however, are not an indiscriminate set. Thus, they always include Pareto candidates and never include non-Pareto candidates as unique winners. But *which* outcome is chosen critically depends on where voters draw the line between acceptable and unacceptable candidates.

Preference-based systems are not responsive to this information and so limit the field of candidates that can win. Despite the bigger menu that AV allows, however, voters are better able to prevent the election of candidates under AV than under other voting systems.

Only under AV are strongly stable outcomes always Condorcet winners when voters choose sincere strategies. By contrast, Condorcet voting systems that purport always to elect Condorcet winners may fail to do so when voters are strategic.

AV allows for other stable outcomes, though not strongly stable ones, such as BC winners and even Condorcet losers. Indeed, we see nothing wrong in such candidates winning if they are the most approved by voters—especially if "majority tyranny" is a concern (Baharad and Nitzan, 2002)—though AV almost always elects Condorcet winners when they exist (Regenwetter, Grofman, Marley, and Tsetlin, 2006).

Beyond the desirability as well as the practicality of AV, it is worth noting that basing social-choice theory on acceptability rather than on preferences is a radical departure from the research program initiated by Borda and Condorcet in late eighteenth-century France (McLean and Urken, 1995). While I do not eschew the election of Borda or Condorcet winners, they should not be the be-all and end-all for judging whether outcomes are acceptable or not.[17] Rather, the pragmatic judgments of sovereign voters about who is acceptable and who is not should be important if not decisive.

This is information that enriches the standard social-choice framework and should, therefore, be incorporated into it.[18] Indeed, in chapter 3 I will consider how both approval and preferences may be incorporated into two different voting systems.

APPENDIX

In this appendix, I give examples that illustrate some difficulties that afflict the Borda count (BC) and the Hare system of single transferable vote (STV)—the two best-known ranking systems—described in section 2.5.

Example A2.1
1. 3 voters: $a\ b\ c$
2. 3 voters: $b\ a\ c$
3. 1 voter: $c\ a\ b$

Under BC, if each class of voters votes sincerely, a will win with 10 points to 9 points for b and 2 points for c. But if the three type 2 voters vote insincerely

---

[17] When they differ, the debate continues on which is the better choice; see, for example, Risse (2005) and Saari (2006). AV, in effect, offers a different criterion for making a choice, which may be neither the Borda nor the Condorcet winner but, instead, a different candidate who is more approved. Whomever AV selects has a strong claim to being the *voters'* choice, though I revisit this issue in chapter 3.

[18] What also enriches the framework is not assuming a fixed set of candidates but, instead, making the field of candidates endogenous, dependent on the voting system, and asking what equilibria are induced (Dellis and Oak, 2006).

by ranking the candidates $b\ c\ a$—dropping a into last place—b wins with 9 points to 7 points for a and 5 points for c.

Such a manipulative strategy would not be hard for voters to determine. The b-voters, knowing (perhaps from a poll) that candidate a is b's closest competitor, can do no better than put this competitor into last place. Curiously, if the a-voters used the same tactic, insincerely ranking the candidates $a\ c\ b$, then a would get 7 points, b would get 6 points—but c would get 8 points and win!

The vulnerability of BC to strategic voting takes a different form in its dependence on the field of candidates (Brams and Fishburn, 1991).

Example A2.2
1. 3 voters: $a\ b\ c$
2. 2 voters: $b\ c\ a$
3. 2 voters: $c\ a\ b$

Under BC, the social ranking of the candidates is

$$a\ (8\ \text{points}) > b\ (7\ \text{points}) > c\ (6\ \text{points}).$$

Now suppose an "irrelevant" candidate x enters; each class of voters, without changing their rankings of candidates a, b, and c, ranks x differently:

Example A2.3
1. 3 voters: $a\ b\ c\ x$
2. 2 voters: $b\ c\ x\ a$
3. 2 voters: $c\ x\ a\ b$

The new social ranking becomes

$$c\ (13\ \text{points}) > b\ (12\ \text{points}) > a\ (11\ \text{points}) > x\ (6\ \text{points}).$$

Although x is indeed irrelevant, coming in last in the field of four, x's presence is not irrelevant to the outcome. In fact, x reverses the social ranking of the three relevant candidates from $a > b > c$ to $c > b > a$.

Clearly, BC is not only vulnerable to strategic voting, but it is also sensitive to the field of candidates. Moreover, if it is not candidates but, say, bills that are being voted on, or propositions in a referendum, then the latter vulnerability shows how, through manipulation of the agenda (Nicholson, 2005), a different winner can be made to win.

The last example illustrates the nonmonotonicity of STV (Brams, 1982a; Brams and Fishburn, 1991).

Example A2.4
1. 7 voters: $a\ b\ c\ d$
2. 6 voters: $b\ a\ c\ d$
3. 5 voters: $c\ b\ a\ d$
4. 3 voters: $d\ c\ b\ a$

Because no candidate has a simple majority of 11 first-place votes, the candidate with the fewest first-choice votes, d, is eliminated first. Then d's 3 votes go to c, giving c 8 votes. Because none of the remaining candidates has a majority at this point, b, with the new lowest total of 6 votes, is eliminated next, and b's second-place votes go to a, who is elected with a total of 13 votes.

Now assume the three type 4 voters raise a from fourth to first place in their rankings without changing their rankings of the other three candidates. Now a has a total of 10 first-place votes, which is not a majority. Hence, the candidate with the fewest first-place votes, c, is eliminated, and his or her 5 votes are given to b, who wins with a total of 11 votes.

This is indeed perverse: a loses when he or she moves up in the rankings of the three type 4 voters from fourth to first place and thereby receives three more first-place votes. Equally strange, candidates may be helped under STV if voters do not show up to vote for them at all, which has been called the "no-show paradox" (Fishburn and Brams, 1983; for later references to this and related monotonicity paradoxes, see Nurmi, 1999, 2002; and Brams and Fishburn, 2002).

The fact that more first-place voters, or even no votes, can hurt rather than help violates what, in my opinion, is a fundamental democratic ethic. Causing a favorite candidate to lose by moving him or her from last to first place, as illustrated by Example A2.4, is probably the most extreme form of this phenomenon.

3

Electing a Single Winner:
Combining Approval and Preference

3.1. INTRODUCTION

As compelling as approval voting (AV) is in theory, and as well as it has worked in practice, other single-winner voting systems may be preferable under certain circumstances. For example, if voters are willing and able to rate candidates on a 3-point scale (good, medium, bad), this system could help voters distinguish acceptable candidates who are good from those who are just medium. In fact, such a refinement of AV has been proposed by Felsenthal (1989), Yilmaz (1999), and Hillinger (2005).

Of course, one could ask voters to make still finer distinctions—say, on a 5-point rating scale—or give them an even greater range of choices (García-LaPresta and Martínez-Panero, 2002; Center for Range Voting, 2007). But are voters sufficiently informed to make such refined judgments? This is not evident. Even if they are, some may have an incentive to misrepresent their sincere choices by rating a favorite candidate far above his or her competitors in order to give that candidate as big a boost as possible. This makes such a system highly vulnerable to strategic voting, though Balinski and Laraki (2007) suggest ways to mitigate this problem in the rating system they propose.

In this chapter, I analyze voting systems that give voters the opportunity to make more refined judgments. Voters are allowed both to rank candidates and, as under AV, to draw the line between those they consider acceptable and those they consider unacceptable.

Rankings and approval, though related, are fundamentally different kinds of information. They cannot necessarily be derived from one another. Both kinds of information are important in the determination of social choices. I propose a way of combining them in two hybrid voting systems, *preference approval voting* (PAV) and *fallback voting* (FV), that have several desirable properties.

Approving of a subset of candidates is generally not difficult, whereas ranking all candidates on a ballot, especially if the list is long, may be arduous.

Note: This chapter is adapted from Brams and Sanver (2008) with permission of Springer Science and Business Media.

PAV asks for both kinds of information, whereas FV asks voters to rank only those candidates they approve of, making it simpler than systems that elicit complete rankings.

I describe and analyze these systems in tandem, starting with definitions and assumptions in section 3.2. In section 3.3 I describe PAV and then analyze which candidates can and cannot win under this system. Although a PAV winner may not be a Condorcet winner or an AV winner, PAV satisfies what I call the *strongest-majority principle for voters*. More specifically, if a majority-approved candidate is preferred by a majority to the AV winner and other majority-approved candidates, PAV "corrects" the AV result by electing this candidate, who is the most preferred of the majority-approved candidates.

A majority-preferred candidate is likely to have a more coherent point of view than a different AV winner, who may be the most popular candidate because he or she is bland or inoffensive—a kind of lowest common denominator who tries to appease everybody. (This problem does not seem to be a common one, as shown by the AV results in section 1.6.) Sometimes *not* choosing such a candidate when two or more candidates receive majority approval makes PAV *coherence-inducing for candidates* by giving an advantage to candidates who are principled but, nevertheless, command broad support.

In section 3.4 I describe FV and compare its properties with those of PAV. Like PAV, FV tends to help those candidates who are relatively highly ranked by a majority of voters. Both systems may give different winners from non-ranking systems like PV and AV, ranking systems like the Borda count and STV, and each other.

In section 3.5 I show that PAV and FV are monotonic in two different senses: voters, by either approving of a candidate or raising him or her in their rankings, can never hurt and may help this candidate get elected. The latter property (rank-monotonicity) is not satisfied by a number of ranking systems, including STV, whereas the former property (approval-monotonicity) is satisfied by AV.

Like practically all voting systems, PAV and FV are manipulable. In section 3.6 I show that voters may induce preferred outcomes either by contracting or expanding their approval sets. Because each voting system may give outcomes in equilibrium when the other does not, neither system is inherently more stable than the other.

In section 3.7 I develop a dynamic model of voter responses to polls in 3-candidate elections, wherein voter preferences are either single-peaked or cyclic. If voters respond to successive polls by adjusting their approval strategies to try to prevent their worst choices from winning, they elect the Condorcet winner, though not necessarily in equilibrium, if their preferences are single-peaked. If their preferences are cyclical, the candidate ranked first or second by the most voters wins after voters respond to several polls. These outcomes are in equilibrium under both PAV and FV.

I conclude in section 3.8 that PAV, and to a less extent FV, subtly interweave two different kinds of information: (i) *approval information* determines those candidates who are above the bar—that is, sufficiently popular to be serious contenders if not outright winners; (ii) *ranking information* enables voters to refine the set of potential winners if more than one candidate receives majority approval.

Together, these two kinds of information facilitate the election of majoritarian candidates who espouse coherent positions. But more than abetting their election, PAV and FV may well have a salutary impact on which candidates choose to run—and how they choose to campaign—encouraging the entry of candidates who appeal to a broad segment of the electorate but do not promise them the moon.

3.2. DEFINITIONS AND ASSUMPTIONS

Consider a set of voters choosing among a set of candidates. As in chapter 1, denote individual candidates by small letters $a, b, c, \ldots$.

Assume that voters strictly rank the candidates from best to worst, so there is no indifference. Thus, for any candidates a and b, either a is preferred to b or b is preferred to a. This assumption simplifies the subsequent analysis but does not in any significant way affect the results, which can readily be extended to the case of nonstrict preferences.

Assume that rankings are transitive, so that for any candidates a, b, and c, a is preferred to c whenever a is preferred to b and b is preferred to c. In addition, assume that a voter evaluates each candidate as either acceptable or unacceptable, as under AV.

The *preference approval* of voters is based on both their rankings and their approval of candidates. Although different, these two types of information exhibit the following consistency: Given two candidates a and b, if a is approved and b is disapproved, then a is ranked above b.

A voter's preference approval is represented by an ordering of candidates from left to right and a vertical bar, to the left of which candidates are approved and to the right of which candidates are disapproved. For example,

$$a \, b \mid c \, d$$

indicates that the voter's two top-ranked candidates, a and b, are approved, and the voter's two bottom-ranked candidates, c and d, are disapproved.

At one extreme, a voter may approve of all candidates; and at the other extreme, of no candidates. While these extreme strategies are dominated strategies in a voting game in which voters have strict preferences—and, therefore, are unlikely to be used—they are not illegal, as such, under PAV or FV.

Some voters will approve of a single favorite candidate, and some will approve of all except a worst choice. Many voters, however, are likely to select some middle ground, approving of two or three candidates in, say, a field of five.

A *preference-approval profile* is a list of preference-approvals of all voters. A *social-choice rule* aggregates preference-approval profiles into social choices, thereby generalizing the standard social-choice model—wherein a voter is characterized simply by his or her ranking of candidates—to one that adds a line in the ranking separating a voter's approvals from disapprovals.

In subsequent sections, I present a number of examples to illustrate results as well as prove some propositions. For simplicity, I assume in the examples that all voters of each type draw the line separating approvals and disapprovals at the same point in their rankings, but none of the results depends on this assumption.

To describe PAV in the next section, define a *voting cycle* among three or more candidates $a, b, c, \ldots$ to occur if $a > b > c > \cdots > a$, where ">" indicates "is preferred by a majority to"; it is a *top cycle* if there are no majority-approved candidates who are preferred to voters in the cycle. (Notice that there never can be an "approval cycle"—approval is strictly ordered from candidates with the most approval to candidates with the least approval.) The majority-preference relation between any two candidates may lead to a tie if and only if there is an even number of voters, which I assume is broken by random tie breaking.

3.3. PREFERENCE APPROVAL VOTING (PAV)

The winner under PAV is determined by two rules, the second comprising two cases:

1. If no candidate, or exactly one candidate, receives a majority of approval votes, the PAV winner is the AV winner—that is, the candidate who receives the most approval votes.
2. If two or more candidates receive a majority of approval votes, then
   (i) if one of these candidates is preferred by a majority to every other majority-approved candidate, he or she is the PAV winner—even if not the AV or Condorcet winner among all candidates;
   (ii) if there is not one majority-preferred candidate because of a top cycle among the majority-approved candidates, the AV winner in the top cycle is the PAV winner—even if not the AV or Condorcet winner among all candidates.

It is rule 2 that distinguishes PAV from AV. It allows for the election of candidates who are not the most approved and, therefore, not AV winners. Indeed, a PAV winner may be the least-approved candidate in a race, as I will show.

Compared with preference-based voting systems, PAV is somewhat more demanding in the information it requires of voters. Besides ranking candidates, voters must indicate where they draw the line between acceptable and unacceptable candidates, which is an issue to which I will return when comparing the complexity of PAV and FV.

In the remainder of this section, I show what kinds of candidates PAV may and may not elect.

Proposition 3.1. *A Condorcet winner may not be a PAV winner under rule 1, rule 2(i), and rule 2(ii).*

Proof. Rule 1. Consider the following 3-voter, 3-candidate example, in which the voters divide into three preference types:

Example 3.1
1. 1 voter: *a b | c*
2. 1 voter: *b | a c*
3. 1 voter: *c | a b*

Candidate *b* is the AV winner, approved of by 2 of the 3 voters, whereas candidates *a* and *c* are approved of by 1 voter each. Because candidate *b* is the only candidate approved of by a majority, *b* is the PAV winner under rule 1. But it is candidate *a*, who is preferred to candidates *b* and *c* by majorities of 2 votes to 1, who is the Condorcet winner.

Rule 2(i). Consider the following 3-voter, 4-candidate example:

Example 3.2
1. 1 voter: *a b c | d*
2. 1 voter: *b c | a d*
3. 1 voter: *d | a c b*

Candidates *b* and *c* tie for AV winner with majorities of 2 votes each. Because candidate *b* is preferred to candidate *c* by 2 votes to 1, *b* is the PAV winner under rule 2(i). But it is candidate *a*, who is preferred to candidates *b*, *c*, and *d* by majorities of 2 votes to 1 (but who is not majority-approved), who is the Condorcet winner.

Rule 2(ii). Consider the following 5-voter, 5-candidate example:

Example 3.3
1. 1 voter: *d a b c | e*
2. 1 voter: *d b c a | e*
3. 1 voter: *e | d c a b*
4. 1 voter: *a b c | d e*
5. 1 voter: *c | b a d e*

Candidates *a* (3 votes), *b* (3 votes), and *c* (4 votes) are all majority approved and in a cycle as well: $a > b > c > a$. Because the Condorcet winner, candidate d (2 votes), is not majority-approved, he or she cannot be the PAV winner. Instead, the most approved candidate in the cycle, c, is the PAV winner. ∎

Not only may PAV fail to elect Condorcet winners when they exist, but it may also fail to elect unanimously approved candidates.

Proposition 3.2. *A unanimously approved AV winner may not be a PAV winner under either rule 2(i) or rule 2(ii).*

Proof. Rule 2(i). Consider the following 3-voter, 3-candidate example:

Example 3.4
1. 2 voters: $a\ b\ |\ c$
2. 1 voter: $b\ c\ |\ a$

Candidate b is approved of by all 3 voters, whereas candidate a is approved of by 2 voters and candidate c by 1 voter. Nevertheless, candidate a is the PAV winner, because under rule 2(i) he or she is preferred by 2 votes to 1 to the other majority-approved candidate, b.

Rule 2(ii). Consider the following 8-voter, 4-candidate example:

Example 3.5
1. 3 voters: $a\ b\ c\ |\ d$
2. 3 voters: $d\ a\ c\ |\ b$
3. 2 voters: $b\ d\ c\ |\ a$

Candidate c is approved of by all 8 voters, whereas candidates a, b, and d are approved of by majorities of either 5 or 6 voters. The latter three candidates are in a top cycle in which $a > b > d > a$; all are preferred by majorities to candidate c, the AV winner. But because candidate a receives more approvals (6) than candidates b and d (5 each), candidate a is the PAV winner under rule 2(ii). ∎

Proposition 3.2 shows how a unanimously approved AV winner may be displaced by a less approved majority winner under PAV. In fact, the conflict between AV and PAV winners may be even more extreme.

Proposition 3.3. *A least-approved candidate may be a PAV winner under rule 2(i).*

Proof. Consider the following 7-voter, 4-candidate example:

Example 3.6
1. 2 voters: $a\ c\ b\ |\ d$
2. 2 voters: $a\ c\ d\ |\ b$
3. 3 voters: $b\ c\ d\ |\ a$

Candidate c is approved of by all 7 voters, candidates b and d by 5 voters each, and candidate a by 4 voters. While all candidates receive majority approval, candidate a is the PAV winner, because he or she is preferred by a majority (type 1 and 2 voters) to the AV winner (candidate *c*), as well as candidates *b* and *d*, under rule 2(i) ∎

When the PAV winner and the AV winner differ, as in Example 3.6, the PAV winner is arguably the more coherent majority choice. Two of the three types of voters rank candidate *a* as their top choice in this example, whereas candidate *c*, the AV winner, is not the top choice of any type of voters.

Finally, I show that PAV may produce winners different from the two best-known ranking systems (for information on several other voting systems besides these, see Brams and Fishburn, 2002).

Proposition 3.4. *A PAV winner may be different from winners under the Borda count (BC) and single transferable vote (STV).*

Proof. If there are *n* candidates, recall (section 2.3) that BC assigns $n-1$ points to the first choice of a voter, $n-2$ points to the second choice, . . . , and 0 points to the last choice; the candidate with the most points wins. In Example 3.6, candidate *c* wins with 14 points (2 points each from all 7 voters), whereas the PAV winner, candidate *a*, receives 12 points (3 points each from 4 voters and 0 points from 3 voters).

Under STV, recall that only first-place votes are counted initially. In Example 3.5, candidates *a*, *d*, and *b* receive 3, 3, and 2 votes, respectively, from the voters who rank them first. Because candidate *b* receives the fewest votes, the votes or his or her supporters are transferred to their second choice, candidate *d*, giving *d* a total of 5 votes, which is a majority and makes candidate *d* the winner. By contrast, candidate *a* is the PAV winner. ∎

In summary, I have shown that PAV may not elect Condorcet winners, or winners under AV, BC, or STV. Nevertheless, PAV winners are strong contenders on grounds of both approval and preference, about which I will say more later.

I turn next to a voting system that asks less information of voters than does PAV, requiring them to rank only those candidates of whom they approve. It shares some properties of PAV but by no means all.

3.4. FALLBACK VOTING (FV)

Fallback voting (FV) proceeds as follows:

1. Voters indicate all candidates of whom they approve, who may range from no candidate (which a voter does by abstaining from voting) to all candidates. Voters rank only those candidates of whom they approve.
2. The highest-ranked candidate of all voters is considered. If a majority of voters agree on one highest-ranked candidate, this candidate is the

FV winner. The procedure stops, and we call this candidate a *level 1 winner*.

3. If there is no level 1 winner, the next-highest-ranked candidate of all voters is considered. If a majority of voters agree on one candidate as *either* their highest or their next-highest-ranked candidate, this candidate is the FV winner. If more than one candidate receives majority approval, then the candidate with the largest majority is the FV winner. The procedure stops, and we call this candidate a *level 2 winner*.

4. If there is no level 2 winner, the voters descend—one level at a time—to lower and lower ranks of *approved* candidates, stopping when, for the first time, one or more candidates are approved of by a majority of voters, or no more candidates are ranked. If exactly one candidate receives majority approval, this candidate is the FV winner. If more than one candidate receives majority approval, then the candidate with the largest majority is the FV winner. If the descent reaches the lowest rank of all voters and no candidate is approved of by a majority of voters, then the candidate with the most approval is the FV winner.

The appellation "fallback" comes from the fact that FV successively falls back on lower-ranked approved candidates if no higher-ranked approved candidate receives majority approval. This nomenclature was first used in Brams and Kilgour (2001b), but it was applied to bargaining rather than voting, in which the decision rule was assumed to be unanimity (the assent of all parties was necessary) rather than a simple majority.

Brams and Kilgour (2001b), in what they called "fallback bargaining with impasse," did not require that the bargainers rank all alternatives. Rather, the bargainers ranked only those they considered better than "impasse," because impasse was preferable to any alternative ranked lower. Bargainers not ranking alternatives below impasse are analogous to voters not approving of candidates below a certain level, whom they do not rank.

Like FV, the majoritarian compromise (MC) proposed by Sertel and his colleagues (Sertel and Yilmaz, 1999; Sertel and Sanver, 1999; Hurwicz and Sertel, 1999) elects the first candidate approved of by a majority in the descent process. However, voters are assumed to rank all candidates—they do not stop their ranking at some point at which they consider candidates they rank lower unacceptable.

As with FV, James W. Bucklin assumed that if a voter did not rank all candidates, he or she disapproved of those not ranked.[1] Thus, when the fallback

---

[1] Bucklin, a lawyer and founder of Grand Junction, Colorado, proposed his system for Grand Junction in the early twentieth century, where it was used from 1909 to 1922—as well as in other cities—but it is no longer used today. See Hoag and Hallett (1926, pp. 485–491), http://www.gjhistory.org/cat/main.htm, http://en.wikipedia.org/wiki/Bucklin_voting, and http://wiki.electorama.com/wiki/ER-Bucklin.

process descends to a level at which a voter no longer ranks candidates, that voter is assumed to approve of no additional candidates should the process continue to descend for other voters because no candidate has yet reached majority approval. Bucklin's system is FV absent the designation of approved candidates, who are implicitly assumed to be only those candidates whom voters rank.

In the analysis of FV that follows, I assume that voters have preferences for all candidates, though they reveal their rankings only for approved candidates. As I will show, the nonrevealed information may lead to the election of different candidates from PAV. First, however, I indicate properties that FV shares with PAV.

Proposition 3.5. *Condorcet winners and unanimous AV winners may not be FV winners, whereas least-approved candidates may be FV winners.*

Proof. In Example 3.1, there is no level 1 winner. Because candidate b is the only candidate approved of by a majority (voters 2 and 3) at level 2, b is the FV winner, whereas candidate a is the Condorcet winner.

In Example 3.4, candidate a is the FV winner at level 1, but candidate b is the unanimous AV winner. In Example 3.6, candidate a is the FV winner at level 1, but a is the least approved of the four candidates. ∎

While FV and PAV share the properties listed in Proposition 3.5, FV, unlike PAV, may fail to elect a majority-preferred candidate among the majority-approved candidates.

Proposition 3.6. *Suppose there are two or more majority-approved candidates. If one is majority-preferred among them, FV may not elect him or her.*

Proof. Consider the following 5-voter, 4-candidate example:

Example 3.7
1. 2 voters: $a\,b\mid c\,d$
2. 1 voter: $d\,c\,a\mid b$
3. 2 voters: $c\,a\mid b\,d$

There is no level 1 majority-approved candidate with at least 3 votes. Because candidate a receives more approval (4 votes) than candidate c (3 votes) at level 2, a is the FV winner. But candidate c is majority-preferred to candidate a by 3 votes to 2. ∎

In fact, candidate c is the Condorcet winner among *all* candidates, defeating candidates b and d as well. PAV, because of rule 2(i), picks candidate c, even though candidate a is more approved at level 2 and is unanimously approved at level 3 (to which FV never descends).

A similar conflict between FV and PAV may occur when there is no Condorcet winner.

Proposition 3.7. *A unanimously approved candidate in a cycle may not be the FV winner.*

Proof. Consider the following 9-voter, 4-candidate example:

Example 3.8
1. 2 voters: $a\ b\ c\mid d$
2. 3 voters: $b\ d\ c\mid a$
3. 4 voters: $c\ a\mid d\ b$

There is a cycle whereby $a > b > c > a$. Candidate c is the only candidate approved of by all 9 voters and so would be the PAV winner under rule 2(ii). Under FV, no candidate is majority-approved at level 1, but at level 2 candidate a receives 6 votes and candidate b receives 5 votes, making a the FV winner. ∎

Proposition 3.8. *FV, PAV, and AV may all give different winners for the same preference-approval profile.*

Proof. Consider the following 9-voter, 4-candidate example:

Example 3.9
1. 4 voters: $a\ b\ c\mid d$
2. 3 voters: $b\ c\mid a\ d$
3. 2 voters: $d\ a\ c\mid b$

There is no level 1 majority-approved candidate, but candidates a and b each receive majority approval (6 and 7 votes, respectively) at level 2. Because candidate b (7 votes) is more approved of than candidate a (6 votes), FV elects candidate b. But candidate c is unanimously approved (9 votes)—at level 3 for the type 1 and 3 voters (to which FV never descends)—so AV elects candidate c. Finally, PAV elects candidate a, who is majority-preferred to the two other majority-approved candidates, b and c. ∎

Note in Example 3.9 that no type of voters ranks the unanimously approved AV winner (candidate c) first, so he or she is likely to be only a lukewarm choice of everybody. Neither FV nor PAV favors such candidates if there are majority-approved candidates ranked higher by the voters.

In Examples 3.7, 3.8, and 3.9, one can determine that candidate a is majority-preferred to candidate b from the rankings of the approved candidates. Thus in Example 3.9, even though the type 2 voters do not indicate that they prefer candidate b to candidate a when they rank their two approved candidates, b and c, the fact these voters do not approve of candidate a implies that candidate b, of whom they do approve, is ranked higher than candidate a. Similarly, one can ascertain from the ranking of the type 3 voters that they prefer candidate a to candidate b.

That PAV would have given a different outcome from FV may not always be revealed.

Proposition 3.9. *Information used to determine an FV winner may not reveal that PAV would have chosen a different winner.*

Proof. Consider the following 3-voter, 4-candidate example:

Example 3.10
1. 1 voter: $a\, b\, c \mid d$
2. 1 voter: $b\, d\, a \mid c$
3. 1 voter: $c \mid a\, b\, d$

There is no level 1 majority-approved candidate, but at level 2 candidate b receives majority approval (2 votes) and is, therefore, the FV winner. Because the type 3 voter does not rank candidates below candidate c under FV, it would not be known whether candidate a would defeat candidate b, or vice versa, in a pairwise contest between these two candidates (while candidate a is preferred by the type 1 voter, candidate b is preferred by the type 2 voter, leaving the contest undecided). But under PAV, wherein voters rank all candidates, the fact that the type 3 voter prefers a to b would not only be revealed but also would render candidate a the winner, because a is majority-preferred to b.[2] ∎

That FV ignores information on the lower-level preferences of voters is one reason why it gives different outcomes from PAV. Although I think information on nonapproved candidates should not be ignored, I recognize that it sometimes may be difficult for voters to provide it.

3.5. MONOTONICITY OF PAV AND FV

Such well-known voting systems as STV are not monotonic, as I showed in the appendix to chapter 2. This renders them vulnerable to "ranking paradoxes" (Brams and Fishburn, 2002, p. 215), such as the no-show paradox that benefits a citizen who does not vote. Because PAV and FV are hybrid voting systems, it is useful to define two kinds of monotonicity.

1. A voting system is *approval-monotonic* if a type of voter, by approving of a new candidate—without changing its approval of other candidates—never hurts and may help this candidate get elected.
2. A voting system is *rank-monotonic* if a type of voter, by raising a candidate in its ranking—without changing its ranking of other candidates—never hurts and may help this candidate get elected.

---

[2] To be sure, if the type 3 voter did not rank any candidates below candidate c, the outcome under PAV would, as under FV, be a tie between candidates a and b. While voters would be encouraged to rank all candidates under PAV, I do not think their ballots should be invalidated if they do not do so.

A *monotonicity paradox* occurs when a voting system is not approval-monotonic or rank-monotonic; violations of rank-monotonicity have been investigated by Fishburn (1982), among others.

Proposition 3.10. *PAV and FV are approval-monotonic.*

Proof. Consider PAV. Under rule 1, a type of voter, by approving of a candidate, helps him or her become the unique AV, and therefore the PAV, winner. Under rule 2(i), a type of voter, by approving of a candidate, helps him or her become one of the majority-approved candidates and, therefore, a possible PAV winner. Under rule 2(ii), a type of voter, by approving of a candidate, helps him or her become the AV, and therefore the PAV, winner among the majority-approved candidates in a cycle. Consider FV. A type of voter, by approving of a candidate, allows him or her to be ranked and receive votes in the descent, thereby helping him or her become the FV winner. ∎

Proposition 3.11. *PAV and FV are rank-monotonic.*

Proof. Consider PAV. Under rule 1, ranks have no effect. Under rule 2(i), a type of voter, by raising a candidate in its ranking, helps that candidate defeat other majority-approved candidates in pairwise contests and thereby become the PAV winner. Under rule 2(ii), a type of voter, by raising a candidate in its ranking, helps that candidate be a member of the cycle—if there is no majority-preferred candidate among the majority-approved candidates—and thereby become a possible PAV winner. Consider FV. A type of voter, by raising a candidate in its ranking, helps that candidate become majority-approved at an earlier level, or receive the largest majority if two or more candidates are majority-approved at the same level, and thereby become the FV winner. ∎

Thus, a type of voters can rest assured that giving either approval or a higher ranking to a candidate can never hurt and may help him or her get elected under PAV and FV. However, this may lead to the defeat of an already approved candidate that one prefers, which is illustrated by the following 7-voter, 4-candidate example:

Example 3.11
1. 1 voter: $a\ b \mid c\ d$
2. 3 voters: $b \mid a\ c\ d$
3. 2 voters: $c\ a \mid b\ d$
4. 1 voter: $d \mid a\ b\ c$

Under PAV, candidate b is the only candidate to be majority-approved (4 votes) and so is the PAV winner under rule 1.

But now assume that the three type 2 voters approve of candidate a as well as candidate b:

2′. 3 voters: $b\ a \mid c\ d$

Candidate a receives 5 votes and candidate b 4 votes, so both are majority-approved. But because candidate a is majority-preferred to candidate b by 4 votes to 3, candidate a is the PAV winner under rule 2(i), contrary to the interests of the type 2 voters who switched from strategy b to strategy ba.

Similarly, for the original approval strategies of the voters in Example 3.11, candidate a is the FV winner, picking up 4 votes at level 2. But when the type 2 voters switch from strategy b to strategy ba, candidate a wins with 5 votes at level 2. As under PAV, the strategy shift by the type 2 voters is detrimental to their interests.

In section 3.7, I will show how information from polls may affect voters' calculations about how many candidates to approve of under PAV, and to approve of and rank under FV. These calculations may or may not result in equilibrium outcomes.

The stability of outcomes under PAV and under FV reflects their robustness against manipulation. Stability may be looked at in either static or dynamic terms. In section 3.6 I view it statically—when will voters be motivated to try or not try to upset an outcome?—whereas in section 3.7 I analyze how unstable outcomes, based on a dynamic poll model, evolve over time.

3.6. NASH EQUILIBRIA UNDER PAV AND FV

Because PAV and FV give the same outcome as AV when either no candidate or one candidate receives the approval of a majority, they share many of the properties of AV. For example, in a field in which at most one candidate is likely to obtain majority approval, PAV and FV, like AV, give candidates an incentive to broaden their appeal to try to maximize their level of approval.

When candidates reach out to try to attract more votes, voters are likely to consider them acceptable and approve of more than one candidate. But if more than one candidate actually receives majority approval, the preferences of voters under PAV and FV matter, so the most-approved candidate may not win, as I showed earlier. Thus, a key question that both PAV and FV raise is how many candidates a voter should approve of if he or she deems more than one acceptable. As I showed in section 3.5, sometimes voting for additional candidates may sabotage the election of a preferred candidate.

In the analysis that follows, assume that voters, in order to try to elect their preferred candidates, choose strategically where to draw the line between approved and disapproved candidates. But assume that they are truthful in their rankings of candidates, which is equivalent to assuming that they choose from among their sincere AV strategies.[3]

---

[3] In section 3.7 I consider the possibility that voters may change their rankings as well as their approval in order to try to manipulate outcomes. For a rigorous study of the manipulability of voting systems that focuses on manipulation through the misrepresentation of rankings, see

Recall from section 2.4 that AV strategies may be unstable, stable, or strongly stable. In the case of PAV and FV, I next consider when strategies are stable, or in equilibrium.

Define an outcome to be *in equilibrium* if the approval strategies of each type of voter that produces it constitute a Nash equilibrium. At such an equilibrium, no type of voter has an incentive to depart unilaterally from its approval strategy, because it would induce no better an outcome, and possibly a worse one, by doing so.

Proposition 3.12. *Truth-telling strategies of voters under PAV and FV may not be in equilibrium. In particular, voters may induce a better outcome either by contracting or expanding their approval sets.*

Proof. I first prove this proposition for PAV using the following 7-voter, 4-candidate example:

Example 3.12
  1. 3 voters: *a b* | *c d*
  2. 2 voters: *c* | *a b d*
  3. 2 voters: *d b* | *a c*

Candidate *b*, approved of by 5 voters, is the only candidate approved of by a majority and so is the PAV winner.

To show the possible effects of contraction, assume that the three type 1 voters *contract* their approval set from strategy *ab* to strategy *a*:

  1′. 3 voters: *a* | *b c d*

Then candidate *a*, who is preferred by the type 1 voters to candidate *b*, will win under PAV rule 1, receiving 3 votes to 2 votes each for candidates *b*, *c*, and *d*.

To show the possible effects of expansion, assume the two type 2 voters in the original Example 3.12 *expand* their approval set from strategy *c* to strategy *ca*:

  2′. 2 voters: *c a* | *b d*

Then candidates *a* and *b* tie with 5 votes each, whereas (candidates *c* and *d* each receive 2 votes). Because candidates *a* and *b* both receive majority approval, PAV rule 2(i) applies. Since candidate *a* is preferred to candidate *b* by a majority of 5 votes to 2, candidate *a*, whom the type 2 voters prefer to candidate *b*, is the winner.

Thereby, both the contraction and the expansion of an approval set by a type of voter may induce a preferred outcome, rendering PAV strategies in

---

Taylor (2005). Excellent case studies of manipulation that are informed by social-choice theory can be found in Riker (1986) and Schofield (2006); how manipulation interacts with rhetoric is analyzed and illustrated in Riker (1996).

Example 3.12 not in equilibrium. It is easy to show that the same contraction and expansion of approval sets induces preferred outcomes under FV (candidate a instead of candidate b in the case of contraction 1′; a tie between candidates a and b in the case of expansion 2′). ∎

I showed earlier that PAV, FV, and AV may lead to three different outcomes for the same preference-approval profile (Proposition 3.8). The fact that an outcome is in equilibrium under one system, however, does not imply that it is in equilibrium under another system.

Proposition 3.13. *When PAV and FV give different outcomes, one may be in equilibrium and the other not.*

Proof. In Example 3.9, I showed that candidate a (the Condorcet winner) wins under PAV and candidate b wins under FV. Candidate a is in equilibrium under PAV, because none of the three types of voters, by switching to a different approval strategy, can induce an outcome they prefer to candidate a. On the other hand, candidate b is not in equilibrium under FV, because the four type 1 voters, by switching from strategy abc to a, can induce the election of candidate a, whom they prefer to candidate b. This example shows that PAV may give an equilibrium outcome when FV does not.

To show that FV may give an equilibrium outcome when PAV does not, consider the following example:

Example 3.13
1. 1 voter: $a\,b\mid c\,d$
2. 1 voter: $c\,a\mid d\,b$
3. 1 voter: $c\mid b\,a\,d$
4. 1 voter: $d\,b\mid a\,c$
5. 1 voter: $d\,b\mid c\,a$

Candidate b is the only candidate approved of by a majority of three voters. No voter, by switching to a different approval strategy under FV, can induce a preferred outcome to candidate b at level 2, making candidate b an equilibrium outcome. Candidate b, being the sole majority-approved candidate, is also the winner under PAV. But voter 2, by switching from strategy ca to cad, can render both candidates d and b majority-approved (3 votes each). Since d is preferred to b by a majority of three voters, including voter 2, voter 2 would have an incentive to induce this tied outcome under PAV, showing that FV may give an equilibrium outcome when PAV does not. ∎

The fact that equilibria under PAV do not imply equilibria under FV, or vice versa, indicates that one system is not inherently more stable than the other.[4]

---

[4] AV yields candidate c in Example 3.9, and candidate b in Example 3.13—but neither in equilibrium—showing that equilibria under PAV and FV are not always the same as under AV.

The *degree* of stability of these systems could more precisely be assessed through computer simulation.

3.7. THE EFFECTS OF POLLS IN 3-CANDIDATE ELECTIONS

In elections for major public office in the United States and other democracies, voters are not in the dark. Polls provide them with information about the relative standing of candidates and may also pinpoint their appeal, or lack thereof, to voters.

In this section, I focus on 3-candidate elections, because they are the simplest example in which information about the relative standing of candidates can affect the strategic choices of voters. Also, such elections are relatively common. I will show how voter responses to a sequence of polls may dynamically change outcomes under PAV and FV.[5]

To assess the effects of polls in 3-candidate elections, I make the following assumptions:

1. *No majority winner.* None of the three candidates, *a*, *b*, or *c*, is the top choice of a majority of voters.
2. *Initial support of only top choice.* Before the poll, each voter approves of only his or her top choice.
3. *Poll information.* The poll indicates the relative standing of the candidates. For example, the ordering $n_a > n_b > n_c$ indicates that candidate *a* receives the most approval votes, candidate *b* the next most, and candidate *c* the fewest (for simplicity, I do not allow for ties).
4. *Strategy shifts.* After the results of the poll are announced, voters may shift strategies by approving of a second choice as well as a top choice. Voters will vote for their two top choices if and only if the poll indicates (i) the about-to-become-winner is their worst choice, and (ii) they can prevent this outcome by approving of a second choice, too, given they did not previously approve of this choice.
5. *Repeated responses.* After voters respond to a poll, they respond to new information that is revealed in subsequent polls, as described in assumption 4 above.
6. *Termination.* Voters cease their strategy shifts when they cannot induce a preferred outcome.

Merrill and Nagel (1987) suggest that outcomes under multistage systems like PAV and FV may be more manipulable than outcomes under single stage systems like AV, but the manipulation of PAV and FV are computationally more demanding and, consequently, probably more impracticable.

[5] The effects of polls under plurality voting and AV were analyzed in Brams (1982b) and Brams and Fishburn (1983, 2007 ch. 7) using a different dynamic model; see also Meirowitz (2004) and citations therein.

Assume that voters truthfully rank the three candidates at the start and do not change these rankings in response to the initial poll or any subsequent poll. I next investigate what outcomes occur in response to polls under PAV for two different kinds of preferences.

1. *Single-peaked preferences.* Voters perceive the candidates to be arrayed along a left-right continuum, with candidate a on the left, candidate b in the middle, and candidate c on the right. Each voter most prefers one of these candidates, next most prefers an adjacent candidate, and least prefers the candidate farthest from his or her most-preferred candidate, who may or may not be adjacent. (Thus, each voter's preferences slope downward from its peak, or most-preferred, candidate.)

More specifically, a-voters on the left with preference ranking $a\ b\ c$ may switch from strategy a to strategy ab, whereas c-voters on the right with preference ranking $c\ b\ a$ may switch from strategy c to strategy cb. The b-voters in the middle split into two groups: one group prefers candidate a over candidate c ($b\ a\ c$); and the other group prefers candidate c over candidate a ($b\ c\ a$). The former group may switch from strategy b to strategy ba, whereas the latter group may switch from strategy b to strategy bc.

Because preferences are single-peaked, the median candidate, b, is the unique Condorcet winner—he or she is preferred by a majority to both candidate a and candidate c. As shown in Table 3.1, there are three qualitatively different poll rankings that the initial poll may give:

$$\text{(i) } n_a > n_b > n_c; \quad \text{(ii) } n_a > n_c > n_b; \quad \text{(iii) } n_b > n_a > n_c,$$

where n_i indicates the number of approval voters of candidate i. If the roles of

TABLE 3.1
Strategy Switches of Voters in Response to a Poll under PAV and FV:
Single-Peaked Preferences with Three Poll Rankings (*b* Condorcet Winner)

Poll Ranking	(i) $n_a > n_b > n_c$	(ii) $n_a > n_c > n_b$	(iii) $n_b > n_a > n_c$
Initial Strategies	$a\mid b\,c$ $b\mid a\,c$ $b\mid c\,a$ $c\mid b\,a$	$a\mid b\,c$ $b\mid a\,c$ $b\mid c\,a$ $c\mid b\,a$	$a\mid b\,c$ $b\mid a\,c$ $b\mid c\,a$ $c\mid b\,a$
Outcome	a	a	b
Shift in Strategies (if any) after Initial Poll	$a\mid b\,c$ $b\mid a\,c$ $b\,c\mid a$ $c\,b\mid a$	$a\mid b\,c$ $b\mid a\,c$ $b\,c\mid a$ $c\,b\mid a$	
Outcome	b	b	

candidates a and c are reversed, there are three analogous rankings, which are not shown in Table 3.1:

$$\text{(iv) } n_c > n_b > n_a; \qquad \text{(v) } n_c > n_a > n_b; \qquad \text{(vi) } n_b > n_c > n_a,$$

For poll ranking (i) in Table 3.1, the voters with preference rankings $b\,c\,a$ and $c\,b\,a$ will switch from strategies b and c, respectively, to strategies bc and cb to try to prevent their worst choice, candidate a, from winning (assumption 4). This results in the election of candidate b, whether candidate b is the unique majority-approved candidate (with approval from three types of voters) or candidate c also wins a majority (with approval from two types of voters); in the latter case, candidate b will defeat candidate c in a pairwise contest. Because no voters can effect a preferred outcome under PAV through any subsequent shifts in their strategies—in response to a poll that shows candidate b to be the unique or largest-majority winner—no voters will have an incentive to make further shifts.

The same shifts will occur for poll ranking (ii), again boosting candidate b to winning status. As for poll ranking (iii), no voters will have an incentive to shift in response to the initial poll, because the plurality winner, candidate b, is not the worst choice of any voters.

Under FV, candidate b will also prevail. In the case of poll rankings (i) and (ii), this occurs because candidate b is the unique or largest-majority winner after the shift. In the case of poll ranking (iii), candidate b is the initial plurality winner, after which the descent of voters ceases because no voter ranks b last.

In summary, whichever of the three qualitatively different poll rankings occurs when voter preferences are single-peaked, the responses of voters to an initial poll leads to the election of Condorcet winner b under both PAV and FV. But when preferences are cyclical and there is no Condorcet winner, the evolution of a winner is more drawn out, requiring up to three shifts rather than just one.

2. *Cyclical preferences.* Consider the simplest case of cyclical preferences, wherein three types of voters, none with a majority of votes initially, have preferences $a\,b\,c$, $b\,c\,a$, and $c\,a\,b$, so $a \succ b \succ c \succ a$. For simplicity, I exclude voters with preferences that do not contribute to the cyclic component of these voters (e.g., $a\,c\,b$).

If, as assumed earlier, voters initially approve of only their top choices, there are two qualitatively different poll rankings that the initial poll may give:

$$\text{(i) } n_a > n_b > n_c; \qquad \text{(ii) } n_a > n_c > n_b.$$

The four other possible rankings are analogous, with candidate b ranked first in two cases and candidate c ranked first in the other two cases:

$$\text{(iii) } n_b > n_a > n_c; \quad \text{(iv) } n_b > n_c > n_a; \quad \text{(v) } n_c > n_a > n_b; \quad \text{(vi) } n_c > n_b > n_a.$$

In Table 3.2, strategy shifts that voters will make in response to poll rankings (i) and (ii) are shown. After an initial poll that shows candidate a to be in

TABLE 3.2
Strategy Switches of Voters in Response to a Poll under PAV and FV: Cyclic Preferences with Two Poll Rankings

Poll Ranking	(i) $n_a > n_b > n_c$	(ii) $n_a > n_c > n_b$
	$a\mid b\,c$	$a\mid b\,c$
Initial Strategies	$b\mid c\,a$	$b\mid c\,a$
	$c\mid a\,b$	$c\mid a\,b$
Outcome	a	a
	$a\mid b\,c$	$a\mid b\,c$
Shift I (Initial Poll)	$b\,c\mid a$	$b\,c\mid a$
	$c\mid a\,b$	$c\mid a\,b$
Outcome	c	c
	$a\,b\mid c$	$a\,b\mid c$
Shift II (2nd Poll)	$b\,c\mid a$	$b\,c\mid a$
	$c\mid a\,b$	$c\mid a\,b$
Outcome	b	b
		$a\,b\mid c$
Shift III (3rd Poll)		$b\,c\mid a$
		$c\,a\mid b$
Outcome		a

first place in each case, there will be one shift by the $b\,c\,a$ voters (Shift I)—and up to two additional shifts (Shift II and Shift III) in response to subsequent polls that show other candidates to be in first place—as voters try to prevent their worst choice from winning.

To illustrate for poll ranking (i), the $b\,c\,a$ voters will switch from strategy b to strategy bc in Shift I to try to prevent candidate a from winning with a plurality of votes. But when this shift leads to candidate c's receiving a majority of votes, the $a\,b\,c$ voters will switch from strategy a to strategy ab in Shift II, giving candidates b and c each a majority.

Under PAV, candidate b will be majority-preferred to candidate c in the contest between these two majority-approved candidates after Shift II. Under FV, candidate b, with approval from both $a\,b\,c$ and $b\,c\,a$ voters at level 2, will receive a larger majority than candidate c—based on the initial poll ranking—with approval from $b\,c\,a$ and $c\,a\,b$ voters.

At this stage, even if the $c\,a\,b$ voters switched from strategy c to strategy ca, they could not induce the election of candidate a, who will get a smaller majority than candidate b, based on the initial poll ranking. Hence, the shifts will terminate after Shift II, resulting in the election of candidate b, the candidate with more first and second-place approval than any other candidate.

For poll ranking (ii), three shifts are required to induce the election of candidate a. In the absence of a Condorcet winner, the most approved candidate in the cycle—when all voters support their two top candidates—emerges as the winner under PAV and FV. In summary, when preferences are cyclical, the candidate who is ranked first or second by the most voters prevails after three shifts under both PAV and FV.

Combining these results with the results on single-peaked preferences gives the following:

Proposition 3.14. *In the poll model for 3-candidate elections under PAV and FV, strategy shifts result in the election of (1) the Condorcet winner if preferences are single-peaked, and (2) the candidate ranked first or second by the most voters if preferences are cyclical.*

Not all these outcomes, however, may be stable.

Proposition 3.15. *In the poll model for 3-candidate elections under PAV and FV, strategy shifts may result in outcomes that are not in equilibrium when there is a Condorcet winner.*

Proof. Assume that voter preferences are single-peaked (Table 3.1), and consider poll ranking (ii) after the shift. Assume that the b c a and c b a voters constitute a majority. Then the c b a voters, by switching from strategy cb to strategy c (a contraction), will induce the election of candidate c, whom they prefer to candidate b. As the sole majority-approved candidate, candidate c wins under both PAV and FV, rendering candidate b not in equilibrium. ∎

Surprisingly, it is not the cyclical preferences of voters (in Table 3.2) that produce instability but the single-peaked preferences of voters (in Table 3.1) for poll ranking (ii)—and poll ranking (i) as well if the b c a and c b a voters constitute a majority in this situation—that produce instability. Thus, the strategy shifts of voters in response to polls, while leading to the outcomes indicated in Proposition 3.14, may not terminate at these outcomes because of the possible nonequilibrium status of candidate b for poll rankings (i) and (ii) in Table 3.1.

This is not to say that the Condorcet winner (in Table 3.1), candidate b, cannot be supported as a Nash equilibrium in this situation. It turns out that the critical strategy profile of candidate b (section 2.4),

$$a\,b\mid c; \qquad b\mid a\,c; \qquad b\mid c\,a; \qquad c\,b\mid a,$$

which maximizes b's approval vis-à-vis the other candidates, supports b as a *strong* Nash equilibrium—no coalition of voter types, by choosing different approval strategies, can induce an outcome they prefer to candidate b. Not only is it impossible for a coalition to displace b with a preferred candidate under PAV and FV, but this is also true of AV. In fact, recall that under AV a nontied candidate is a strong Nash equilibrium at his or her critical strategy profile if and only if he or she is a unique Condorcet winner (Propostion 2.11).

I have assumed until now that while voters may change their levels of approval in order to try to induce preferred outcomes, they are steadfast in their rankings of candidates, which I assumed are truthful. But what if they can falsify their rankings? Then the candidates will be more vulnerable. But falsifying rankings, especially if information is incomplete, is a risky strategy that many voters are likely to shun.[6]

3.8. SUMMARY AND CONCLUSIONS

It is worth emphasizing that PAV and FV duplicate AV when at most one candidate receives a majority of approval votes. In such a situation, there seems good reason to elect the AV winner, because if there is a different Condorcet winner, he or she would not be majority-approved. If the AV winner also is not majority-approved, his or her election seems even more compelling, because this is the most acceptable candidate in a field in which nobody is approved of by a majority.

When two or more candidates are majority-approved, PAV and FV may elect different winners from AV, the Borda count, STV, and one another. PAV chooses the majority-preferred candidate, if there is one, among those who are majority-approved, whereas FV chooses the first candidate to receive a unique or largest majority in the descent.[7]

If there is no majority-preferred candidate among the majority-approved candidates, PAV chooses the most approved candidate in the cycle. FV does the same if this candidate is in the first set of candidates to receive majority approval in the descent; if not, a majority-approved candidate with less approval—but received earlier—will be the FV winner. PAV and FV winners, if different

---

[6] AV, of course, does not permit such falsification since voters do not rank candidates. While AV leads to the same outcomes as PAV and FV in the poll model, it may give very different outcomes in other situations, as I showed earlier.

[7] Majority approval may be too high a bar to impose if the field of candidates is large. This bar has been lowered in some plurality elections, wherein a candidate can win outright if he or she obtains at least 40 percent of the vote; otherwise, there is a runoff election between the two highest vote getters. I am quite open about the amount of approval (i) that two or more candidates must receive in order that rule 2 take effect under PAV, or (ii) that one candidate must receive for the descent to stop under FV. Perhaps a simple majority should not be the *sine qua non*. A lower threshold may be appropriate in elections in which at most one candidate is likely to receive majority approval and, therefore, the winner will always be the AV winner, obviating the need for PAV or FV. For a discussion of "flexible majority rules," see Gersbach (2005).

from the AV winner, are likely to have more coherent majoritarian positions, not just be the lukewarm choices of most voters.

Candidates with coherent positions are more likely to run if they believe, without egregious pandering, that they can win. Consequently, PAV and FV may well encourage candidates to enter the fray who might otherwise be deterred because they are unwilling to sacrifice their fundamental tenets in order to win.

PAV and FV afford voters the opportunity to approve of lower-ranked candidates without necessarily helping them to win. Unlike AV, in which voting for a less-preferred candidate can cause the displacement of a more-preferred candidate, PAV and FV impede this event, though they do not rule it out entirely.

PAV, for example, takes into account voter preferences, which can override the greater approval a less-preferred candidate receives. Both PAV and FV are approval-monotonic and rank-monotonic, so approving of a candidate or ranking him or her higher never hurts, and may help, this candidate to get elected.

PAV is more information demanding than FV, which asks voters to rank only their approved candidates. Without complete information on preference rankings, FV is less able to ensure the election of a majority-preferred—or the most approved if there is no majority-preferred—candidate among the majority-approved candidates.

PAV and FV may elect different candidates in equilibrium if voters contract or expand their approval sets; neither system is inherently more stable than the other. In the 3-candidate dynamic poll model, Condorcet winners are elected after one shift when voter preferences are single-peaked—though not always in equilibrium—whereas candidates ranked first or second by the most voters are equilibrium choices after several shifts when voter preferences are cyclic.

By combining information on approval and preferences, PAV and FV may yield outcomes that neither kind of information, by itself, produces. Although PAV is more likely to lead to majority-preferred winners among the majority-approved, its greater information demands of voters may make FV a better practical choice. Such trade-offs require careful consideration, as do other ways of mixing approval and preferences to coax better social choices out of a voting system.[8]

---

[8] Ossipoff and Smith (2005) survey a number of such voting methods, several of which disqualify candidates if another candidate is ranked above them on more than half the ballots. Thus, if there is a Condorcet winner, this candidate will disqualify all others and will, therefore, be elected, independent of how approved he or she is. In my view, a Condorcet winner who receives less than majority approval—as I showed can happen under very different circumstances in Examples 3.1, 3.2, and 3.3—should *not* be elected when there are other candidates who receive majority approval. Both PAV and FV give precedence to majority-approved candidates over Condorcet winners when there is a conflict. But *among* majority-approved candidates, Condorcet winners take precedence under PAV.

Finally, it is worth mentioning a situation in which PAV was recently adopted by the New York University politics department because of a failure, at least initially, of plurality voting (PV) to choose a candidate for a faculty position. Two candidates, A and B, were vying for that position, with almost two-thirds of department members favoring one or the other.

But the department split almost evenly over which candidate members preferred. Because the more than one-third who favored neither candidate won under PV, it seemed that neither candidate would be hired, though a substantial majority preferred either A or B over no hire. In the end, however, the majority prevailed in a second vote over hiring one or the other, with a third vote showing which one of the two candidates was preferred.

Under PAV, there would have been three options: hire A, hire B, or hire neither (the position did not have to be filled). The nearly two-thirds who favored either A or B over no hire presumably would have approved of both, at which point their preferences for either A or B would have elected one of the two candidates (except in the case of a tie).

Note that AV might not have succeeded, because some of the A and B supporters might have approved of only their favorite, which could have prevented either from winning. But under PAV, there is no good strategic reason for A and B supporters not to approve of both, knowing that their preferences will determine a winner between the two if both are majority-approved. Thus, PAV mitigates, if not prevents, certain kinds of strategizing to which AV may be vulnerable, including what Nagel (2006, 2007) calls the "Burr dilemma."[9]

[9] This dilemma occurred in the U.S. presidential election of 1800, in which Republicans Thomas Jefferson and Aaron Burr tied in electoral votes when Republican electors voted for both of them. (AV was used at that time: electors could vote for more than one candidate; the candidate with the most votes became president, and the candidate with the second-most votes became vice-president.) The tie in the Electoral College sent the election to the House of Representatives, wherein each state had one vote. After thirty-five ballots cast over six days, Jefferson finally won. But if PAV had been used in the Electoral College, Republican electors could, as they did, approve of both Jefferson and Burr with impunity and thereby defeat the opposition candidates. In addition, by being able to express their preferences for either Jefferson or Burr, the Republican electors would have elected their preferred candidate as president and their nonpreferred candidate as vice-president, except in the unlikely event of a tie.

4

Electing Multiple Winners:
Constrained Approval Voting

4.1. INTRODUCTION

In electing a committee, a council, or a legislature comprising more than one person, the usual rationale is to afford different factions or interests the opportunity to gain representation in proportion to their numbers, which is referred to as *proportional representation* (PR).[1] In this chapter, I focus on one procedure for doing so that uses an approval ballot. As with AV, voters can approve or disapprove of each candidate, but who is elected will be constrained in certain ways, which is why I call the procedure *constrained approval voting* (CAV).[2]

CAV was inspired by the request of a professional association to advise it on a voting procedure to use in electing its governing board. Although I will not reveal the association's identity, I have altered no essential facts of the study that it commissioned, which largely concerned how a fair division of seats on its governing board might be achieved through a PR election procedure.

The procedure I recommended is applicable when members of an electorate can be classified in more than one way. For the association, the classification was by region and specialty.

The association was reviewing its election procedures in response to pressures for reform. In particular, some members thought that certain types of members were underrepresented on the board—and other types overrepresented—creating biases that affected the association's policies. A different voting system was seen as a possible way to address this perceived misrepresentation.

Note: This chapter is adapted from Brams (1990) with permission of the Institute for Operations Research and Management Sciences; see also Brams (1975, 2003, ch. 3), Brams and Fishburn (1984b, 1984c), and Brams and Taylor (1996, ch. 10).

[1] Other methods for achieving PR are analyzed in Rapoport, Felsenthal, and Maoz (1988a, 1988b). Paradoxes to which *party-list systems* of PR—in which voters can vote for one party, which receives representation in parliament proportional to its number of votes—are described in Van Deemen (1993). I do not analyze such systems here, because they achieve PR directly, at least to the degree that there is agreement on a method of rounding. See Balinski and Young (1982, 2001) and Young (1994, ch. 3) for an analysis of different rounding methods.

[2] Baharad and Nitzan (2005) propose different constraints for what they call "restricted approval voting," which restricts the minimal and maximal number of candidates that voters can approve of but not, as here, the number or kinds of candidates that can be elected in different categories.

After considering several different PR systems, the association had found each wanting in ensuring the kind of fair representation that it desired. For one thing, the leadership of the association did not want to encourage groups within the association to act like political parties that blatantly campaign for seats to increase their shares of votes on the governing board. Indeed, the leadership viewed with repugnance the factionalism that such campaigns might induce in the organization.[3]

At the same time, the leadership wished to consider the possibility of a PR system that would smooth over existing divisions within the association by giving to aggrieved groups what they considered their fair shares of seats on the board. The representation problem was complicated, however, by the fact that there was a desire not only that groups representing different specialties be represented but also that different regional interests be represented, too (the association had offices worldwide). This problem is illustrated by an example in section 4.2.

After much discussion, the system that attracted the most interest was CAV, because it placed constraints on the number of candidates that could be elected in different categories. It ensured that no category got too few or too many seats according to its size. These constraints, and the problems they give rise to, are analyzed in sections 4.3 and 4.4.

The winners under CAV are the candidates most approved of by all the voters, subject to the constraints. In the example analyzed in section 4.5, the constraints mandate the election of one-half the candidates to be elected; the other half are the candidates who are most approved.

CAV significantly modifies the purpose for which AV was originally designed. It also raises a number of questions about the properties of the constraints, and their likely effects on the representation of different interests on the board, which I explore in section 4.5.

In section 4.6 I briefly analyze two voting systems, cumulative voting and additional-member voting systems, in which voters are not restricted to approving or disapproving of candidates but can express themselves in other ways. These systems, which are used in different jurisdictions, offer alternative ways of achieving PR. I summarize the results and draw some conclusions in section 4.7.

4.2. BACKGROUND

Prior to its consideration of CAV, members of the association that I advised had voted in previous elections for a slate of candidates prepared by a nominating

---

[3] For a detailed examination of a committee (the Federal Open Market Committee of the U.S. Federal Reserve System) in which factionalism is muted, see Chappell, McGregor, and Vermilyea (2005).

committee. The committee prepared a slate of about twice as many candidates as there were seats to be filled; members could vote only for as many candidates as there were seats to be filled—no more and no less. Those candidates with the most votes were elected to fill the open seats over a multiyear cycle. If the cycle were two years, for example, board members would serve two years, with half the seats being filled in each annual election.

No change was contemplated in the size of the board or in the terms of office of its members. The issue was whether to elect board members by constituencies representing different interests of the association. What made the problem unusual was that a constituency was defined by two dimensions—region and specialty—rendering the problem of assigning seat shares more difficult than if only one dimension (e.g., a left-right policy continuum, along which political parties may take positions) needed to be represented.[4]

To illustrate the two-dimensional problem, assume that there are two *regional* divisions of the association (A and B) and three *specialty* divisions (X, Y, and Z). The percentages of members that fall into each category can be shown in a 2 × 3 matrix, where the rows indicate the regional divisions and the columns the specialty divisions (see Matrix 4.1).

MATRIX 4.1
Specialty

Region	X	Y	Z	Row total
A	27	16	17	60
B	21	9	10	40
Column total	48	25	27	100

These percentages may be interpreted as targets for the composition of the board.

The targets in any election will depend not only on the number of members that fall into each category but also on the number of board members continuing in each category. For example, if elections are held over a two-year cycle, and cell AY already contains 24 percent of the continuing board members, the 16 percent shown in the table should be reduced to 8 percent as a target—and underrepresented cells increased accordingly—to ensure that members from

---

[4] The mathematical problem, nonetheless, of assigning an integer number of seats in a parliament to parties, based on the votes they receive, is not a trivial one. Balinski and Young (1982, 2001) offer a thorough analysis of this problem—mainly in the context of assigning seats to congressional districts based on their populations—and also provide an engrossing history of this problem; for a critical discussion of their proposed solution to the one-dimensional apportionment problem, see Brams and Straffin (1982). See also Woodall (1986a), Young (1994, ch. 3), and Edelman (2006a), who proposes a new method that he argues better meets the stipulations of the U.S. Supreme Court.

AY do not continue to be overrepresented but instead are properly represented at the 16 percent level (the average of 24 percent and 8 percent) on the next board.

The percentages shown in the table are an ideal: given that each board member has one vote, no allocation of board members (and therefore votes) to each category may mirror the percentages perfectly. Although a system of weighted voting could lead to a better fit, no consideration was given to endowing different board members with different numbers of votes that mirror the above percentages in the matrix.

For the next step, assume that the percentages are the targets, and six new members are to be elected to the board. (By coincidence, this number exactly matches the number of cells.) Multiplying the target percentages by 6 gives the *target election figures* (TEFs) shown in Matrix 4.2.

MATRIX 4.2
Specialty

Region	X	Y	Z	Row total
A	1.62	0.96	1.02	3.60
B	1.26	0.54	0.60	2.40
Column total	2.88	1.50	1.62	6.00

Manifestly, the remainders of each of the TEFs preclude a perfect matching of (whole) representatives to the cells. But this does not prevent narrowing the possibilities to those that are, in some sense, best-fitting.

4.3. CONTROLLED ROUNDINGS

To determine what integer representations are best-fitting, consider the following constraints:

1. *Row and column minima.* Rounded down, the TEF column totals are 2, 1, and 1; the row totals are 3 and 2. Assuming these as *minima* for the totals of regional and specialty representatives, respectively, there are 65 distinct cases that satisfy these constraints and whose cell entries sum to 6, as exhaustively enumerated in Table 4.1.

There are systematic procedures, but there seem to be no efficient algorithms, for generating all these cases. For a 2×3 matrix, a hand calculation is feasible; for larger matrices, the association used computer spreadsheets to find integer allocations, reflecting finer breakdowns of the association.

2. *Row and column maxima.* Rounded up, the TEF column totals are 3, 2, and 2; the row totals are 4 and 3. Assuming these as *maxima* for the totals of

TABLE 4.1

Sixty-Five Cases That Satisfy Constraint 1

Sum of First Row = 4 (30 Cases)

4 0 0	3 1 0	3 1 0	3 1 0	3 0 1	3 0 1	3 0 1	2 2 0	2 2 0
0 1 1	1 0 1	0 1 1	0 0 2	1 1 0	0 2 0	0 1 1	1 0 1	0 1 1
2 2 0	2 1 1	2 1 1	2 1 1	2 1 1	2 1 1	2 1 1	2 0 2	2 0 2
0 0 2	2 0 0	1 1 0	1 0 1	0 2 0	0 1 1	0 0 2	1 1 0	0 2 0
2 0 2	1 3 0	1 2 1	1 2 1	1 2 1	1 1 2	1 1 2	1 1 2	1 0 3
0 1 1	1 0 1	2 0 0	1 1 0	1 0 1	2 0 0	1 1 0	1 0 1	1 1 0
0 3 1	0 2 2	0 1 3						
2 0 0	2 0 0	2 0 0						

Sum of First Row = 3 (35 Cases)

3 0 0	3 0 0	3 0 0	2 1 0	2 1 0	2 1 0	2 1 0	2 1 0	2 1 0
1 1 1	0 2 1	0 1 2	2 0 1	1 1 1	1 0 2	0 2 1	0 1 2	0 0 3
2 0 1	2 0 1	2 0 1	2 0 1	2 0 1	2 0 1	1 2 0	1 2 0	1 2 0
2 1 0	1 2 0	1 1 1	0 3 0	0 2 1	0 1 2	2 0 1	1 1 1	1 0 2
1 1 1	1 1 1	1 1 1	1 1 1	1 1 1	1 1 1	1 0 2	1 0 2	1 0 2
3 0 0	2 1 0	2 0 1	1 2 0	1 1 1	1 0 2	2 1 0	1 2 0	1 1 1
0 3 0	0 2 1	0 2 1	0 2 1	0 1 2	0 1 2	0 1 2	0 0 3	
2 0 1	3 0 0	2 1 0	2 0 1	3 0 0	2 1 0	2 0 1	2 1 0	

Note: Row and column sums are no less than the row and column TEFs rounded down (minima).

regional and specialty representatives, respectively, 30 of the 65 cases satisfying constraint 1 are excluded. Specifically, the row maxima exclude none of the 65 cases, but the first, second, and third column maxima exclude 8, 11, and 11 cases, as shown in Table 4.2.

These column-maxima constraints are mutually exclusive: the cases excluded by each one are not excluded by either of the other two. Thirty-five cases remain admissible.

3. *Cell minima and maxima.* Rounding down and up the TEFs of all the cells gives a minimum and a maximum for each cell. Satisfying these minimal and maximal cell constraints reduces the 35 admissible cases meeting constraints 1 and 2 to just 10 "controlled roundings," as shown in Table 4.3. Note that the sum of the first row in the first five cases is 4, whereas this sum in the last five cases is 3.

The three constraints, applied progressively, have reduced the number of admissible cases from 65 (constraint 1) to 35 (constraints 1 and 2) to 10

TABLE 4.2

Sixty-Five Cases That Satisfy Constraint 1 and Are Either Excluded or Included by Constraint 2

30 Cases Excluded

  8 Cases in Which the *First* Column Sums to More Than 3

4 0 0	3 1 0	3 0 1	2 1 1	3 0 0	2 1 0	2 0 1	1 1 1
0 1 1	1 0 1	1 1 0	2 0 0	1 1 1	2 0 1	2 1 0	3 0 0

  11 Cases in Which the *Second* Column Sums to More Than 2

0 2 2	2 1 1	1 3 0	1 2 1	0 3 1	2 1 0	2 0 1	1 2 0	1 1 1
0 1 1	0 2 0	1 0 1	1 1 0	2 0 0	0 2 1	0 3 0	1 1 1	1 2 0

0 3 0	0 2 1
2 0 1	2 1 0

  11 Cases in Which the *Third* Column Sums to More than 2

2 1 1	2 0 2	1 1 2	1 0 3	0 1 3	2 1 0	2 0 1	1 1 1	1 0 2
0 0 2	0 1 1	1 0 1	1 1 0	2 0 0	0 0 3	0 1 2	1 0 2	1 1 1

0 1 2	0 0 3
2 0 1	2 1 0

35 Cases Included

  16 Cases in Which the First Row Sums to 4

3 1 1	3 1 0	3 0 1	3 0 1	2 2 0	2 2 0	2 1 1	2 1 1	2 1 1
0 1 1	0 0 2	0 2 0	0 1 1	1 0 1	0 0 2	1 1 0	1 0 1	0 1 0

2 0 2	2 0 2	1 2 1	1 2 1	1 1 2	1 1 2	0 2 2
1 1 0	0 2 0	2 0 0	1 0 1	2 0 0	1 1 0	2 0 0

  19 Cases in Which the First Row Sums to 3

3 0 0	3 0 0	2 1 0	2 1 0	2 1 0	2 0 1	2 0 1	2 0 1	1 2 0
0 2 1	0 1 2	1 1 1	1 0 2	0 1 2	1 2 0	1 1 1	0 2 1	2 0 1

1 2 0	1 1 1	1 1 1	1 1 1	1 0 2	1 0 2	0 2 1	0 2 1	0 1 2
1 0 2	2 1 0	2 0 1	1 1 1	2 1 0	1 2 0	3 0 0	2 0 1	3 0 0

0 1 2
2 1 0

Note: Row and column sums are no more than the row and column TEFs rounded up (maxima).

TABLE 4.3

Ten Controlled Roundings That Satisfy Constraints 1 and 2 and, in Addition,
Constraint 3

#1	#2	#3	#4	#5	#6	#7	#8	#9	#10
2 1 1	2 1 1	1 1 2	1 1 2	2 0 2	2 0 1	1 1 1	1 1 1	1 1 1	1 0 2
1 1 0	1 0 1	2 0 0	1 1 0	1 1 0	1 1 1	2 1 0	2 0 1	1 1 1	2 1 0

Note: Cell entries are no less than the cell TEFs rounded down (minima), and no more than the
   cell TEFs rounded up (maxima).

(constraints 1, 2, and 3) whose cell entries, and column and row totals, sum to
the grand total of 6. Satisfying these three constraints results in what is called a
controlled rounding, which can always be found for any matrix (Cox and Ernst,
1982). For larger arrays (i.e., three dimensions or more), however, a controlled
rounding may not exist (Fagan, Greenberg, and Hemmig, 1988).

   A controlled rounding can be defined more straightforwardly as one in
which, for every column and row, the sum of its cell TEFs, rounded down or
up, equals the column or row (total) TEF, rounded down or up, with the roundings summing to some grand total. Constraints 1 and 2 give all possible cases
that are roundings of the column and row TEFs and sum to 6; constraint 3
limits these to those that are also roundings of the cell TEFs.

4.4. FURTHER NARROWING: THE SEARCH MAY BE FUTILE

One could reduce the number of cases still further by invoking various criteria. For example, define an integer representation to be *cell-consistent* if the
TEF of a cell that is assigned a larger integer is at least as great as one that is
assigned a smaller integer. In controlled-rounding case #1, the TEF of cell BY
is 0.54 and that of BZ is 0.60; yet BY is assigned a 1 and BZ a 0, which makes
this representation cell-inconsistent.

   In fact, the only two cases that are cell-consistent are #2 and #9. Allocation
#2 is cell-consistent because cell AX is the largest TEF (1.62) and receives the
only two seats that are assigned to a cell; BY is the smallest TEF (0.54) and
receives the only zero. Allocation #9 is cell-consistent because 1's are assigned to all cells, so no smaller TEF receives a larger assignment than a
larger TEF. When the integer assignments to the column and row sums are
also consistent, the allocation is said to be *consistent*.

   Another criterion for reducing the number of integer representations—in
this case, to exactly one—is the Hamilton method of rounding (Balinski and
Young, 1982, 2001), which has two steps:

1. Allocate to each category—both the six cells and the column and row sums—the integer portion of its TEF (i.e., its number to the left of the decimal point).
2. Of those seats remaining (out of the six to be allocated in the example), allocate them to the TEFs with the largest remainders—starting with the TEF with the biggest remainder—until the six seats are exhausted.

To illustrate the Hamilton method for the TEFs given in Matrix 4.2, the integer allocations according to step 1 are shown in Matrix 3.3 (note that the column sums total 4 and the row sums total 5).

MATRIX 4.3

$$
\begin{array}{ccc|c}
1 & 0 & 1 & 3 \\
1 & 0 & 0 & 2 \\
\hline
2 & 1 & 1 & 4/5
\end{array}
$$

The remaining seats are now allocated, according to step 2, on the basis of the TEFs having the largest remainders:

- three seats to cells in the 2×3 matrix (to which three seats have already been allocated)
- two seats to the column sums (to which four seats have already been allocated)
- one seat to the row sums (to which five seats have already been allocated)

These assignments give as a final allocation the figures shown in Matrix 4.4.

MATRIX 4.4

$$
\begin{array}{ccc|c}
2 & 1 & 1 & 4 \\
1 & 0 & 1 & 2 \\
\hline
3 & 1 & 2 & 6
\end{array}
$$

This allocation, called a *Hamilton allocation*, is the same as (consistent) allocation #2 in Table 4.3. Allocation #9, on the other hand, is consistent but not Hamilton, which illustrates

Proposition 4.1. *Hamilton allocations are always consistent, but consistent allocations are not always Hamilton.*

Proof. The second part of this proposition is proved by allocation #9. The first part follows from the fact that, by step 1 of the Hamilton method, TEFs with larger integer portions never receive fewer seats than TEFs with smaller integer portions; by step 2, TEFs with larger remainders never receive fewer seats than TEFs with smaller remainders. Hence, larger TEFs can never be assigned fewer seats than smaller TEFs. ■

Controlled roundings #3 and #5, in addition to the Hamilton allocation (#2), have column and row sums identical to that of the Hamilton allocation. However, these allocations are cell-inconsistent and hence inconsistent. By Proposition 4.1, they cannot be Hamilton, because Hamilton allocations are a subset of consistent allocations. (Except for possible ties, in which a seat might be randomly assigned at step 1 or step 2, Hamilton allocations are unique.)

So far, it would appear, a Hamilton allocation is the most sensible of the (consistent) controlled rounding allocations. But there is a rub: one may not exist.

For example, the percentages and TEFs shown in Matrix 4.5 and Matrix 4.6 for the allocation of six seats differ very little from those used in the earlier example (see Matrix 4.1 and Matrix 4.2).

MATRIX 4.5

Percentages

30	10	19	59
18	12	11	41
48	22	30	100

MATRIX 4.6

Target Election Figures

1.80	0.60	1.14	3.54
1.08	0.72	0.66	2.46
2.88	1.32	1.80	6.00

Now applying the Hamilton method to the TEFs—both the cells and the column and row sums—one obtains the allocations for each shown in Matrix 4.7.

MATRIX 4.7

2	0	1	4
1	1	1	2
3	1	2	6

Although the column allocations sum to their Hamilton allocations, the first row sums to 3 (not 4), and the second row sums to 3 (not 2). This example proves that Hamilton allocations to the cells may not agree with Hamilton allocations to the column or row sums.

The fact that the Hamilton allocations in this example are unique, but not a controlled rounding, immediately implies

Proposition 4.2. *A controlled rounding that is Hamilton may not exist.*

A requirement of a controlled rounding is that the sums of the rounded cell TEFs for each column and row equal the corresponding rounded column and row TEFs.

Finally, to settle the question of the existence of a consistent controlled rounding, consider the percentages shown in Matrix 4.8 for a 2×2 matrix. If three seats are to be filled, the TEFs are given in Matrix 4.9.

MATRIX 4.8
Percentages

30	22	52
21	27	48
51	49	100

MATRIX 4.9
Target Election Figures

0.90	0.66	1.56
0.63	0.81	1.44
1.53	1.47	3.00

The only cell-consistent allocation is to assign one seat to all entries except the lowest (0.63). But the consistency of the column sums demands that the first column receive two seats, when in fact the sum of its cell-consistent entries $(0 + 1)$ is 1. This example proves the following:

Proposition 4.3. *A controlled rounding that is consistent may not exist.*

There is a final difficulty with cell-consistent controlled roundings, illustrated by the percentages and TEFs shown in Matrix 4.10 and Matrix 4.11.

MATRIX 4.10
Percentages

14.6	9.4	18.0	42.0
9.6	9.6	0.4	19.6
19.8	9.2	9.4	38.4
44.0	28.2	27.0	100.0

MATRIX 4.11

Target Election Figures

0.73	0.47	0.90	2.10
0.48	0.48	0.02	0.98
0.99	0.46	0.47	1.92
2.20	1.41	1.39	5.00

The cell-consistent allocation of seats shown in Matrix 4.12 is not consistent, because the first two columns and the second two rows do not sum to values consistent with their column and row TEFs:

MATRIX 4.12

1	0	1	2
1	1	0	1
1	0	0	2
2	2	1	5

However, a new difficulty arises in this example: the second row is entitled to only 0.98 seats, but the cell-consistent assignments for this row sum to 2.

When the discrepancy between the cell-consistent sum of a column or row and its TEF is greater than 1, it does not satisfy *quota* (Balinski and Young, 1982, 2001). Put another way, the column or row TEF, rounded either up or down, is not equal to the cell-consistent sum. In the example, 0.98 rounded up is 1, but the cell-consistent sum of the second row is 2, which proves the following:

Proposition 4.4. *A cell-consistent rounding may not satisfy quota.*

An assignment of seats that violates quota, of course, is not a controlled rounding.

Because the narrowing-down criteria I have discussed—consistent allocations and Hamilton allocations—may be incompatible with all controlled roundings, they cannot reliably be used to distinguish either a very few or a single best allocation. In addition, a cell-consistent allocation not only may be inconsistent but also may fail to satisfy quota—and therefore not be a controlled rounding.

Other criteria have been proposed for filtering out the best-fitting controlled roundings. For example, Balinski and Demange (1989) and Gassner (1991) provide analyses of technical criteria for finding best-fitting biproportional allocations (that is, those proportional to the TEFs in two dimensions, as here). In addition, they suggest their application in a political context, in which biproportionality might be based on political parties and geographical constituencies rather than specialties and regions.

L. Cox (1987) argues for unbiased controlled roundings and provides a computationally efficient procedure for finding them. Other approaches and algorithms have been proposed by Kelly, Golden, and Assad (1989), but they are not tied to fundamental principles of fair representation. These are the focus of concern of the apportionment methods discussed by Balinski (2006), Maier (2006), Pukelsheim (2006), and other contributors to Simeone and Pukelsheim (2006).

For the purpose of choosing an elected board that is a reasonable approximation of the TEFs, the fact that no controlled rounding may satisfy a requirement as weak as consistency—not to mention give allocations compatible with a specific apportionment method like Hamilton—casts doubt on the recommendation of exactly one allocation on purely theoretical grounds. Indeed, a case can be made that any of the controlled roundings is good enough: each cell receives representation within one seat of what it is entitled to, and so does each geographical region and functional category.

In the absence of compelling criteria for singling out a best controlled rounding in an election, I recommended an empirical solution to the association. Underlying it, however, was a theoretical rationale tied to the notion of voter sovereignty: let the voters themselves choose the outcome they most favor.

4.5. CONSTRAINED APPROVAL VOTING (CAV)

The association considered whether to require that members vote for a preset number of candidates, such as six—the number to be elected—in the example. But because newer members did not have sufficient knowledge to make more than two or three informed choices, it was felt that they were likely to be unduly influenced by the casual advice of more senior members. To prevent this, it was decided that all members should be permitted the more flexible option of voting for as many candidates as they liked. Hence, AV ballots would be used, but votes would not be aggregated in the usual way.

Instead, it was decided to restrict the domain of possible outcomes to counteract a possible bias that AV might introduce. Specifically, if members of the largest categories, like AX with 27 percent of the members in the example, tended to concentrate their votes on candidates in this category, they could inordinately affect the election outcome.

In fact, even under the extant system, in which voters were required to vote for an entire slate, the nominating committee "engineered" the slate to thwart voters from electing "too many" members of one type. (This was referred to as *slate engineering*, which might be roughly defined as rigging an election to produce certain desired results.) For example, if AX members were overrepresented on the continuing board, the nominating committee might propose AX candidates who were not well known in order to diminish their chances of election.

This form of manipulation is well known to political scientists (Riker, 1986). But it is an informal device that on occasion had not worked as planned, which is one reason why the association wanted to explore alternatives that offered more formal protection. Presumably, an election system that explicitly ensured the fair representation of different interests would also gain the confidence of voters. Consequently, whatever outcome it produced would be considered more legitimate.

If the admissible outcomes are restricted to the set of controlled roundings and not a particular one, then voters would decide not only who is elected from each cell but also, within limits, how many. Of course, limiting voters to outcomes in the set of controlled roundings is drastically different from using AV to elect single winners in multicandidate elections, with no restrictions on who can be elected.

Indeed, one might argue that the ten controlled roundings in the example are too restrictive a set. The 35 (or even 65) cases available if constraint 3 (or 2 as well) were lifted would give the voters more control in the choice of a board and hence greater sovereignty.

Thereby, the board's composition would be more responsive to their voting. Whereas the controlled roundings guarantee that an integer representation is no more than one seat from the TEFs, the less restrictive set of, say, 35 cases would permit 25 additional outcomes—each of which leads to the election of at least one different candidate—and still guarantee column and row sums within one seat of their TEFs.

The acceptability of this set versus the ten controlled roundings depends upon the importance one attaches to the principle that the number of cell seats—compared with the regional and specialty totals—should all be within one seat of the TEFs. To put this matter somewhat differently, the designation of what outcomes are admissible depends on whether one thinks more popular (i.e., approved) candidates should be permitted to win at the price of causing deviations from the cell TEFs greater than one seat.

Once the voters have chosen the set of outcomes they deem admissible, the outcome they select under CAV will be that with the greatest total number of approval votes. For example, if the admissible outcomes are the ten controlled roundings in the earlier example, they have in common the certain election of exactly one candidate from the three cells AX, AZ, and BX, as shown in Matrix 4.13.

MATRIX 4.13

$$\begin{vmatrix} 1 & 0 & 1 \\ 1 & 0 & 0 \end{vmatrix}$$

This means that the biggest vote-getters in each of these cells will be guaranteed election, whichever of the ten controlled roundings wins. The votes for these candidates can then be set aside.

The particular controlled rounding that wins will be the one in which the total approval vote for the three remaining "discretionary" choices is greatest. In case #1, for example, the runner-up in cell AX, the winner in cell AY, and the winner in cell BY would complete the six choices. The total of the votes for these three candidates would be compared with analogous totals in the nine other cases—given by the sums of the votes of the three best-performing candidates in the appropriate cells who were not certain winners—to determine the winning controlled rounding.[5]

If one admits the 35 cases that meet constraints 1 and 2, what they have in common is that no candidates are guaranteed election from any cell (see Figure 4.2). In this set of cases, all six choices are discretionary, although the constraints that these cases—as well as the ten controlled roundings—must satisfy still impose restrictions on the possible outcomes (e.g., four discretionary choices cannot all be chosen from one cell). At the least stringent level of criterion 1 alone, the election of four candidates from cell AX is permitted in one case (see Figure 4.1), which is more than a two-seat (122 percent) deviation from its TEF of 1.80.

In my opinion, this deviation is too large, and I therefore recommend tighter restrictions. Both the ten controlled roundings and the 35 cases that ensure the row and column totals will be within one seat of the TEFs seem acceptable to me. Making this choice, the association could determine how much leeway it wants to permit the voters.

Clearly, AV, by allowing voters to vote for as many candidates as they like, gives voters greater discretion than does restricting their votes to a fixed number. But where one draws the line to preclude the election of candidates who would not form a representative board is a value judgment. This analysis, I believe, clarifies the trade-offs that this judgment entails.

4.6. UNCONSTRAINING VOTES: TWO ALTERNATIVES TO CAV

There are other ways of achieving PR besides putting voters into categories, which I will explore in later chapters. Before turning to these, however, I briefly discuss two voting systems that achieve PR by giving voters greater freedom to express themselves than does AV or CAV. In one system, voters can allocate different numbers of votes to candidates, whereas in the other they can cast a vote not only for a candidate but also for a party.

---

[5] Integer programs that efficiently accomplish this calculation are given in Potthoff (1990) and Straszak, Libura, Sikorski, and Wagner (1993).

Cumulative Voting

Cumulative voting is a voting system in which each voter is given a fixed number of votes to distribute among one or more candidates; the candidates with the most votes win. This system allows voters to express their intensities of preference rather than simply to rank candidates, as under STV and the Borda count. It is a PR system in which minorities can ensure their approximate proportional representation by concentrating their votes on a subset of candidates commensurate with their size in the electorate.

To illustrate this system and the calculation of optimal strategies under it, assume there is a single minority position among the electorate favored by one-third of the voters and a majority position favored by the remaining two-thirds. Assume, further, that the electorate comprises 300 voters, and a six-member governing body is to be elected. The six candidates elected will be those with the greatest numbers of votes.

If each voter has 6 votes to cast for as many as six candidates, and if each of the 100 voters in the minority casts 3 votes each for only two candidates, these voters can ensure the election of these two candidates no matter what the 200 voters in the majority do. To see this, notice that each of the two minority candidates will get a total of 300 (100 × 3) votes, whereas the two-thirds majority, with a total of 1,200 (200 × 6) votes to allocate, can at best match this number for four candidates (1,200/4 = 300).

If the two-thirds majority instructs its supporters to distribute their votes equally among five candidates (1,200/5 = 240), it will not match the vote totals of the two minority candidates (300 each) but can still ensure the election of four (of its five) candidates—and possibly get its fifth candidate elected if the minority puts up three candidates and instructs its supporters to distribute their votes equally among the three (giving each 600/3 = 200 votes).

Against these strategies of either majority (support five candidates) or the minority (support two candidates), it is easy to show that neither side can improve its position. To elect five (instead of four) candidates with 301 votes each, the majority would need 1,505 instead of 1,200 votes, holding constant the 600 votes of the minority. Similarly, for the minority to elect three (instead of two) candidates with 241 votes each, it would need 723 instead of 600 votes, holding constant the 1,200 votes of the majority.

It is evident that the optimal strategy for the leaders of both the majority and minority is to instruct their members to allocate their votes as equally as possible among certain numbers of candidates. The number of candidates they should support will be proportionally about equal to the number of their supporters in the electorate (if known).

Any deviation from this strategy—for example, by putting up a full slate of candidates and not instructing supporters to vote for only some on this slate—offers the other side an opportunity to capture more than its proportional share of the seats. Patently, good planning and disciplined supporters are required in order to be effective under this system. In the face of uncertainty about their level of support in the electorate, party leaders may well make nonoptimal choices about how many candidates their supporters should concentrate their votes on, which weakens the argument that cumulative voting can in practice guarantee PR.

A systematic analysis of optimal strategies under cumulative voting is given in Brams (1975), including the role that uncertainty may play. These strategies are compared with strategies actually adopted by the Democratic and Republican parties in elections for the Illinois General Assembly, where cumulative voting was used until 1982.

Cumulative voting has also been used in elections for some boards of directors, which has enabled shareholders representing minority positions to gain seats on these boards. In 1987 cumulative voting was adopted by two cities in the United States (Alamorgordo, New Mexico, and Peoria, Illinois)—and other small cities more recently—to satisfy court requirements of minority representation in municipal elections. Cumulative voting has most often been used as a remedy in cities in which minority voters are widely scattered, rendering impractical the creation of so-called minority districts under a districting system; experience with its use is analyzed in Bowler, Donovan, and Brockington (2003).

Additional-Member Voting Systems

In most parliamentary democracies, it is not candidates who run for office but political parties that put up lists of candidates. Under party-list voting, voters vote for parties, which receive representation in a parliament proportional to the number of votes they receive. Usually there is a threshold, such as 5 percent, which a party must exceed in order to gain any seats in the parliament.

This is a rather straightforward means of ensuring the proportional representation of parties that attain or surpass the threshold. More interesting are *additional-member voting systems*, in which some legislators are elected from districts, but additional members may be added to the legislature to ensure, insofar as possible, that the parties underrepresented on the basis of their national-vote proportions gain additional seats.

Denmark and Sweden, for example, use total votes, summed over each party's district candidates, as the basis for allocating additional seats. In elections to Germany's Bundestag and Iceland's Parliament, voters vote twice, once for district representatives and once for a party. Half of the Bundestag is chosen from party lists, on the basis of the national party vote, with adjustments made to the district results so as to ensure the approximate proportional representation of parties. In 1993 Italy adopted a similar system for its Chamber of Deputies, except that three-quarters rather than one-half of the seats are filled by deputies from the districts; the remaining one-quarter is used to approximate PR to the extent possible. Other additional-member voting systems, sometimes referred to as mixed-member proportional systems, include those of New Zealand, Scotland, and Wales.

In Puerto Rico, no fixed number of seats is added unless the largest party in one house of its bicameral legislature wins more than two-thirds of the seats in district elections. When this happens, that house can be increased by as much as one-third to ameliorate the possible underrepresentation of minority parties.

To offer some insight into an important strategic feature of additional-member systems, assume, as in Puerto Rico, that additional members can be added to a legislature to adjust for underrepresentation, but this number is variable. More specifically, consider a voting system, called *adjusted district voting*, or ADV (Brams and Fishburn, 1984b, 1984c), that has the following attributes:

1. There is a jurisdiction divided into equal-size districts, each of which elects a single representative to a legislature.
2. There are two main factions in the jurisdiction, one majority and one minority, whose size can be determined. For example, if the factions are represented by political parties, their respective sizes can be determined by the votes that each party's candidates, summed across all districts, receive in the jurisdiction.
3. The legislature consists of all representatives who win in the districts *plus* the the largest vote getters among the losers—necessary to achieve PR—if it is not realized in the district elections. Typically, this adjustment will involve adding minority-faction candidates, who lose in the district races, to the legislature, so that it mirrors the majority-minority breakdown in the electorate as closely as possible.
4. The size of the legislature is *variable*, with a lower bound equal to the numbers of districts (if no adjustment is necessary to achieve PR), and an upper bound equal to twice the number of districts (if a nearly 50 percent minority wins no seats).

As an example of ADV, suppose that there are eight districts in a jurisdiction. If there is an 80 percent majority and a 20 percent minority, the majority is likely to win all the seats unless there is an extreme concentration of the minority in one or two districts.

Suppose the minority wins no seats. Then its two biggest vote getters could be given two "extra" seats to provide it with representation of 20 percent in a body of ten members, exactly its proportion in the electorate.

Now suppose the minority wins one seat, which would provide it with representation of $1/8 \approx 13$ percent. If it were given an extra seat, its representation would rise to $2/9 \approx 22$ percent, which would be closer to its 20 percent proportion in the electorate. However, assume that the addition of extra seats can never make the minority's proportion in the legislature exceed its proportion in the electorate, which I call the *proportionality constraint.*

Paradoxically, the minority would benefit by winning no seats and then being granted two extra seats to bring its proportion up to exactly 20 percent. To prevent a minority from benefiting by *losing* in district elections, assume the following *no-benefit constraint*: the allocation of extra seats to the minority can never give it a greater proportion in the legislature than it would obtain had it won more district elections.

How would this constraint work in the example? If the minority won no seats in the district elections, then the addition of two extra seats would give it $2/10 = 1/5$ representation in the legislature, exactly its proportion in the electorate. But I just showed that if the minority had won one seat, it would not be entitled to an extra seat—and $2/9$ representation in the legislature—because of the proportionality constraint. Hence, its representation would have to remain at $1/8$ if it won in exactly one district.

Because $1/5 > 1/8$, the no-benefit constraint prevents the minority from gaining two extra seats if it wins no district seats initially. Instead, it would be entitled in this case to only one extra seat, giving it $1/9$ rather than $1/5$ representation; since $1/9 < 1/8$, the no-benefit constraint is satisfied.

But $1/9 \approx 11$ percent is only about half the minority's 20 percent proportion in the electorate. In fact, one can prove in the general case that the no-benefit constraint may prevent a minority from receiving up to about half of the extra seats it would be entitled to—on the basis of its total vote—were the constraint not operative and it could therefore get up to this proportion (e.g., two out of ten seats in the example) in the legislature (Brams and Fishburn, 1984a).

The no-benefit constraint may be interpreted as a kind of "strategy-proofness" feature of ADV: it makes it unprofitable for a minority party deliberately to lose in a district election in order to do better after the adjustment that gives it extra seats. But strategy-proofness, in precluding any possible advantage that might accrue to the minority from throwing a district election, has a price.

As the example demonstrates, it may severely restrict the ability of ADV to satisfy PR (11 percent versus 20 percent for the minority), giving rise to the following dilemma. Under ADV, one cannot guarantee PR if one wishes to satisfy both the proportionality constraint and the no-benefit constraint, which allows the minority only 11 percent representation in the example. But dropping the no-benefit constraint allows the minority to obtain its full proportional representation (two out of ten seats in the augmented legislature). However, this gives the minority an incentive deliberately to lose in the district contests in order to do better after the adjustment (i.e., get its full 20 percent by winning no district seats).

It is worth pointing out that the "second chance" for minority candidates afforded by ADV would encourage them to run in the first place, because even if most or all of them are defeated in the district races, their biggest vote getters would still get a chance at the (possibly) extra seats in the second stage. But these extra seats might be cut by a factor of up to two from the minority's proportion in the electorate, assuming one wants to motivate district races with the no-benefit constraint.

Consider what might have happened under ADV after the June 1983 British general election. The Alliance (comprising the new Social Democratic Party and the old Liberal Party) got more than 25 percent of the national vote but less than 4 percent of the seats in Parliament (by coming in second in most districts behind the Conservative or Labor Party candidates). ADV would have assigned the Alliance 20 percent representation in an augmented 817-seat Parliament— 26 percent larger than the present 650-seat Parliament[6]—which is somewhat less than the Alliance's proportional 25 percent share that the no-benefit constraint precludes (Brams and Fishburn 1984b).

4.7. SUMMARY AND CONCLUSIONS

There is no perfect voting procedure—at least perfect for all purposes. Because different procedures satisfy different desiderata, one must decide on those desiderata one considers important in order to make an intelligent choice.

In this chapter, I focused on a new system, constrained approval voting (CAV), that is designed to ensure PR within certain constraints. In the example I discussed, there were ten controlled roundings that satisfied these constraints, which reflected the division of a society into three specialties and two regions. One-half of its six representatives were mandated by this rounding— they had to be the biggest vote getters in three of the six categories—but the other three representatives were the most approved candidates from one of the ten controlled roundings.

Did the professional association for which I proposed CAV adopt it? A majority of the committee that had been formed to consider election-reform proposals—and make recommendations to the board—thought that breaking the association down into specialty and regional categories would violate its unitary philosophy. Instead, they wanted its members to view it as a single entity.

Consequently, the association decided to continue to use slate engineering to ensure, insofar as possible, a representative board. And once it had rejected categorizing candidates, it saw AV as a secondary issue and did not deem it desirable, even with constraints.

---

[6] Under ADV, it might be advisable to cut considerably the number of districts in Britain so that, after adjustments are made, Parliament would generally be close to its present size and therefore not too unwieldy.

The failure of the professional association to adopt CAV seems also to have been a function of another factor. Most of the work done on the design of CAV was with a subcommittee of five people, but the full committee that made the final decision on adoption had ten additional members. Although these people received extensive written and oral reports summarizing the work of the subcommittee, they probably did not fully appreciate CAV's advantages over slate engineering, which had worked fairly well in the past.

The options (including CAV) presented to the full committee when it convened were phrased neutrally; moreover, ample time was given to discuss the arguments for CAV. However, despite rumblings of discontent about the makeup of the board, there was no crisis at the time. Furthermore, because the committee had previously considered and rejected other voting systems, including ranking systems like BC and STV, most members felt that they had given due consideration to possible alternatives. In short, there seemed no overriding reason to make a change.

Sometime later, the association went through a crisis and split in two. One of the two entities that formed experienced a second crisis and no longer exists today.

Cumulative voting and additional-member voting systems drop the constraint of approving or disapproving of candidates and allow voters to express themselves in different ways. Cumulative voting offers a means for parties (or other groups) to guarantee their proportional representation, whatever the strategies of other parties, by concentrating their multiple votes on a number of candidates proportional to the party's size in the electorate. However, its effective use requires considerable organizational efforts on the part of parties—and the disciplined behavior of their members—especially in the face of uncertainty about the parties' level of support in the electorate before the election.

Additional-member voting systems, and specifically adjusted district voting (ADV) that results in a variable-size legislature, provide a mechanism for approximating PR in a legislature without the organizational efforts or discipline required of cumulative voting. But imposing the no-benefit constraint on the allocation of additional seats—in order to eliminate the incentive of parties to throw district races—also vitiates fully satisfying PR, underscoring the difficulties of satisfying a number of desiderata.

In subsequent chapters, I explore other ways of electing a representative committee or council, or achieving PR in a legislature, with and without approval balloting. An understanding of difficulties each procedure encounters, and the possible trade-offs they entail, facilitates the selection of a procedure to meet the needs one considers most pressing.

5

Electing Multiple Winners:
The Minimax Procedure

5.1. INTRODUCTION

In this chapter I analyze a voting procedure, called the *minimax procedure*, which is designed to elect committees whose members are representative of the electorate as a whole. It is based on approval balloting—whereby voters approve of as many candidates as they like, as under approval voting (AV) and constrained approval voting (CAV)—but votes are not aggregated in the usual manner.[1]

Instead of selecting the candidates that receive the most votes, the minimax procedure selects the set of candidates that minimizes the maximum "Hamming distance" to voters' ballots, where these ballots are weighted by their proximity to other voters' ballots. This set of candidates constitutes the *minimax outcome*. I define and illustrate Hamming distance in section 5.2 and show how the "proximity weighting" of this distance is determined. I also offer a geometric interpretation of minimax outcomes.

In contrast to the minimax outcome, the set of candidates that minimizes the sum of the Hamming distances to all voters is the *minisum outcome*. In fact, this is the set of majority winners under AV, but if fewer or more candidates are to be elected, the minisum outcome is the set of candidates receiving the most votes (whether a majority or not). After defining and illustrating minimax and minisum outcomes, I give examples in section 5.3 in which tied and nontied minimax and minisum outcomes may be diametrically opposed.

When committees of two or more candidates are to be elected, there are good reasons for preferring a minimax over a minisum outcome. A minimax outcome ensures that no voter is "too far away" from the committee that is elected—based on proximity-weighted Hamming distances—whereas minisum outcomes

---

Note: This chapter is adapted from Brams, Kilgour, and Sanver (2007) with permission of Springer Science and Business Media.

[1] Merrill and Nagel (1987) distinguish between a balloting method and a procedure for aggregating voter choices on the ballot, which need not be a numerical threshold (e.g., a plurality or majority of votes). Throughout I assume the balloting method is approval balloting and compare different ways of *aggregating* approval votes (minimax versus minisum).

ensure that voters will, on average, be closer to the committee, even though a few voters may be far away.

In section 5.4 I discuss the applicability of the minimax and minisum procedures when there are restrictions on the possible committees to be elected, either in size or in composition. In section 5.5 I show that while the minisum procedure is not manipulable, the minimax procedure is (when preferences are based on Hamming distance), though in practice the minimax procedure is probably almost as invulnerable as the minisum procedure because of its computational complexity.

In section 5.6 I analyze the 2003 Game Theory Society (GTS) election of twelve new members to the GTS council from a list of twenty-four candidates. There were $2^{24} \approx 16.8$ million possible ballots under approval balloting, because each voter could approve, or not, each of the twenty-four candidates. Given this huge number of possible ballots, it is hardly surprising that all but two of the 161 GTS members who voted in this election cast different ballots.[2]

In section 5.7 I conclude that the minimax procedure is a viable alternative to the minisum procedure for electing committees. Besides professional societies like the GTS, colleges, universities, and other organizations that rely substantially on committees to make recommendations and decisions seem good candidates for the minimax procedure.

In other arenas, such as faction-ridden countries like Afghanistan and Iraq, the minimax procedure could facilitate the choice of councils and cabinets that mirror the diversity of interests in the electorate. (There is an ongoing debate on which electoral systems best promote the consolidation of democracy, especially in ethnically divided societies; it is joined in Diamond and Plattner [2006], whose authors compare the experiences of a variety of countries.) The minimax procedure could also be used to resolve multi-issue disputes. In fact, a simplified version of this procedure would have led to a different outcome

---

[2] If all ballots are assumed equiprobable, the probability that no two (of the 161) voters cast identical ballots is $[(s)(s-1) \ldots (s-159)(s-160)]/s^{161}$, where s is the number of possible ballots (16,774,216 in this case). This follows from the fact that the first voter can cast one of s different ballots; for each of these, there are $(s-1)$ ways for the second voter to cast a different ballot; and so on to the 161st voter. The product of these numbers, divided by the number of possible ballots, s^{161}, gives the probability that no two voters cast identical ballots; the complement of this probability is the probability that at least two voters cast the same ballot. In the GTS election, the latter probability was only 0.000768, or less than 1 in 1,000, indicating that it was highly improbable that two or more voters would cast the same ballot, given all ballots are equiprobable (also highly unlikely). These calculations are similar to those used to solve the "birthday problem" in probability theory, which asks how many people must be in a room to make the probability greater than one-half that at least two people have the same birthday (the answer is twenty-three or more). In section 5.5 I define a more "empirical" probability, based on the number of voters voting for different numbers of candidates, which suggests that the probability that some ballots are identical is much higher.

from that achieved in oil-pollution treaty negotiations of thirty-two countries in 1954 (Brams, Kilgour, and Sanver, 2004, 2007).

5.2. MINISUM AND MINIMAX OUTCOMES

Assume there are n voters and k candidates. Under approval balloting, a *ballot* is a binary k-vector, $(p_1, p_2, \ldots, p_k)$, where p_i equals 0 or 1. These binary vectors indicate the approval or disapproval of each candidate by a voter.

To simplify notation, I write ballots such as (1, 1, 0) as 110, which indicates that the voter approves of candidates 1 and 2 but disapproves of candidate 3. (Vectors such as 110 are also used to represent election outcomes—that is, the committees that are chosen by the voters.) The number of distinct ballots, or possible election outcomes, is 2^k.

To illustrate the selection of representative committees based on the minisum and minimax criteria, consider the following example, in which 4 voters cast three distinct ballots for $k = 3$ candidates:

1 voter: 100
1 voter: 110
2 voters: 101

Under the standard method of aggregating approval votes, it is usual to ask which candidate receives the most votes.

Observe that candidate 1 receives approval from all 4 voters, candidate 2 from 1 voter, and candidate 3 from 2 voters, so candidate 1 is elected. If a committee is being chosen and the criterion for being elected is to receive a majority of votes, normally one would say that candidate 3, who is approved of by exactly half the voters, would not be elected, but the version of majority voting I use here allows for candidate 3 to be elected or not. That is, outcomes 101 and 100 are both majority voting outcomes.[3]

The *Hamming distance* between two ballots, p and q, is $d(p, q)$, the number of components on which they differ. For example, if $k = 3$ and a voter's ballot is 110, the distances, d, between it and the eight binary 3-vectors (including itself) are shown below:

Ballot	$d = 0$	$d = 1$	$d = 2$	$d = 3$
110	110	100	000	001
		010	101	
		111	011	

[3] In general, if there is a tie between the yes (1) and no (0) votes for a candidate, then there are multiple majority-voting outcomes, both including and excluding this candidate. Defining majority-voting outcomes in this way makes them coincide with minisum outcomes (more on this later).

Observe that there are three ballots at Hamming distance $d = 1$, and three more at $d = 2$; ballot 110 is at distance $d = 0$ from itself, and its *antipode*, the ballot on which all components differ, is at $d = 3$.

Define a *majority-voting* (MV) committee to be any subset of candidates that includes all candidates who receive more than $n/2$ voters and none that receive less than than $n/2$ votes, where n is the number of voters. Brams, Kilgour, and Sanver (2004, 2007, Proposition 4) proved that a committee is an MV committee if and only if the sum of the Hamming distances between all voters and the committee is a minimum. For this reason, I refer to MV committees as *minisum committees*.

As the 4-voter example demonstrates, there may be more than one MV committee (100 and 101). In general, an MV committee is not unique if and only if n is even and at least one candidate receives exactly $n/2$ votes. (If n is odd, MV committees will always be unique since no candidate can receive exactly half the votes.)

Minisum Committees with Count Weights

Following Kilgour, Brams, and Sanver (2006), I focus not on the individual ballots but on the distinct ballots, and the number of times that each is cast. For instance, committees 100 and 101 minimize the sum of the Hamming distances to all voters in the 4-voter example—or, equivalently, the sum of the Hamming distances to all distinct ballots weighted by the numbers of voters who cast each one. This is shown by the weighted Hamming distances to the

TABLE 5.1
Derivation of Minisum and Minimax Committees Based on Count Weights
(4-Voter Example)

Ballot:	100	110	101	Sum	Max
Count Weight:	1	1	2		
1. 000	1	2	4	7	4
2. 100	0	1	2	3[a]	2[a]
3. 010	2	1	6	9	6
4. 001	2	3	2	7	3
5. 110	1	0	4	5	4
6. 101	1	2	0	3[a]	2[a]
7. 011	3	2	4	9	4
8. 111	2	1	2	5	2[a]

[a]Minimum in column.

eight possible committees in Table 5.1. Call the weights *count weights*, because they count the numbers of voters who cast each ballot.

The sums of the entries in each row are shown in the Sum column of Table 5.1. Clearly, the two MV committees, 100 and 101, whose sums of 3 are noted, minimize the sum of the weighted Hamming distances. Choosing a committee that minimizes the sum of the weighted Hamming distances, based on count weights, is equivalent to choosing an MV committee. In the example, this committee always includes candidate 1 and may or may not include candidate 3.

Minimax Committees

Following Brams, Kilgour, and Sanver (2004, 2007) and Kilgour, Brams, and Sanver (2006), note that there are other ways to define the most representative committee. Instead of finding a committee that minimizes the sum of the Hamming distances to all ballots, find the committee(s) that minimize the maximum Hamming distance, weighted in each case by count weights. In the example, these are the three committees that tie with values of 2, which are noted, in the Max (for "maximum") column of Table 5.1.[4]

Note that a third committee, 111, ties with minisum committees 100 and 101 as most representative, based on count weights. Because 111 is not a minisum committee, however, it is arguably an inferior choice to 100 and 101. But there is a more fundamental issue regarding minimax committees: the minimax procedure, based on count weights, does not seem as compelling as the same procedure based on a different weighting scheme that I describe next.

Minimax Committees with Proximity Weights

Proximity weights, like count weights, reflect the number of voters who cast each of the distinct ballots. But they also incorporate information about the closeness of a ballot to all other ballots, based on Hamming distances.

The closer a ballot is to all other ballots, and the more voters who cast it, the more influence it should have on the determination of a committee. The minisum procedure with proximity weights works in this way. The proximity weight of ballot q^j is

$$w_j = \frac{m_j}{\sum\limits_{h=1}^{t} m_h d(q^j, q^h)}, \tag{5.1}$$

---

[4] These committees are also the product of fallback bargaining (FB) and fallback voting (FV), where the descent stops when at least one candidate receives a majority of votes; for more on this connection and references, see note 5.

where m_j is the number of voters who cast ballot $q^j = (q_1^j, q_2^j, \ldots, q_k^j)$ for k candidates and t is the number of distinct ballots cast. The denominator of the fraction is the sum of the Hamming distances from ballot j to all ballots (including ballot j), weighted by the number of voters who cast each ballot.

To illustrate in the example, the Hamming distances of ballot 100 to itself, 110, and 101 are 0, 1, and 1, respectively. Because these three ballots are cast by 1, 1, and 2 voters, respectively, ballot 100 has weight $1/[(1 \times 0) + (1 \times 1) + (2 \times 1)] = 1/3$, with the numerator reflecting the fact that one voter cast this ballot.

Similarly, ballots 110 and 101 have weights of 1/5 and 2/3. As shown in Kilgour, Brams, and Sanver (2006), it is the relative sizes of the weights that matter, so, for convenience, I multiply the fractions by 15 to clear denominators. This yields weights of 5, 3, and 10 for ballots 100, 110, and 101, respectively, as shown in Table 5.2. This table is the same as Table 5.1 except that it is based on proximity weights rather than count weights.

Notice that only committee 101 minimizes both the sum and the maximum of weighted Hamming distances, based on proximity weights. While committee 101 is also one of the committees singled out by the minisum and minimax criteria, based on count weights, this coincidence will not necessarily be the norm. In fact, I show in section 5.3 that the minisum outcome, based on count weights, and the minimax outcome, based on proximity weights, may be antipodes.

A Geometric Interpretation of Minimax Outcomes

Minimax outcomes may be interpreted geometrically, which I illustrate next. Represent the eight possible ballots for three candidates as the vertices of the

TABLE 5.2
Derivation of Minisum and Minimax Committees Based on Proximity Weights (4-Voter Example)

Ballot:	100	110	101	Sum	Max
Proximity Weight:	5	3	10		
1. 000	5	6	20	31	20
2. 100	0	3	10	13	10
3. 010	10	3	30	43	30
4. 001	10	9	10	29	10
5. 110	5	0	20	25	20
6. 101	5	6	0	11[a]	6[a]
7. 011	15	6	20	41	20
8. 111	10	3	10	23	10

[a] Minimum in column.

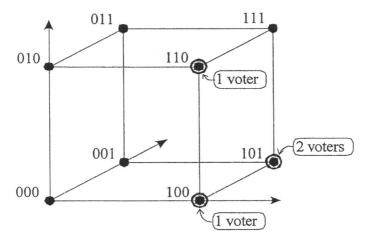

Figure 5.1. Geometric Representation of Ballots in 4-Voter Example (Voted-for Ballots Circled)

cube in Figure 5.1, in which approval (1) or disapproval (0) of each candidate is represented on a different axis (the first candidate on the horizontal axis, the second candidate on the vertical axis, and the third candidate on the planar axis). The three distinct ballots in the example (100, 110, and 101) are circled in Figure 5.1.

The proximity weights of (5, 3, 10) for ballots (100, 110, 101) can be thought of as the *inertias* of these ballots. Voters who cast them would depart from them, moving outward along the edges of the cube toward other vertices that they find acceptable, at velocities inversely proportional to these inertias.[5] Thus, after ten units of time, the two voters who cast ballot 101 would move distance 1 (i.e., traverse one edge from their node); the one voter who casts ballot 100 would move distance 2 (i.e., traverse $10/5 = 2$

[5] This interpretation is inspired by fallback bargaining (FB) (Brams and Kilgour, 2001b), which can be applied to approval balloting (Brams, Kilgour, and Sanver, 2004, 2007; Kilgour, Brams, and Sanver, 2006), as I did with fallback voting (FV) in chapter 3. Technically, Brams and Kilgour (2001b) define FB only when preferences form a linear order over all alternatives (i.e., voters' preferences are strict). Here I make a straightforward extension of this procedure to allow for nonstrict preferences. Under this procedure, voters fall back, or descend lower and lower, in their preferences until they reach an alternative on which all agree. This alternative minimizes the maximum distance they must traverse to make their agreement unanimous; for results on its computational complexity, see LeGrand, Markakis, and Mehta (2006). The innovation here is that voters may descend at different rates, depending on the weighting scheme used; a proof that this descent minimizes the maximum *weighted* Hamming distance is given in Kilgour, Brams, and Sanver (2006). If the requirement is that only a majority, not all, voters must agree, the FB outcome is essentially the "majoritarian compromise," described in Hurwicz and Sertel (1999), Sertel and Sanver (1999), and Sertel and Yilmaz (1999); see sections 2.3 and 3.4.

edges from its node); and the one voter who casts ballot 110 would move distance $10/3 = 3\frac{1}{3}$ (i.e., traverse $10/3 = 3\frac{1}{3}$ edges from its node). Moving at these relative rates, it is easy to see that the first committee that *all* voters would reach would be 101 in six units of time: The 110 voter would find 101 acceptable at time 6; the other voters would find it acceptable sooner (the 100 voter at time 5, and the two 101 voters at time 0, because the latter voters start out at this node).

If count weights rather than proximity weights are used (see Table 5.1), an analogous argument shows why there are three tied minimax outcomes. The two voters who cast ballots 100 and 110 traverse edges twice as vast as the two voters who cast ballot 101. From Figure 5.1, it is apparent that the first outcomes on which all four voters will agree will be within one edge of 101, and within two edges of 100 and 110, which are outcomes 100, 101, and 111. These are precisely the minimax outcomes, based on count weights, shown in Table 5.1.[6]

Henceforth I will use proximity weights, not count weights, to define minimax outcomes. Count weights reflect only the number of voters who cast a particular ballot but not how close this ballot is to other ballots, whereas proximity weights take into account both factors.

Thus in the example, with count weights the two 101 voters have twice the inertia of each of voters 100 and 110, even though the 110 voter is not as close to the two 101 voters as the 100 voter is (see Figure 5.1). But with proximity weights, the greater closeness of the 100 voter to the two 101 voters increases the 100 voter's inertia, and therefore influence, compared to the 110 voter, in the ratio 5:3.

More generally, with proximity weights voters whose ballots are close, but not necessarily identical, to the ballots of other voters add weight to these ballots (i.e., give them greater inertia). Likewise, extreme voters—outliers who are far from other voters—have reduced influence on the outcome.[7]

A minimax outcome can be visualized as the first outcome that all voters will converge upon as they move along the edges of a hypercube—in all directions from their ballots—at speeds inversely proportional to their proximity weights. Not only may this outcome differ radically from the minisum outcome (based on count weights), as I show in section 5.3, but this difference raises the question of under what circumstances a minimax outcome is preferable to a minisum outcome in the selection of a committee.

---

[6] It is worth noting that if the count weights were all 1 (if there were one 101 voter rather than two), the minimax outcome would be 100, which is the node exactly "between," and one edge distant from, 110 and 101 (see Figure 5.1).

[7] Other weighting schemes, of course, are possible, but proximity weights seem to balance the need to give representation to outliers, but downgrade this representation according to how far away (disconnected from other voters) they are.

5.3. MINIMAX VERSUS MINISUM OUTCOMES:
THEY MAY BE ANTIPODES

Minimax and minisum outcomes may be identical or overlap, as I showed in the previous example. But they may also diverge maximally, as I show next. In each case, I ask which committee—minimax or minisum—better represents the electorate. As I will demonstrate, the answer depends on which candidates, based on their patterns of support, one thinks should be on the committee.

Proposition 5.1. *If there are two or more candidates,* tied *minisum and* tied *minimax outcomes may include antipodes.*

Proof. Assume there are $n = 2$ voters who cast ballots 00 and 11 for $k = 2$ candidates. (Geometrically, the four possible committees (00, 10, 01, 11) can be represented by a square.) The minimax outcomes, 01 and 10, are antipodes, each lying at distance 1 from both ballots. These outcomes, as well as outcomes 00 and 11 that are also antipodes, are all minisum outcomes, whose Hamming distances to the two ballots all sum to 2. The 2-voter example can easily be extended to any larger number of voters or candidates. ∎

Outcomes 01 and 10 lead to the election of just one person. This is not a committee as this term is usually used, but Proposition 5.1 holds for larger tied minimax and minisum committees.[8] These examples illustrate not only that minisum and minimax may produce antipodes but also that each voting system, by itself, may produce them as well.

Note that there are half as many minimax outcomes as minisum outcomes in the 2-voter example. Whereas the minimax outcomes, 10 and 01, seem reasonable compromises, the additional minisum outcomes, 00 and 11, entirely favor one voter or the other. Manifestly, neither of the latter outcomes represents *both* voters well.

In the examples that follow, I will, for reasons of exposition, use antipodes like 0000 (no candidate elected) and 1111 (all candidates elected). These outcomes can readily be converted into antipodes, like 1100 and 0011, that more plausibly reflect real-world election possibilities.

---

[8] Consider the following example comprising 4 voters and 3 candidates: (1) 110; (2) 101; (3) 010; (4) 001. By constructing a table analogous to Tables 5.1 and 5.2, it is not difficult to show that there are four minimax outcomes, {000, 100, 011, 111}, which include two antipodal pairs; all eight possible outcomes are minisum. Notice that a larger minimax or minisum committee may not include a smaller committee; for example, 011 does not include 100. This failure of monotonicity—larger committees may not include smaller committees as subsets—is shared with other voting procedures, like the Kemeny rule, that have also been proposed to elect committees (Ratliff, 2003). Ratliff (2006) surveys other properties of various committee-selection procedures that have been proposed, including a procedure used to select committees at Wheaton College that he had proposed and that was used.

The next two propositions show that minimax and minisum outcomes may be antipodes when there are as few as four candidates (with ties) and five candidates (without ties).

Proposition 5.2. *If there are four or more candidates, a* nonunique *minimax and a* unique *minisum outcome may be antipodes.*

Proof. Consider the following example, in which there $n = 11$ voters and $k = 4$ candidates:

1. 3 voters: 0000
2. 2 voters: 0111
3. 2 voters: 1011
4. 2 voters: 1101
5. 2 voters: 1110

Applying equation (5.1), the proximity weight of ballot 1 is

$$3/[(3 \times 0) + (2 \times 3) + (2 \times 3) + (2 \times 3) + (2 \times 3)] = 3/24 = 1/8.$$

The proximity weight of ballot 2—and, by symmetry, ballots 3, 4, and 5—is

$$2/[(3 \times 3) + (2 \times 0) + (2 \times 2) + (2 \times 2) + (2 \times 2)] = 2/21.$$

Multiplying the weights by a factor $(8 \times 21 = 168)$ that clears denominators produces a proximity weight of 21 for ballot 1, and a proximity weight of 16 for each of ballots 2, 3, 4, and 5. Thus, the voters who cast ballot 1 are more influential than all other voters under the minimax procedure.

Note that one of the seven tied minimax outcomes in Table 5.3 is ballot 1 (0000), whereas the unique minisum (or MV) outcome is the antipode, ballot 16 (1111), as can be calculated directly: six of the eleven voters approve of each candidate. This 4-candidate example of antipodal minisum and minimax outcomes can easily be extended to any larger number of candidates. ∎

In the example in the proof of Proposition 5.1, there were more minisum outcomes than minimax outcomes (4 minisum and 2 minimax), whereas the opposite is true for the example in the proof of Proposition 5.2 (1 minisum and 7 minimax). Observe that the 3 voters who cast ballot 0000 in the latter example will be totally dissatisfied by minisum outcome 1111, a Hamming distance of 4 away. This seems a good argument for a minimax outcome, which is at maximum distance of 3 from the ballot of any voter.

The most stark clash of minimax and minisum outcomes occurs when they are both unique and antipodal.

Proposition 5.3. *If there are five or more candidates, a* unique *minimax and a* unique *minisum outcome may be antipodes.*

Proof. Consider the following example, in which there are $n = 11$ voters and $k = 5$ candidates:

TABLE 5.3
Derivation of Minimax Committees Based on Proximity Weights (11-Voter Example)

Ballot:	0000	0111	1011	1101	1110	Max
No. of Voters:	3	2	2	2	2	
Proximity Weight:	21	16	16	16	16	
1. 0000	0	48	48	48	48	48*
2. 1000	21	64	32	32	32	64
3. 0100	21	32	64	32	32	64
4. 0010	21	32	32	64	32	64
5. 0001	21	32	32	32	64	64
6. 1100	42	48	48	16	16	48[a]
7. 1010	42	48	16	48	16	48[a]
8. 1001	42	48	16	16	48	48[a]
9. 0110	42	16	48	48	16	48[a]
10. 0101	42	16	48	16	48	48[a]
11. 0011	42	16	16	48	48	48[a]
12. 1110	63	32	32	32	0	63
13. 1101	63	32	32	0	32	63
14. 1011	63	32	0	32	32	63
15. 0111	63	0	32	32	32	63
16. 1111	84	32	32	32	32	84

[a] Minimum of column.

1. 11100
2. 11010
3. 11001
4. 10110
5. 10101
6. 10011
7. 01110
8. 01101
9. 01011
10. 00111
11. 00000

Instead of constructing a table like Table 5.3, with a row for each of the 32 possible committees, I exploit the example's symmetry by noting that 10 voters approve of exactly three candidates in the $\binom{5}{3} = 10$ different ways that this is possible; voter 11 approves of no candidates.

Applying equation (5.1), the proximity weight of ballot 1 is

$$1/[0 + 2 + 2 + 2 + 2 + 4 + 2 + 2 + 4 + 4 + 3] = 1/27;$$

by symmetry, it is the same for ballots 2 through 10. The proximity weight of ballot 11 is

$$1/[(10 \times 3) + (1 \times 0)] = 1/30.$$

Clearing denominators, the proximity weight of the first 10 ballots is 10, and the proximity weight of ballot 11 is 9. Thus, the voter who casts ballot 11 is slightly less influential than the voters who cast the other 10 ballots.

Because the maximum Hamming distance between any two of the first 10 ballots is 4, the maximum weighted Hamming distance of one of these ballots is $4 \times 10 = 40$. By contrast, the maximum weighted distance of ballot 11 is $3 \times 9 = 27$, because this ballot is a Hamming distance of 3 from each of the 10 other ballots (and 0 from itself).

To show that every one of the $32 - 11 = 21$ other committees (ballots) has a greater maximum weighted Hamming distance than 27, consider (i) the one committee with 5 members (maximum weighted distance of 5×9 from 00000), (ii) the 5 different committees with 4 members (maximum weighted distance of 4×9 from 00000), (iii) the ten committees with 2 members (maximum weighted distance of 5×10 from one of the 3-member committees), and (iv) the five committees with 1 member (maximum weighted distance of 4×10 from one of the 3-member committees). In all these cases, the maximum weighted distances exceed the maximum weighted distance of $3 \times 9 = 27$ of ballot 11 from all others, so the latter distance is minimal. Therefore, ballot 11 is the minimax outcome, whereas 11111 is the minisum outcome because the 1's are majority winners. This five-candidate example of antipodal minisum and minimax outcomes can easily be extended to any larger number of candidates. ∎

Once again, a minimax committee (00000) seems better to represent *all* voters than a minisum committee (11111). (Recall that these antipodes might be more plausible 2-member and 3-member committees, such as 11000 and 00111.) Whereas the voter casting ballot 11 would be completely dissatisfied by 11111, the other 10 voters would mildly prefer 11111 to 00000.

These results for antipodes suggest that minimax committees may be more representative of all voters than minisum committees, because they leave no voter too aggrieved, especially not voters whose ballots are relatively close to those of many other voters. To be sure, if the aggrieved voters are only an isolated minority, like voter 11 in the foregoing example, it may be preferable to give better representation to the large majority than to appease the minority.

Our main purpose in this section has been to highlight such a trade-off by posing minimax outcomes as an alternative to minisum outcomes. Whether or not the minimax procedure or the minisum procedure should be used depends on the importance one attaches to the Rawlsian criterion (Rawls, 1971) of making the worst-off voter as well off as possible. Computer simulation, based on a spatial model, could help answer the question of how different, on average, minimax and minisum outcomes are, and under what conditions the minimax procedure might be better suited than the minisum procedure at finding representative committees.

In section 5.5 I show that the divergence between minisum and minimax outcomes is not purely theoretical but actually occurred in a real-life election that used approval balloting to elect a committee of twelve members. But first I discuss elections in which not every subset of candidates is a possible outcome.

5.4. ENDOGENOUS VERSUS RESTRICTED OUTCOMES

So far I have assumed that any subset of the candidates can be the minisum or minimax committee elected, whereas it is commonplace to put restrictions on the outcome. For example, one may want to specify the size of the committee to be elected (to ensure that it is neither too small nor too large to function efficiently) or its composition (to ensure that certain groups are at least minimally represented).

Call elections *endogenous* if all outcomes are possible winners; otherwise, they are *restricted*. As shown in Kilgour, Brams, and Sanver (2006), both minimax and minisum procedures apply equally well to restricted and endogenous elections. In a restricted election, one constructs tables, like Table 5.1, in which only rows representing eligible committees—that is, those not disqualified by the restrictions—appear.

In the election of a committee restricted according to size, the minisum procedure is equivalent to a more familiar procedure—namely AV—as shown by the next proposition.

Proposition 5.4. *When the size of a committee is restricted to c members, the minisum outcomes are the sets of c candidates receiving the most votes.*

Proof. See appendix.

The idea behind the proof is the following. When there is no restriction, the minisum outcome is the set of candidates that win a majority of votes (Brams, Kilgour, and Sanver, 2004, 2007 Proposition 4). Assume that the number of majority winners is less than the desired committee size c. Then adding to the majority winners those nonmajority candidates with the most approval until the committee size is exactly c minimizes the sum of the weighted distances to these members and, therefore, the sum of weighted distances to these members plus the majority winners. Likewise, if the number of original majority winners is greater than the desired committee size c, subtracting the candidates with the fewest approvals until the committee size is exactly c minimizes the sum of the weighted distances of the candidates who remain.

Unlike the minisum case, I know of no algorithm to find minimax outcomes—short of constructing tables, like Table 5.2, which is a "brute-force" calculation. If outcomes are endogenous, I previously showed that minisum and minimax outcomes may be antipodes. Restricting outcomes will not necessarily lead to a common minisum and minimax outcome, as the next example with $n = 4$ voters and $k = 4$ candidates illustrates:

1. 1100
2. 1010
3. 1001
4. 0111

Assume a committee of size $c = 1$ is to be chosen. It is easy to see that 1000 is the unique minisum outcome, because candidate 1 receives 3 votes when the three other candidates receive 2 votes each.[9]

To find the minimax outcome, use equation (5.1) to calculate the proximity weights, which for ballot 1100 is

$$1/[0 + 2 + 2 + 3] = 1/7,$$

and, by symmetry, is the same for ballots 1010 and 1001. Similarly, the proximity weight of ballot 0111 is

$$1/[3 + 3 + 3 + 0] = 1/9.$$

Clearing denominators, the proximity weights of the first three ballots are 9 each, and the proximity weight of ballot 0111 is 7. Thus, the voter who casts ballot 0111 is less influential than the voters who cast the other three ballots.

As shown in Table 5.4, the unique minimax outcome is 1111, which does not satisfy the restriction of the committee to one member.[10] (Note that 1111 is not the ballot of any voter, nor are some minisum outcomes, such as 1000.)[11] Surprisingly, of the four possible committees that include one member (see committees 2–5 in Table 5.4), the three tied minimax outcomes—0100, 0010,

---

[9] If there were no single-winner restriction, the election of candidate 1 and any 1, 2, or all 3 of the other candidates (i.e., outcomes 1100, 1010, 1001, 1110, 1101, 1011, and 1111) are tied minimax outcomes that are, like outcome 1000, also minisum.

[10] I show the sixteen possible outcomes in Table 5.4 to illustrate how the restriction to $c = 1$ may alter minimax outcomes, making them, as in this example, disjoint from the unrestricted outcome.

[11] Brams, Kilgour, and Zwicker (1997, 1998) were the first to show that the minisum outcomes need not correspond to the ballot of any voter, which they called the "paradox of multiple elections" and which is analyzed in chapter 7. Özkal-Sanver and Sanver (2005) show that a voting rule ensures a Pareto-optimal outcome if and only if it never exhibits this paradox.

TABLE 5.4
Derivation of Minimax Committees Based on Proximity Weights (4-Voter Example)

Ballot:	1100	1010	1001	0111	Max
No. of Voters:	1	1	1	1	
Proximity Weight:	9	9	9	7	
1. 0000	18	18	18	21	21
2. 1000	9	9	9	28	28
3. 0100	9	27	27	14	27
4. 0010	27	9	27	14	27
5. 0001	27	27	9	14	27
6. 1100	0	18	18	21	21
7. 1010	18	0	18	21	21
8. 1001	18	18	0	21	21
9. 0110	18	18	36	7	36
10. 0101	18	36	18	7	36
11. 0011	36	18	18	7	36
12. 1110	9	9	27	14	27
13. 1101	9	27	9	14	27
14. 1011	27	9	9	14	27
15. 0111	27	27	27	0	27
16. 1111	18	18	18	7	18[a]

[a] Minimum of column.

and 0001, which are committees 3, 4, and 5—do not include the minisum outcome, 1000.

It may seem bizarre not to elect the most approved candidate, especially one approved of by a majority, in a single-winner election. I will revisit this issue in the concluding section, asking whether the minimax criterion is reasonable, especially in single-winner elections.

5.5. MANIPULABILITY

A voting procedure is *manipulable* if it is possible for a voter, by misrepresenting his or her preferences, to obtain a preferred outcome. To define "preferred," I relate Hamming distance to preferences.

Specifically, assume that a voter's ballot indicates his or her most-preferred committee, or *top preference*. In addition, assume that the voter's preference is *spatial* in the sense that outcomes that are farther (as measured by Hamming distance) from his or her top preference are less preferred. Thus, if p^i is voter i's top preference and p and q are any outcomes such that $d(p^i, p) < d(p^i, q)$, then voter i strictly prefers p to q. In other words, as distance increases, a voter's preference falls off, reaching a minimum at the antipode of the top preference; note that I make no assumption about the voter's preference among ballots that are at equal Hamming distance from the top preference.[12]

I make an additional assumption about preferences over sets. If outcome a is preferred to outcome b, then $\{a\}$ is preferred to $\{a, b\}$, which can be interpreted as a tie in which each of the two outcomes occurs with positive probability. This assumption, which is used in the proof of Proposition 5.5, seems eminently reasonable.

Proposition 5.5. *The minimax procedure is manipulable, whereas the minisum procedure is not.*

Proof. First consider the minimax procedure. In the example in Table 5.2, I showed that the unique minimax outcome is 101, which is a Hamming distance of 2 from the ballot of the voter who casts ballot 110. But if this voter falsely indicates his or her ballot to be 100, then the situation would appear as follows:

2 voters: 100
2 voters: 101

It is easy to see that the proximity weights according to equation (5.1) are now equal, so one need only ask which outcome minimizes the maximum Hamming distance of the voters from their ballots. Clearly, the ballots themselves do this, so the minimax outcome is $\{100, 101\}$.

The 110 voter who falsely indicated a top preference of 100 prefers this tied outcome, because he or she prefers 100 to 101, and, by the assumption I made earlier, prefers $\{100, 101\}$ to 101. Hence, the minimax procedure is manipula-

---

[12]Like both the minimax and minisum procedures, spatial models of preference could be based on other metrics, such as "root-mean-square," which is essentially Euclidean distance. The Hamming metric is particularly well suited for measuring the distance of a voter from an outcome, because it reflects equally the voter's disagreements with the candidates elected and with those not elected. However, spatial models based on this metric cannot mirror well the preference of a voter who wants a "balanced" committee—say, with an equal number of men and women. For example, assume that eight male (M) and female (F) candidates are listed as follows, FFFFMMMM, and the committee is to have four members. The "balanced" voter's two most-preferred committees might be 11001100 and 00110011, which are antipodes, so voting for either will work to rule out the other, especially under minimax. Of course, the worst case for this voter, 11110000 or 00001111 (all women or all men), can be precluded if it is mandated that the committee must have equal numbers of men and women. Then the male chauvinist who votes for the four males (00001111) will never get his favorite committee but will, instead, support equally 11000011 and 00111100—in fact, *all* balanced committees. Thus, if this voter wants to have some effect on the outcome, it behooves him to vote for some women! For a review of the literature on ranking sets of items, see Barberà, Bossert, and Pattanaik (1998).

ble. The proof that the minisum procedure is not manipulable is given in the appendix. ■

The idea of the proof for the minisum procedure is easy to describe. Because a voter's choices are binary on each candidate, it is always in his or her interest to support those, and only those, candidates of whom he or she approves. Moreover, the voter's decision on each candidate does not affect which other candidates are elected, so each voter cannot be worse off as a consequence of voting truthfully on *all* candidates.

Although the minimax procedure is vulnerable to manipulation in theory, in practice it is probably almost as resilient to manipulation as the minisum procedure. To exploit it would require a manipulative voter to have virtually complete information about the voting intentions of other voters, which is unlikely in most real-world situations. Indeed, merely finding the truthful minimax outcome is computationally hard, as I indicated earlier, reflecting the fact that the number of possible outcomes increases exponentially with the number of candidates.[13]

I next turn to a real-world election. This election renders concrete some of the theoretical and practical issues I have discussed and raises some new questions as well.

5.6. THE GAME THEORY SOCIETY ELECTION

In 2003, the Game Theory Society (GTS) used AV for the first time to elect 12 new council members from a list of 24 candidates. (The council comprises 36 members, with 12 elected each year to serve 3-year terms.)[14] I give below the numbers of members who voted for from 1 to all 24 candidates (no voters voted for between 19 and 23 candidates).

Votes cast	1	2	3	4	5	6	7	8	9	10	11	12	13	14	15	16	17	18	24
# of voters	3	2	3	10	8	6	13	12	21	14	9	25	10	7	6	5	3	3	1

Casting a total of 1,574 votes, the 161 voters, who constitute 45 percent of the GTS membership, approved, on average, $1574/161 \approx 9.8$ candidates; the median number of candidates approved of, 10, is almost the same.

---

[13] Although this might be seen as a disadvantage of the minimax procedure, computers make the calculation of minimax outcomes feasible for 30 or more candidates (in section 5.5 I analyze an election with 24 candidates). For more on the computability of minimax, see Kilgour, Brams, and Sanver (2006).

[14] The fact that there is exit from the council after three years makes the voting incentives different from a society in which members, once elected, do not leave and are the ones who decide who is admitted (Barberà, Maschler, and Shalev, 2001).

The modal number of candidates approved of is 12 (by 25 voters), echoing the ballot instructions that 12 of the 24 candidates were to be elected. The approval of candidates ranged from a high of 110 votes (68.3 percent approval) to a low of 31 votes (19.3 percent approval). The average approval received by a candidate was 40.7 percent, though only candidates who received at least 69 votes (42.9 percent approval) were elected.

In the GTS election, there were $2^{24} \approx 16.8$ million possible ballots. It turned out that 2 of the 161 voters voted identically. As one might expect, the identical ballot, 111100011001101000000111, was cast by 2 of the 25 modal voters who voted for 12 candidates. If all ballots approving of 12 candidates are assumed equiprobable, the probability that no two of 25 ballots are identical is

$$[t(t-1)(t-2)\ldots(t-24)]/t^{25} \approx 0.999889,$$

where $t = \binom{24}{12} = 2,704,156$, based on reasoning given in note 2. The complement of this probability, 0.000111, is the probability that at least two voters cast identical ballots.[15]

If there had been no restriction in the GTS election, the 5 candidates approved of by a majority—at least 81 of the 161 voters—would have been elected. Adding the next 7 biggest vote getters gives the minisum outcome under the restriction that 12 candidates are to be elected.

Do these candidates best represent the electorate? In fact, 4 of the 12 minimax winners differ from the minisum winners. Each set of winners is given below—ordered from most popular on the left to least popular on the right—with differences between those elected to each council underscored.[16]

---

[15] I made this calculation for each category of voter—from those who cast 1 vote to those who cast 18 votes—excluding only the category containing the one voter who voted for all 24 candidates (because there is only one such ballot). The voters most likely to cast an identical ballot are the three who vote for one candidate; the probability that at least two of them cast the same ballot is 0.121528. To generalize for all voters, let p_i be the probability that no two voters who cast i votes chose an identical ballot. It follows that the probability that no two voters in any category cast an identical ballot is $p_1 p_2 \ldots p_{17} p_{18}$, so the complement of this probability is the probability that at least two voters in one or more categories cast identical ballots. The latter probability in the GTS election is 0.131009; it is far greater than the probability that I indicated in note 2 (0.000768), which did not take into account the eighteen categories into which voters sorted themselves empirically. Even this greater probability is likely an underestimate, because it does not reflect the fact that some candidates were far more approved of than others, rendering dubious the assumption that all ballots in each category are equiprobable.

[16] It is worth pointing out that minisum outcomes are always Pareto-optimal; if this were not the case, then there would be some other outcome such that some voter is less distant and no voter is more distant, contradicting the defining property of minisum. By contrast, minimax outcomes need not be Pareto-optimal. To illustrate, I revisit the example in note 8, in which the top preferences of 4 voters for 3 candidates are as follows: (1) 110; (2) 101; (3) 010; (4) 001. There are four minimax outcomes: (a) 000; (b) 100; (c) 011; (d) 111. Because outcome (c) is at least as good as outcome (a) for all voters, and better for voters (3) and (4), and outcome (b) is at least as

Council Restricted to Twelve Members
Minisum: 1 1 1 1 1 1 1 1 1 1 1 1 1 0 0 0 0 0 0 0 0 0 0 0
Minimax: 1 1 1 1 1 1 0 1 1 0 0 0 1 1 1 0 0 0 0 0 0 0 0 1

Observe that four of the minisum winners would have been displaced by candidates who received fewer votes, one of whom was the candidate who received the fewest votes.[17]

To exclude this candidate would have put some voters at a greater weighted distance than including him or her. Thereby, minimax gives voice to voters who approve of unpopular candidates if they make the council more representative by not leaving some voters "out in the cold."

If the size of the council had not been restricted to 12 winners but instead had been endogenous, the minisum and minimax councils would have differed substantially.

Unrestricted Council (Minisum, Five Members; Minimax, Eight Members)
Minisum: 1 1 1 1 1 0 0 0 0 0 0 0 0 0 0 0 0 0 0 0 0 0 0 0
Minimax: 1 1 1 1 0 0 0 0 1 0 1 0 1 0 1 0 0 0 0 0 0 0 0 0

As noted earlier, the minisum council would have comprised only the 5 majority winners. By contrast, the minimax outcome includes 8 candidates; four of these candidates came in 9th, 11th, 13th, and 15th.

I showed in section 5.5 that it was possible, in theory, for a voter successfully to manipulate a minimax outcome, but I contended that this would be well-nigh impossible in most elections. As a case in point, consider the single voter who voted for all 24 candidates in the GTS election and who, presumably, was indifferent toward all the candidates.[18]

Might this voter have influenced the outcome if the size of the council had been endogenous? In fact, if minisum had been the procedure, the 5 biggest vote getters would still have been elected had this voter not voted, and the eight minimax winners also would have won.

Likewise, this voter's absence would have had no effect on the composition of the 12-member minimax council given earlier. This suggests that single outliers are unlikely to be consequential, whether or not the size of a committee is fixed, especially in an election with 100 or more voters.

Making the number of candidates to be elected endogenous, and using the minisum procedure, is tantamount to electing only candidates approved of by

---

good as outcome (d) for all voters, and better for voters (1) and (2), only outcomes (b) and (c) are Pareto-optimal.

[17] Fishburn (2004) shows that the 12 minisum winners tended to be supported somewhat more strongly by voters who voted for few candidates, whereas the reverse was true for the losers. In effect, voters who approved of few candidates were more discriminating, helping to put the minisum winners over the top.

[18] Of course, this voter might simply have relished the role of being an outlier by approving of everybody, even though he or she had no effect on the actual (minisum) outcome under the GTS rules.

a majority (5 candidates in the case of the GTS council). In fact, this rule is used to determine who is admitted to certain societies, though the quota or threshold for entry is not always a simple majority.[19]

In the concluding section, I summarize the results and comment on the feasibility of the minimax procedure in different kinds of elections. Not only may this procedure give a dramatically different outcome from the minisum procedure, but this outcome may better reflect diverse views within the electorate.

5.7. SUMMARY AND CONCLUSIONS

Under approval balloting, each voter approves of a subset of candidates. The minisum and minimax procedures find subsets that are as close as possible to the ballots of all voters, but according to two different notions of "closeness." Whereas the minisum procedure selects the outcome that minimizes the sum of Hamming distances to all voters—or, equivalently, the average Hamming distance—and uses count weights, the minimax procedure finds the outcome that minimizes the maximum weighted Hamming distance and uses proximity weights.

Geometrically, the latter can be visualized in terms of voters moving from their vertices of a hypercube, which represent their ballots, along the edges at speeds inversely proportional to their proximity weights (or inertias). The node or nodes reached first by *all* voters are the minimax outcomes.

Minimax and minisum may yield diametrically opposed outcomes, or antipodes, if there are as few as four candidates (with ties), or five candidates (without ties). If "representation" means not antagonizing any voters—especially those with similar or identical preferences—too much, then minimax outcomes seem more representative of the entire electorate than minisum outcomes.

My analysis of the 2003 election by the Game Theory Society (GTS) of 12 new members to its council showed that the minimax procedure would have given 4 different winners from the minisum procedure, which was the procedure actually used by the GTS. There would have been a greater difference if the number of candidates to be elected had been endogenous. The minisum procedure would have elected only the 5 candidates who were approved of by a majority, whereas the minimax procedure would have elected 8 candidates, including 4 relatively unpopular candidates that better represented certain sets of voters.

In single-winner elections, the AV winner (minisum outcome) would seem the normatively most desirable choice. But as I showed in an example in section 5.4, a different candidate may less antagonize a minority (1 of the 4 voters in

---

[19] Using the minisum procedure in this manner, which is effectively AV, is what Barberà, Sonnenschein, and Zhou (1991) call "voting by quota" or, more generally, "voting by committee." They show that such a scheme is nonmanipulable if voter preferences are separable, about which I will say more in chapter 7.

this example), so it is not apparent—even in single-winner elections—that the minisum winner should always triumph.

Whereas the minisum procedure is not manipulable, the minimax procedure is. In practice, however, it would be virtually impossible for a voter to induce a preferred outcome because of (i) a lack of information about other voters' ballots, and (ii) the computational complexity of processing such information, even if it were available.

In the GTS election, the absence of the outlier who voted for all 24 candidates would not have changed the minimax outcome. This suggests that single outliers cannot easily manipulate the minimax outcome, because their "pull" on an outcome is small.

Whether the size of a committee should be fixed or endogenous (perhaps within a range) will depend, I think, on the importance of electing a committee whose size significantly affects its ability to function. Even if size is made endogenous, voters should probably be given some guidance as to roughly what size would be appropriate. Without this information, it may be hard for them to gauge how many candidates to approve of in an election.

These practical considerations aside, I believe that more theoretical research on the properties of the minimax procedure is needed. For example, if this procedure is used, is it appropriate to break ties among the minimax winners using minisum? Are there other ways of combining criteria? What effects do the correlated preferences of voters, or perceived similarities in candidates, have on the minimax and minisum outcomes, or on the likelihood of antipodes? How might information (e.g., from polls) affect the manipulability of the procedure?

In addition to these questions, other procedures, especially those that allow for proportional representation, such as those I present in chapter 6, should be considered. Just as AV in single-winner elections stimulated considerable theoretical and empirical research beginning a generation ago (Brams and Fishburn, 1983; Weber, 1995), I hope that the minimax procedure generates new research on using approval balloting to elect committees.

APPENDIX

Proposition 5.4. *When the size of a committee is restricted to c members, the minisum outcomes are the sets of c candidates receiving the most votes.*

Proof. Assume that there are n voters and k candidates, and that $m_h > 0$ voters cast the ballot $q^h = (q_1^h, q_2^h, \ldots, q_k^h)$, where $h = 1, 2, \ldots, t$. Note that $\sum_{h=1}^{t} m_h = n$. For an arbitrary binary k-vector $x = (x_1, x_2, \ldots, x_k)$, define

$$d_j(x, q^h) = \begin{cases} 0 & \text{if } x_j = q_j^h \\ 1 & \text{if } x_j \neq q_j^h \end{cases}$$

for $h = 1, 2, \ldots, t$ and $j = 1, 2, \ldots, k$. Then it is clear that the Hamming distance from x to q^h is given by $d(x, q^h) = \sum_{j=1}^{k} d_j(x, q^h)$. The endogenous minisum winner is any x that minimizes $D(x) = \sum_{h=1}^{t} m_h d(x, q^h)$, whereas the restricted minisum winner is any x that minimizes $D(x)$ among all x's containing exactly c 1's.

To find an equivalent representation for $D(x)$, define $S_j(x) = \sum_{h=1}^{t} m_h d_j(x, q^h)$ for any x and j. $S_j(x)$ represents the number of voters who disagree with k-vector x on candidate j. Note that $D(x) = \sum_{j=1}^{k} S_j(x)$, and that $S_j(x)$ depends only on x_j and not on the other $k - 1$ components of x. Therefore, x represents a minisum winner if and only if x minimizes

$$D(x) = \sum_{h=1}^{t} m_h d(x, q^h) = \sum_{j=1}^{k} S_j(x). \tag{A5.1}$$

With equation (A5.1) all minisum committees can be characterized. Define $K_j = \sum_{h=1}^{t} m_h q_j^h$, so K_j is the number of voters who vote for candidate j; clearly, $n - K_j$ is the number of voters who vote against j. Now[20]

$$S_j(x) = \begin{cases} n - K_j & \text{if } x_j = 1 \\ K_j & \text{if } x_j = 0 \end{cases} \tag{A5.2}$$

Next consider how to choose $x = (x_1, x_2, \ldots, x_k)$ so that exactly c of the x_j's equal 1 and $D(x)$ is minimized. By equation (A5.2), $D(x)$ will be the sum of c values of $n - K_j$ (corresponding to the members of the winning committee) and $k - c$ values of K_j (corresponding to the unsuccessful candidates). Clearly, putting the c candidates with the largest values of K_j on the committee minimizes $D(x)$. In other words, the minisum winners under the restriction that a committee of size c is to be elected must correspond to a vector x such that $x_j = 1$ if and only if j belongs to some subset of c candidates that receives the most votes. ∎

Proposition 5.5. *The minimax procedure is manipulable, whereas the minisum procedure is not.*

Proof that minisum is nonmanipulable. To show that a voter is best served by voting for his or her top preference, apply the preference model introduced in the text to equations (A5.1) and (A5.2). Assume that voter i is one of the m_h voters whose top preference is $q_h = p^i$.

To show that i cannot do better than to cast ballot $q^h = p^i$, suppose that i's top preference, x_j, satisfies $x_j = 1$ for some j. Then the preference assumptions

---

[20] From equation (A5.2) it follows that among all possible committees, $x = (x_1, \ldots, x_k)$ minimizes $D(x)$ if and only if, for each j, $x_j = 1$ if $n - K_j < K_j$ and $x_j = 0$ if $K_j < n - K_j$. This minimization proof is different from, and more general than, the proof given in Brams, Kilgour, and Sanver (2004, 2007).

imply that voter i prefers any committee with $x_j = 1$ to the committee that is otherwise the same but has $x_j = 0$.

Consider i's decision to vote truthfully ($p_j^i = 1$) or untruthfully ($p_j^i = 0$) on candidate j. According to equation (A5.2), selecting $p_j^i = 0$ rather than $p_j^i = 1$ reduces the value of $S_j(x)$ from K_j to $K_j - 1$. The four possibilities implied by equation (A5.1) for whether candidate j belongs to the minisum winner(s) are set forth in Table A5.1.

TABLE A5.1

	j elected if $p_j^i = 1$?	j elected if $p_j^i = 0$?
$K_j < n - K_j$	Never	Never
$K_j = n - K_j$	Sometimes	Never
$K_j = n - K_j + 1$	Always	Sometimes
$K_j > n - K_j + 1$	Always	Always

In the table, "sometimes" means that the total votes for and against candidate j are equal; if such a tie occurs, the set of minisum committees consists of one or more pairs of committees that differ only in that one includes candidate j and one does not.

Because the voter always prefers that candidate j be a member of the committee, it follows from the table that voter i is never worse off by voting truthfully (i.e., for candidate j) and may be better off. The argument is analogous if i's top preference satisfies $x_j = 0$. ∎

6

Electing Multiple Winners: Minimizing Misrepresentation

6.1. INTRODUCTION

In this chapter, I extend and generalize proportional-representation (PR) systems that were first proposed by Monroe (1995). Monroe's systems select the winning candidates for a representative body that minimize the sum of "misrepresentation" values of voters.[1] These values are based on information that the voters provide on their ballots, such as candidate rankings or approval votes.

Suppose, for example, that one wishes to choose a set of candidates so that as many voters as possible approve of at least one candidate who is elected. Or if the voters rank candidates, suppose that one wishes to choose candidates who are as highly ranked, on average, as possible. These criteria of selection minimize, in some sense, the misrepresentation of voters by ensuring that the candidates elected are approved of, or preferred to, the maximum degree possible.

Monroe's voting systems are innovative, but his analysis of them has gaps. For example, he offered no mathematically explicit and verifiable method for achieving the intended minimization. Without such a method, there is a danger of selecting a set of winners that fails to minimize total misrepresentation; in fact, this deficiency plagued some of Monroe's own calculations. I will suggest various ways that his systems can be extended and his notion of misrepresentation generalized.

I begin by showing how integer programming provides a mathematical procedure for selecting winners so as to minimize the sum of misrepresentation values of voters. It provides a way of assigning one or more candidates to each voter, up to the total number of candidates, in order to determine the set of candidates that minimizes total misrepresentation.

Note: This chapter is adapted from Potthoff and Brams (1998) with permission of Sage Publications Ltd.

[1] To be consistent with Monroe's terminology, I use *misrepresentation* rather than terms such as *nonrepresentation* or *unrepresentativeness*. As used in this chapter, misrepresentation has nothing to do with voters' misrepresentation of their preferences, as in strategic voting. Also excluded from consideration are party-list systems, in which parties obtain seats in parliament in proportion to the votes they receive, which is the most common form of PR in democracies today.

I explore in more detail than did Monroe the ramifications of using approval voting (AV) to achieve PR. Of the PR systems analyzed, one involves the use of AV in three-member districts, which is a competitor to cumulative voting that I briefly discussed in section 4.6.

Monroe's systems and their extensions raise several questions in large electorates, where computational streamlining is desirable, and in small electorates, where the chance of ties is the greatest. For example, one question I analyze is how to fill vacancies that arise after an election. Another is making fractional assignments of voters to candidates so that all candidates have exactly the same number of voters assigned to them—rather than numbers that differ by 1, which is the only difference that Monroe permitted.

The integer-programming approach equalizes the number of *voters* assigned to winning candidates. This allows for restrictions that may be more or less stringent than those Monroe imposed. In fact, imposing no restrictions gives a system that was originally proposed by Chamberlin and Courant (1983), which Monroe discusses (and rejects).

If one lifts restrictions on the number of *candidates* to be elected, Monroe's system is equivalent to one proposed by Tullock (1967, ch. 10), which is mentioned in the Chamberlin-Courant article but not in the Monroe article. Both the Chamberlin-Courant system and the Tullock system, which use weighted voting, are subject to criticisms that have been made of weighted-voting systems.

Although integer programming offers a powerful technique for implementing PR systems that have not yet been tried, which system is best is by no means obvious. I consider the pros and cons of both weighted-voting and nonweighted-voting PR systems.

In addition, I briefly discuss the degree of nonmanipulability and the degree of representativeness of PR systems, which are important criteria in the assessment of any voting system. Finally, I propose a new system of "hierarchical PR" that offers voters representation at different levels (e.g., representative, senator, president). Because it requires only one election to do so, it effectively enables the electorate to choose, all at once, the most qualified candidates at each level.

Throughout I present a number of examples that highlight anomalies and pitfalls to which Monroe's systems and their extensions are vulnerable. While they warn us about possible defects that can arise, these systems are superior in many ways to extant PR systems. Despite their complexity, they provide representation to voters that standard PR systems often fail to achieve.

6.2. OBSTACLES TO THE IMPLEMENTATION OF PROPORTIONAL REPRESENTATION (PR)

Before embarking on a study of new PR systems, it behooves one to ask the following question: Do they stand any chance of being used in actual elections?

Clearly, several obstacles exist. A major one is the inertia of overcoming the status quo, which is probably less of an obstacle for small electorates than for large ones, and for new legislative bodies than for established ones. Countries that have had little experience with PR, like the United States, will probably be more resistant than countries, like most democracies in Europe, that have used PR for many decades. Another obstacle to the use of PR in the United States is a 1967 federal law that requires states to use single-member districts to elect members to the House of Representatives.

Other obstacles include the complexity and the computational problems of PR systems that use approval votes or candidate rankings; these problems will surely be ameliorated as computer capabilities continue to advance. Not to be ignored is the problem of educating voters about new PR systems and their possible advantages in different situations.

There is one obstacle to the implementation of any PR system that has not been generally recognized. Since *Baker v. Carr* (1962) and subsequent U.S. Supreme Court apportionment cases, the division of a jurisdiction into single-member districts has required that each district have almost the same total population, where population means citizens of all ages. Although one might question whether districting should instead be based instead on the total adult population, or perhaps even on the total number of registered voters or actual voters, the current requirement is well established.

By contrast, PR representation in legislatures is based on actual votes, which determines the number of seats a party wins, rather than on population. A switch from single-member districts to PR is likely to lead to reduced voting strength for ethnic and other groups that have low ratios of (i) adult to total population, (ii) registered voters to all adults, or (iii) actual voters to registered voters.

Of course, there might be offsetting features of PR that could operate so as to increase the voting strength of an affected group. But surely some groups will view PR, based as it is on actual votes cast, as likely to undermine their legislative strength—and hence will resist it as a threat to their interests.

Such resistance may be reduced, and the support for PR systems enhanced, as a consequence of recent court decisions, beginning with *Shaw v. Reno* (1993), that have disallowed bizarre-looking districts for the purpose of increasing the representation of minorities. Indeed, such gerrymandered districts, which traditionally have been created to entrench politicians, have caused outrage since the nineteenth century. This outrage may encourage the adoption of PR systems that more accurately reflect citizen preferences, and make representatives more responsive to voters, than do artificially constructed districts.[2]

---

[2] Term limits in several states now provide a device to prevent politicians from staying in office indefinitely. But this reform has yet to be adopted in U.S. federal elections for the House of Representatives or the Senate, though the president is limited to two terms (eight years).

Trying to implement these PR systems at the state or national level is probably premature. I indicate later that they might first be tried out in city councils or private organizations that seek to mirror different interests in proportion to their size in the electorate. The key idea behind minimizing misrepresentation of these interests is to find representatives who as many voters as possible rank highly, or at least approve of, that support these interests.

6.3. INTEGER PROGRAMMING

Because an integer program is a special type of linear program, I begin by defining the latter. In *linear programming,* one wants to find nonnegative values of the t variables, $x_1, x_2, \ldots, x_t$, that maximize or minimize the linear function,

$$z = c_1 x_1 + c_2 x_2 + \cdots + c_t x_t,$$

subject to a set of s linear constraints that can be written as

$$a_h x_1 + a_h x_2 + \cdots + a_h x_t \ [\leq, =, \text{or} \geq] \ b_h,$$

where $h = 1, 2, \ldots s$. The function z is called the *objective function.*

Integer programming, or *integer linear programming,* is a standard tool in certain disciplines, such as management science and operations research (see, e.g., Davis and McKeown, 1984, ch. 8; Gould, Eppen, and Schmidt, 1993, ch. 9), but it has rarely been used in political science.[3] An integer program is simply a linear program with an additional set of constraints: the t x-variables must all be integers.

A *zero-one integer program* is an integer program in which each x-variable is constrained to be either 0 or 1. In a *mixed-integer program,* or *mixed-integer linear program,* some but not all of the x-variables are constrained to be integers.

Computer software packages are available to solve integer-programming problems. Although the mathematical techniques used by the packages are complex, one does not generally need to understand these techniques in order to use the software and obtain solutions.

Before describing how integer programming can be applied to PR problems, consider a simple non-PR example that illustrates the general kind of problem

---

[3] The only political-science applications of integer programming that I am aware of are to constrained approval voting (CAV), as discussed in chapter 4; see Potthoff (1990) and Straszak, Libura, Sikorski, and Wagner (1993). While the problem of apportioning representatives to the U.S. House of Representatives and similar bodies can be formulated as an integer program, integer programming apparently has not been used, though other mathematical approaches have been (Balinski and Young, 1982, 2001; Ernst, 1994), which I discuss in chapter 9. Apart from voting and apportionment, Herlihy (1981) applied mixed-integer programming to a districting problem in the Republic of Ireland.

integer programming is designed to solve. Suppose a merchant advertises bags of emeralds for sale at a certain price per bag, with the stipulation that each bag will contain no more than four emeralds totaling at least 65 units in weight. Suppose a customer arrives at a time when the merchant's inventory consists of seven emeralds weighing 8, 8, 12, 15, 22, 32, and 41 units. Which emeralds does the merchant include in the bag for this customer so that the advertised conditions are observed yet the total weight is as low as possible?

The problem can be set up and solved as a zero-one integer program. There are $t = 7$ x_i-variables to be solved for, where x_i $(i = 1, 2, \ldots, 7)$ is 1 if the ith emerald is to be included in the bag and 0 if it is to be left out. The objective function, here to be minimized, is

$$z = 8x_1 + 8x_2 + 12x_3 + 15x_4 + 22x_5 + 32x_6 + 41x_7.$$

Other than the requirements for each x_i to be either 0 or 1, there are only two constraints:

$$8x_1 + 8x_2 + 12x_3 + 15x_4 + 22x_5 + 32x_6 + 41x_7 \geq 65$$

$$x_1 + x_2 + x_3 + x_4 + x_5 + x_6 + x_7 \leq 4.$$

(Although the first constraint has the same coefficients as the objective function, this will not generally be the case in an integer program.) The solution turns out to be $x_3 = x_5 = x_6 = 1$ and $x_1 = x_2 = x_4 = x_7 = 0$. That is, the emeralds selected are those with weights of 12, 22, and 32. Their total weight, $z = 66$, is the minmum weight satisfying these conditions. This weight, 66, is not below 65; and the number of emeralds, three, does not exceed four.

6.4. MONROE'S SYSTEM

Let there be M candidates, m of whom are to be elected, and n voters. I use i $(i = 1, 2, \ldots, M)$ to index the candidates and j $(j = 1, 2, \ldots, n^*)$ to index the voters, where for now $n^* = n$. For later purposes (see Appendix A), n^* will be less than n when j indexes not just single voters but *groups* of voters who all vote the same way.

Misrepresentation value μ_{ij} measures the extent to which voter j would be misrepresented by candidate i. As Monroe notes, if each voter is asked to rank the candidates from 1 (most preferred) to M (least preferred),[4] it is reasonable to define μ_{ij} to be voter j's rank of candidate i minus 1. For example, if voter

---

[4] An unranked candidate receives the average of the voter's unused ranks. For example, if a voter ranks only his or her two top choices out of $M = 6$ candidates, the ranks are (1, 2, 4.5, 4.5, 4.5, 4.5).

5 ranks candidate 4 in third place, $\mu_{45} = 3 - 1 = 2$; the highest-ranked candidate of each voter will have a value of 0. For AV, μ_{ij} is defined to be 0 if voter j approves of candidate i, 1 otherwise, so μ_{ij} will be either 0 or 1.

For any subset of m of the M candidates, Monroe assigns one of these m candidates to each voter in such a way that the number of voters connected to each candidate is n/m, or a number that differs from n/m by less than 1 if n/m is not an integer. If a unique set of m candidates produces an assignment that minimizes the sum of the misrepresentation values across all voters, it is declared to be the set of m winners. Although this criterion is well defined, Monroe did not provide a mathematically proven way to find the m winners so as to minimize the total misrepresentation of voters.[5]

Setting up the problem as an integer program does provide such a way, which I next describe. Let x_i be 1 if candidate i is a winner, 0 otherwise. Let x_{ij} be 1 if candidate i is assigned to voter j, 0 otherwise.

Note that the x-variables are now of two types: M of them have one subscript, and $Mn*$ have two. Let the objective function be

$$z = \sum_i \sum_j \mu_{ij},$$

which is the sum of the misrepresentation values. It is to be minimized subject to the following constraints (L and U are defined below):

$$\sum_i x_i = m \tag{6.1}$$

$$\sum_i x_{ij} = 1 \quad \text{for each } j \, (j = 1, 2, \ldots, n*) \tag{6.2}$$

$$-Lx_i + \sum_j x_{ij} \geq 0 \quad \text{for each } i \, (i = 1, 2, \ldots, M) \tag{6.3}$$

$$-Ux_i + \sum_j x_{ij} \leq 0 \quad \text{for each } i \tag{6.4}$$

---

[5] Comparing the results in the first column of Table 6.1 with those of Monroe (1995, Table 2) reveals disagreements in both total misrepresentation and winning candidates. The rule used for assigning ranks to unranked candidates (note 4) was the same reasonable rule that Monroe (1995, p. 928) said he used. It seems, though, that Monroe, in his calculations, did not use this rule; rather, if a voter left, for example, four of the six candidates unranked, then Monroe's ranks were (1, 2, 3, 3, 3, 3) instead of (1, 2, 4.5, 4.5, 4.5, 4.5). Because such a discrepancy would not in itself prove Monroe's solution method (which was not fully explained) to be faulty, a rerun of the integer programs using what appeared to be the rule that Monroe used showed agreement with Monroe's results for $m = 1$, 2, and 3, but not for $m = 4$. For $m = 4$ with the altered rule, the (unique) set of winners was {F, H, I, T} compared with Monroe's answer of {D, F, H, I} tied with {F, G, I, T}; total misrepresentation was 12 compared with 13.

$$x_i \text{ is an integer less than or equal to 1 for each } i \qquad (6.5)$$

$$x_{ij} \text{ is an integer less than or equal to 1 for each } (i, j) \text{ combination.} \quad (6.6)$$

Constraint (6.1) simply says that there are to be m winners out of the M candidates. Constraint (6.2) specifies that one candidate is to be assigned to each voter.

Constraints (6.3) and (6.4) say that the number of voters to whom a candidate is assigned must be greater than or equal to L (lower bound) but less than or equal to U (upper bound) for a winning candidate ($x_i = 1$). For a losing candidate ($x_i = 0$), constraint (6.4) forces this number to be 0.

If n/m is not an integer, then L is set equal to the largest integer less than n/m, and U to the smallest integer greater than n/m. If n/m is an integer, then constraints (6.3)–(6.4) can still be used, with $L = U = n/m$; or constraints (6.3)–(6.4) can be replaced by the single set of constraints,

$$-Bx_i + \sum_j x_{ij} = 0 \quad \text{for each } i, \qquad (6.7)$$

where $B = n/m$.

There are ways to create the misrepresentation values μ_{ij} other than basing them on rankings or approval votes. For example, each voter could rate each candidate on a $(T + 1)$-point scale from 0 to T, with μ_{ij} set equal to T minus the rating. AV is a special case of such a system, with $T = 1$.

Another possibility would be to give each voter a fixed number of votes to spread across the candidates in any way desired, including giving more than one vote to the same candidate. Then μ_{ij} could be the negative of the number of votes that voter j gives to candidate i, or perhaps some other decreasing function of this number. The voting rules would be like those for cumulative voting (see section 4.6), but the rules for determining the winners would not, because the latter rules would be based on misrepresentation values rather than on total votes.

For $m = 1, 2, 3,$ and 4 winners, Monroe (1995) applied his voting system to an election in which the $M = 6$ candidates were actually newspapers; they were ranked by $n = 33$ voters. Integer programs in SAS® were used to solve for the winning sets of candidates for up to $m = 4$ winners (following Monroe) as well as for $m = 5$.

These solutions appear in the first column of Table 6.1, which shows the winning candidates along with the associated total misrepresentation values. (The remaining columns in Table 6.1, as well as all of Table 6.2, will be explained shortly.) These results do not fully agree with those of Monroe, as I indicated in footnote 5. Fortunately, integer programming offers a foolproof technique for obtaining a correct answer.

TABLE 6.1
Integer-Programming Solutions for m Winners, and Their Total Misrepresentation Based on Ranks (k candidates are assigned to each voter using Monroe's data)

	$k = 1$	$k = 2$	$k = 3$	$k = 4$	$k = 5$
$m = 1$	I; 48				
$m = 2$	I,T; 27	I,T; 114.5			
$m = 3$	F,I,T; 16	F,I,T; 84.5	G,I,T; 198		
$m = 4$	F,G,I,T; 13	F,G,I,T; 70	F,G,I,T; 161	F,G,I,T; 290	
$m = 5$	D,F,H,I,T and F,G,H,I,T (tie);9	D,F,H,I,T; 59.5	F,G,H,I,T; 142	F,G,H,I,T; 250	F,G,H,I,T; 391

Note: The entry in a cell for any m and k shows the set(s) of winners, followed by the value of total misrepresentation.

TABLE 6.2
Integer-Programming Solutions for m Winners, and Their Total Misrepresentation Based on AV (k candidates are assigned to each voter using Monroe's data)

	$k = 1$	$k = 2$	$k = 3$	$k = 4$	$k = 5$
$m = 1$	I; 5 ($p = 3$)				
$m = 2$	F,I and I,T (tie); 7 ($p = 2$)	I,T; 13 ($p = 4$)			
$m = 3$	F,I,T; 9 ($p = 1$)	G,I,T; 8 ($p = 3$)	G,I,T; 23 ($p = 4$)		
$m = 4$	F,H,I,T; 7 ($p = 1$)	D,F,I,T and F,H,I,T (tie); 15 ($p = 2$)	F,G,I,T; 26 ($p = 3$)	F,G,I,T and G,H,I,T (tie); 43 ($p = 4$)	
$m = 5$	F,G,H,I,T; 7 ($p = 1$)	D,F,H,I,T; 12 ($p = 2$)	D,F,H,I,T; 40 ($p = 2$)	F,G,H,I,T; 47 ($p = 3$)	F,G,H,I,T; 63 ($p = 4$)

Note: The entry in a cell for any m and k shows the set(s) of winners, followed by the value of total misrepresentation and the value of p that was used. An approval vote was counted for the p top-ranked candidates on each ballot, or for all the ranked candidates if fewer than p were ranked.

6.5. ASSIGNING MORE THAN ONE CANDIDATE TO A VOTER

Monroe's system can be extended in various ways. Although some extensions may appear abstruse, they broaden the menu of PR options, one of which provides an alternative to Illinois' former use of cumulative voting for three-member districts (Brams, 1975, 2003, ch. 3).

Monroe's system can be generalized to allow for k candidates ($1 \leq k \leq m$), rather than just one, to be assigned to each voter for the purpose of determining the set of winners that minimizes total misrepresentation. Integer programming can easily accommodate such a generalization. One simply replaces constraint (6.2) with

$$\sum_i x_{ij} = k \quad \text{for each } j. \tag{6.8}$$

Then for constraints (6.3)–(6.4), L will now be the largest integer less than or equal to kn/m, and U will be the smallest integer greater than or equal to kn/m. If kn/m is an integer, constraints (6.3)–(6.4) can be replaced by constraint (6.7) with $B = kn/m$.

A value of k greater than 1 may be useful with AV, as I will show shortly. But for AV as well as other voting systems, one must guard against making k too high. If k is equal to its maximal value of m, then by constraint (6.7) each winner is assigned to all n voters, making the winning candidates simply those whose μ_{ij} totals across all voters are lowest. (Two examples follow.) As a consequence, minorities, even large ones, may receive underrepresentation, even zero representation, if a disciplined majority votes in ways that crowd out the opposition. Thus, one may want to avoid $k = m$ to ensure minority representation.

As an example, consider an at-large election for an m-member legislative body, where each voter is allowed to vote for up to m candidates. With $k = m$, define μ_{ij} to be 0 if voter j votes for candidate i, 1 otherwise. The integer program need not actually be run, because the winners will simply be the m candidates with the most votes. Thus, a 51 percent majority, all voting for the same m candidates, will win all the seats, however the 49 percent minority votes.

If the μ_{ij} are based on rankings and $k = m$, Monroe's system picks as the m winning candidates those with the m highest Borda-count scores. Monroe (1995, p. 928) noted this result for the case $k = m = 1$.

For Monroe's voting data that used rankings, it is instructive to run integer programs with the constraints (6.1), (6.3)–(6.6), and (6.8) for all values of m from 1 to 5, and all values of k from 1 to m. The results appear in Table 6.1. For a given m, the sets of winners in a few cases change slightly as k increases, but mostly they stay the same.

To explain how integer programming produced the results in Table 6.1, consider the cell at the intersection of $m = 4$ and $k = 2$. It shows F, G, I, and T to be the four winners. The integer-programming solution yielded $x_2 = x_3 = x_5 = x_6 = 1$ and $x_1 = x_4 = 0$, where the subscripts $i = 1, 2, 3, 4, 5,$ and 6 refer, respectively, to candidates D, F, G, H, I, and T. The solution for the 198 x_{ij}'s assigns the candidate pairs (F, G), (F, I), (F, T), (G, I), (G, T), and (I, T), respectively, to 5, 3, 8, 8, 3, and 6 voters. For example, F is assigned to 16, G to 16, I to 17, and T to 17 voters.

The total misrepresentation, 70 units, is the result of summing the 66 μ_{ij}'s that correspond to each voter's assigned candidates. For example, voter 3, who is assigned to candidates G and I, ranked the candidates in the order (I, T, G), with no preference shown among the other three candidates. Thereby this voter produced values of (4, 4, 2, 4, 0, 1) for μ_{13} through μ_{63}. The contribution of voter 3 to the total misrepresentation is 2 + 0, or 2.

Note that total misrepresentation in this case has to be at least 33, because each voter will contribute at least a value of 1. More generally, total misrepresentation when rankings are used with Monroe's system cannot be less than $nk(k-1)/2$.

6.6. APPROVAL VOTING

Monroe gave only limited consideration to using AV with his system, but he was optimistic about its prospects. To infer approvals from Monroe's ranking data, one must make some assumption about how voters would have voted had they used AV.

Assume that the p top-ranked candidates on each ballot, or each of the ranked candidates if fewer than p were ranked, would have been approved. Because casting approval votes for about half the candidates is often a good strategy (Brams and Fishburn, 1983, 2007, ch. 5), start by setting $p = 3$ (recall that $M = 6$ newspapers were ranked).

For this value of p, however, several sets of m candidates all have the same total misrepresentation—namely, zero. To illustrate, for $k = 1$ and $p = 3$, 9 of the 15 possible sets of $m = 4$ candidates, and 5 of the 6 possible sets of $m = 5$ candidates, give a total misrepresentation of zero, reflecting the fact that an approved candidate can be assigned to every voter.

This embarrassment of riches—too many subsets of candidates tied for winning—is hardly acceptable. Its likelihood increases the higher m, the lower k, or the higher p is.

Possible remedies include lowering p or raising k. To be sure, in a real-world AV election, one cannot reduce p, because this entails altering the ballots. On the other hand, voters faced with a high m or a low k might well realize that their ballots will have a better chance of being decisive if they approve of fewer candidates, which corresponds to reducing p.

Table 6.2 gives integer-programming solutions under AV when p varies according to m and k. This table resembles Table 6.1, except that each cell shows the value of p in addition to the set(s) of m winners and the total misrepresentation. The method used to select p, based on m and k, is designed to attain some consistency with respect to average misrepresentation.[6]

[6] Briefly, for each (m, k), choose $p = p(m, k)$ to be the value of $p(1, 2, 3, 4, \text{or } 5)$ for which $E(m, k, p)$ is closest to $E_0(m, k)$ (using the higher value of p in case of a tie). $E(m, k, p)$ denotes the average misrepresentation per voter for a given (m, k, p); $E_0(m, k)$ denotes the "ideal" misrepresentation.

Although the sets of winners in Table 6.2 are similar to those in Table 6.1, they are not identical. As for ties, there is little difficulty. Table 6.2 shows only three, all of them two-way. The lower left-hand corner of Table 6.2 does not show values of total misrepresentation equal to or even close to zero, as would have been the case had the table been based on $p = 3$ throughout. Thus, lowering p removed the problem of a large number of m-candidate sets tied at zero.

An awareness of possible problems with large m and $k = 1$, however, may not be sufficient to deter voters from casting approval votes for "too many" candidates. To reduce the danger that many subsets of candidates could be tied at zero, one can set $k > 1$.

In Table 6.2, the total misrepresentation generally increases as one moves across each row from left to right—in the direction of increasing k—despite the fact that p is also increasing. Setting k equal to some value greater than 1 pushes total misrepresentation well above zero, thereby removing the danger of many ties and making AV more suitable for use with Monroe's system.

The larger the electorate, however, the less will be the danger of having many subsets of candidates tied at zero when $k = 1$. This is because a larger number of voters increases the probability that, for any given subset, there will be at least one voter who does not approve of any candidate and who, therefore, will necessarily be misrepresented. But even though no subsets of candidates have a misrepresentation value of zero, there can still be a number whose misrepresentation totals are closely bunched slightly above zero—a result that may be almost as unsatisfactory as having many subsets tied at zero.

There is a second potential difficulty illustrated by the following example. Set $k = 1$, and suppose that $n = 87$ voters, 15 from party A and 72 from party B, are to choose $m = 3$ winners from $M = 4$ candidates, of whom one is from A and three are from B. Furthermore, suppose that all 15 A-voters approve of only the single candidate of their party, and all 72 B-voters approve of all three of their candidates.

Choose $E_0(m, k) = E_0(k) = (k + 1)/4$, which is a 50–50 mixture of $E_{01}(k) = 1/2$ and $E_{02}(k) = k/2$. Assume, for simplicity, that the p candidates whom a voter approves of are a random draw from the $M = 6$ candidates on the ballot, and let r (a random variable) denote how many of these p candidates are winners. Define $C(N, R) = N!/[R!(N - R)!]$ if $0 \leq R \leq N$, and 0 otherwise. The average misrepresentation is then

$$E(m, k, p) = \sum_{r=0}^{k} (k - r)P(m, p; r),$$

where $P(m, p; r) = C(6 - m, p - r)\, C(m, r)/C(6, p)$ is the probability that exactly r of the p approved candidates are winners. Thus, for $m = 5$ and $k = 3$, $E(5, 3, p)$ is 13/6, 8/6, 3/6, 0, 0 for $p = 1, 2, 3, 4, 5$. Consequently, $p(5, 3) = 2$, because $E(5, 3, 2) = 8/6$ is closer to $E_0(3) = 1$ than $E(5, 3, 3) = 3/6$ is.

These voting strategies, one would think, would be the natural ones. Total misrepresentation, however, is 14 for any set of three potential winners that includes the A-candidate, but it is 15 for the set of all three B-candidates. Consequently, the A-candidate is a winner even though the strength of party A in the electorate is only about 17 percent—far below 25 percent, the threshold required (Brams, 1975, 2003, ch. 3) for a party to win a seat in the case of cumulative voting with two parties and three winners.

Suppose now that the B-voters somehow split themselves into three groups of 24 voters each, with those in each group voting to approve of a different one of the three B-candidates and no one else. Because total misrepresentation remains at 15 for the subset comprising all three B-candidates but rises to 24 for any subset containing the A-candidate, the A-candidate is no longer a winner.

Thus, the result that is consistent with the cumulative-voting threshold can be achieved, but only if the B-voters adopt a voting strategy that requires coordination among them.[7] Although this result constitutes an equilibrium, the complexity required to achieve the coordination is anything but desirable.

In summary, when Monroe's system is used with AV and $k = 1$, there is a potential problem with (i) ties and (ii) unduly low thresholds for minority candidates. More technical problems that may arise with Monroe-type systems generally—not just using AV—are discussed in three appendixes: Appendix A deals with easing the computational burdens for large electorates, B with ties, and C with the filling of vacancies.

I turn next to major extensions of Monroe's system. Because there is no inherent reason to assume that assignments of voters to representatives must be integers, I begin by considering the possibility of fractional assignments.

6.7. FRACTIONAL ASSIGNMENTS

Instead of restricting the x_{ij}'s to integers, as in constraint (6.6), assume they are allowed to be fractions. Thus, when $m = 4$ and $k = 1$, instead of assigning, as Monroe did, one candidate to 9 voters and the other three to 8 voters, one could assign the four candidates to 8.25 voters. When $m = 2$ and $k = 1$, instead of splitting the voters 17 and 16 between the two winners, one could assign 16.5 voters to each. Although Monroe considered the possibility of fractional

---

[7] For the general case of AV under Monroe's system with $k = 1$, two parties, m winners, one A-candidate, and m B-candidates, party A will win one seat if its strength exceeds the threshold of $1/2m$ of the electorate (equal to $1/6$ for $m = 3$) when all A-voters vote only for the A-candidate and all B-voters vote for all m of the B-candidates. If the B-voters coordinate their strategies so that only $1/m$ of them vote for each of their m candidates, then the threshold for A to win one seat becomes $1/(m + 1)$, the same as the cumulative-voting threshold. It turns out that $1/(m + 1)$ is a threshold not only for cumulative voting but also for the Hare system of single transferable vote (STV), so it appears to be a natural threshold.

assignments in a brief footnote, he did not pursue the idea of assigning exactly kn/m voters to each winning candidate when kn/m is not an integer.

Mathematically and computationally, fractional assignments are easily handled through mixed-integer programming. The x_i's are still constrained to be integers, but the x_{ij}'s are not. Thus,

$$x_{ij} \leq 1 \quad \text{for each } (i, j) \text{ combination} \tag{6.9}$$

replaces constraint (6.6). In addition, constraint (6.7) is used with $B = kn/m$, regardless of whether or not kn/m is an integer, so constraints (6.3)–(6.4) are never used. The full set of constraints is, therefore, (6.1), (6.8), (6.7), (6.5), and (6.9).

Although the provision for fractional assignments may seem like a minor emendation, it can result in significant differences, especially in small electorates. First, fractional and integer assignment can yield different sets of winners. For example, for $m = 5$ and $k = 2$ with voting based on rankings, the winning set (unique in both cases) changes from {D, F, H, I, T} in Table 6.1 to {F, G, H, I, T} under fractional assignments, so G replaces D. Second, if each voter were cloned $(m - 1)$ times, thereby producing an electorate m times as large, one would think that the set of winners ought not to change. This property is always satisfied under fractional assignments but not under integer assignments.[8]

Even with the provision for fractional assignments, not many of the x_{ij}'s will actually be fractional. For example, if $m = 4$ and $k = 1$ for either AV (with $p = 1$) or voting based on rankings, just three of Monroe's 33 voters have fractional x_{ij}'s. (For both types of voting, the solution is a 3/4-assignment of one winner to each of these three voters, together with a 1/4-assignment of each of the other three winners to one of these same voters.) In the case where kn/m is an integer, there will always be a solution that minimizes total misrepresentation and has no fractional x_{ij}'s at all, even though the constraints permit fractional x_{ij}'s.[9]

---

[8] Mathematically, the minimization problem is virtually the same after the cloning as before when fractional assignments but not integer assignments are made. In the example earlier in the paragraph, if Monroe's 33 voters are each cloned four times to produce an electorate of size $n = 165$, then (i) kn/m becomes an integer, 66; (ii) the winning set is (F, G, H, I, T) not only under fractional assignments but also under integer assignments; and (iii) total misrepresentation is 305 under both types of assignment, as compared with 61 under fractional assignments and 59.5 under integer assignment before the cloning.

[9] The proof of this last statement is based on two observations. First, if the x_i's are held fixed so as to correspond to any specific potential set of m winners, then the minimization problem for the specific set—with no integer restrictions on the x_i's—is a special case of what is known as the *capacitated transportation problem*. Second, there exists an all-integer optimal solution to a capacitated transportation problem provided that all the constraint constants [B and the right-hand sides of (6.8) and (6.9) in the present case] are integers. See Dantzig (1963, pp. 377–380).

6.8. NONINTEGER k

Once provision is made for fractional assignments, it becomes possible to use a noninteger value for k. But when would one want to assign a noninteger number of candidates to each voter?

Consider the following example under AV. There are

- two parties, A and B
- $M = M_A + M_B$ candidates, M_A from A and M_B from B
- n voters, Pn from A (where $P < 1/2$) and $(1 - P)n$ from B
- m voters

Assume that the n voters vote for all the candidates of their own party and no others.

Suppose first that $k = 1$. As in the example discussed at the end of section 6.5 and in note 7, A will win a seat even if P is below the cumulative-voting threshold of 1/4, provided that P exceeds 1/6. This leads to the difficulties noted earlier (namely, the B-voters can stop A from winning a seat only by carefully coordinating their votes).

But now let $k = 3/2$ rather than 1, and let all voters continue to approve of all candidates of their own party and no others. Then the threshold for party A to win a seat increases from 1/6 to 1/4, the cumulative-voting threshold.[10] Thus, the threshold problem disappears.

If there are to be $m = 2$ winners rather than 3, the threshold for A to win a seat is $P = 1/4$ if $k = 1$, but it increases to the cumulative-voting threshold of $P = 1/3$ if k is chosen to be 4/3. Thus, one avoids the threshold problem by setting $k = 4/3$ rather than 1.

Unfortunately, the cumulative-voting thresholds cannot be duplicated exactly (through the choice of k) when $m > 3$. This is because there is more than one threshold to contend with in this situation.

For many years, Illinois voters used cumulative voting to elect the lower house of their legislature from three-member districts. A disadvantage was that parties were uncertain as to how many candidates they should run in a district; they could suffer by running either too many or too few and, in fact, did on occasion (Brams, 1975, 2003, ch. 3).

In the present model with $m = 3$ and $k = 3/2$ under AV, however, the threshold for party A to win a seat is $P = 1/4$, the cumulative-voting threshold,

---

[10] For any permissible values of P, m, and k, the minimized total misrepresentation, after division by n, is $(k - 1)P + (k/m - P)$ if A receives one seat (provided that $P > k/m$), and it is kP if A receives no seats. Equating these two expressions and solving for P yields $P = k/2m$, the threshold for A to win one seat. If one equates this expression for P with $P = 1/(m + 1)$, the cumulative-voting threshold, and then solves for k, the result is $k = 2m/(m + 1)$, which is the value of k required to duplicate the cumulative-voting threshold. For $m = 3$ winners, $k = 3/2$; and for $m = 2$ winners, $k = 4/3$.

regardless of what M_A and M_B, the numbers of candidates, are. In this respect, therefore, the system is better than cumulative voting, at least for the case of $m = 3$ (and $m = 2$ as well).

6.9. THE CHAMBERLIN-COURANT SYSTEM

As Monroe (1995, p. 936) noted, Chamberlin and Courant (1983) proposed a PR voting system that is effectively the same as his, except that

- it does not require constituency sizes to be equal or nearly equal; and
- it uses weighted voting in the legislature to reflect constituency size.

Thereby the Chamberlin-Courant system allows the number of voters assigned to a winning candidate to vary across different winners. In turn, the winners cast these numbers of votes in the legislature.

Solutions for sets of winners under the Chamberlin-Courant system can be found through integer programming. The objective function and the constraints are the same as under Monroe's system, except that constraint (6.3) is dropped (which is equivalent to retaining it with $L = 0$) and constraint (6.4) is applied with $U = n$. Although Chamberlin and Courant defined their system only for $k = 1$, it can be used for any k.

The constraint set then comprises (6.1), (6.8), (6.4) with $U = n$, (6.5), and either (6.6) or (6.9). If k is an integer, using constraint (6.9) rather than constraint (6.6) does not alter any solutions for winners, but (6.9) must be used in case k is not an integer. The grouping of like ballots, ties, and the filling of vacancies are discussed in Appendixes A, B, and C.

If $k = m$, the Monroe and Chamberlin-Courant systems are equivalent, whatever the μ_{ij}'s. This is because their sets of constraints are the same and thus lead to the same solutions.

Although Chamberlin and Courant did not mention it, their system, like Monroe's, is applicable not just to voting based on rankings but also to voting based on AV and, in fact, to any kind of voting from which misrepresentation values, μ_{ij}, can be derived. Under either their system or Monroe's, a set of winners that minimizes the total misrepresentation of voters can, of course, always be determined from a solution to the integer program. But the Chamberlin-Courant system also requires the determination of the weights that each winner has in the legislature.

It might appear that these weights could be obtained from the integer-programming results simply by taking $\sum_j x_{ij}$, the number of voters to whom candidate i is assigned, as the weight for candidate i. Doing this will work in some cases but not in others. It will always give correct weights if voting is based on rankings and every voter ranks all M candidates, or if the μ_{ij}'s are such that no two of them are the same for the same voter. In these cases, ties with differing sets of x_{ij}'s for a given set of winners are not possible.

As shown in Appendix B, ties of this type are inconsequential under Monroe's system, because the x_{ij}'s are not used for anything under his system. Under the Chamberlin-Courant system, by comparison, multiple optimal solutions with different values of the x_{ij}'s can cause a problem. One cannot just use $\sum_j x_{ij}$ as the weight for candidate i, because the x_{ij}'s will differ among different solutions. Instead, one must make a simple supplemental calculation once the set of m winners has been obtained from the integer program.

To illustrate how this is done—and the problem corrected—consider first an example with voting based on rankings. Let $m = 3$ and $k = 1$, and suppose that the integer program yields candidates A, B, and C as the winners. For a voter who leaves all three winners unranked, 1/3 of a unit of weight (or possibly 0 units) could be contributed to each of A, B, and C; otherwise, 1 unit could accrue to the winner to whom the voter gives the top rank.

As a second example, let $m = 4$ and $k = 2$ under AV, and let A, B, C, and D be the winners. A voter who approves of all winners except C will contribute 2/3 of a unit to each of A, B, and D; a voter who approves of only C and D will contribute 1 unit to each; and a voter who approves of only B will contribute 1 unit to B and 1/3 of a unit (or possibly 0) to each of A, C, and D.

One can conceive of a spectrum of PR voting systems, with the Chamberlin-Courant system at one extreme and the fractional-voting system at the other. In between, there will be a continuum of voting systems, based on the degree to which the number of voters assigned to a candidate is forced to be uniform among winners.

Mathematically, the continuum can be defined in terms of a variable q that runs from 0 to 1. The constraints are then (6.1), (6.8), (6.3) with $L = qkn/m$, (6.4) with $U = qkn/m + (1 - q)n$, (6.5), and (6.9). The Chamberlin-Courant system has $q = 0$; the fractional-voting system, $q = 1$. Although the Monroe system cannot be placed exactly along this continuum, it falls, roughly speaking, at a q-value close, but not equal to, 1 unless kn/m is an integer.

The matter of weighted-versus-unweighted voting need not be tied to the value of q. Although Chamberlin and Courant advocated weighted voting for $q = 0$, and Monroe favored unweighted voting for q at or near 1, either option could be used anywhere along the continuum. If unweighted voting is used and q is not close to 1, however, the representation might be far from proportional. At $q = 1$, of course, weighted voting with weights equal to the constituency sizes is equivalent to unweighted voting.

6.10. TULLOCK'S SYSTEM

Tullock (1967, ch. 10) proposed a PR voting system under which each voter votes for just one candidate, every candidate who receives at least one vote is a winner (so that *all* the candidates are assured of being winners if they vote for themselves), and weighted voting is used in the legislature.[11] The weights of the legislators are equal to the numbers of votes cast for them. Candidates who receive just one vote, though they can join the legislature, may be restricted with respect to legislative activities other than voting.

From a mathematical standpoint, Tullock's system can be subsumed within the integer-programming framework as a special case of the Chamberlin-Courant system with $m = M$ and $k = 1$. One can define μ_{ij} to be 0 if candidate i is the top choice of voter j, and 1 otherwise. Although the integer program will yield the correct solution, there is no need to use it since obviously the solution will give $x_i = 1$ for each i and $x_{ij} = 1 - p$ each μ_{ij}.[12]

Tullock's system has several virtues. It is conceptually simple and easily understood. It is nonmanipulable, because no voters have an incentive to vote insincerely for strategic reasons (e.g., to prevent a candidate from being a representative, or to support a second choice rather than a first choice who has little chance of winning).

Furthermore, Tullock's system is a perfect reflection of the electorate, at least insofar as voters have complete information about the candidates' positions and insofar as they vote for themselves in cases where no other candidate perfectly represents their views. On the other hand, two significant objections can be raised against Tullock's system.

First, the size of the legislature may be enormous. Tullock anticipated this objection. He offered some suggestions for operating a large legislature and some others for holding down its size. The latter included imposing a maximum on the number of members, and a minimum on the number of votes for a candidate to be elected. But, these restrictions would make his system vulnerable to strategic voting, undermining its aforementioned nonmanipulability.

To hold down the legislature's size, one could use some form of the Chamberlin-Courant system with a value of m that is less than M but still quite high. Voters would then need to indicate more than just a single first choice; that would seem desirable even under the pure Tullock system (where $m = M$), because otherwise the system would have no ready way to handle a vacancy.

The second objection is with weighted voting itself, which Tullock's system uses and which I discuss in the next section. The same objection can also be raised, as Monroe has done, against the Chamberlin-Courant system.

---

[11] Alger (2006) revised Tullock's system to allow for "ancillary roles," which he calls "voting by proxy," but here I focus on Tullock's system.

[12] Consider a system in which each district has n voters. Assume that n_0 of the n ballots are drawn randomly without replacement. The candidates marked on these ballots are declared elected, with weights in the legislature equal to their numbers of drawn ballots. Then Tullock's system, which has no random element, is a special case in which $n_0 = n$; a lottery system proposed by Amar (1984) is a special case in which $n_0 = 1$.

6.11. WEIGHTED VOTING

The weights of players do not necessarily reflect their power, as measured by their pivotalness or criticalness in winning coalitions. For example, analyses based on the so-called power indices, especially those proposed by Shapley-Shubik and Banzhaf (Shapley and Shubik, 1954; Banzhaf, 1965; Riker and Shapley, 1968; Brams, 1975, 2003, ch. 5; Felsenthal and Machover, 1998), show that legislators with more weight—or parties able to discipline their members to vote as a bloc—usually have disproportionately more voting power than is warranted by their size.[13] However, in PR systems of the kind discussed here, some of the liabilities of weighted voting may be mitigated by several factors:

1. Constituencies are self-selecting under Tullock's system and may be largely so under the Chamberlin-Courant system. The aforementioned analyses of weighted voting do not assume self-selected constituencies—based on voters' choices—but instead that weights have been fixed, usually based on the population of a constituency.
2. The constituencies will change with every election, as will the weights. Thus, any adverse effects of weighting need not be long lasting. In particular, voters will have frequent opportunities to reduce the weights of heavily weighted legislators.
3. Self-selected constituencies can be expected to be quite homogeneous. By contrast, a geographical constituency may well be so heterogeneous that its legislator will often cast a vote that is opposed by almost half the constituency; the resulting misrepresentation is aggravated if that legislator also carries a high weight. But a legislator with a homogeneous self-selected constituency, whether the constituency's weight is high or low, is not likely to cast a vote that is opposed by more than a small number of his or her constituents.
4. Arguments that voter power is proportional to the square root of constituency size, which led to the conclusion that disparate constituency sizes should be avoided (Banzhaf, 1966, 1968), rest explicitly on the assumption of heterogeneous constituencies.

[13] For example, consider a weighted voting body in which players (A, B, C) have weights (50, 49, 1), and the decision rule is a simple majority of 51 votes. The weightiest player, A, has a power value equal to 2/3 according to the Banzhaf index and 3/5 according to the Shapley-Shubik index, though it has only 50 percent of the votes. This is because A is a necessary member of every winning coalition—such a coalition would not be winning without A, which gives A a veto. By comparison, because B and C are interchangeable in any winning coalition, they have exactly the same power values (1/6 according to the Banzhaf index, 1/5 according to the Shapley-Shubik index). This gives the least weighty player, C, a power value far out of proportion to its weight. Such anomalies in the measurement of voting power are common.

5. The acceptability of weighted voting may depend partly on whether the legislative body is large or small. As Banzhaf (1965) and Riker and Shapley (1968) indicated, the adverse effects of weighted voting tend to diminish as the size of the legislature increases.

6. The acceptability of weighted voting may vary according to the degree to which a legislative body serves as a vehicle for mirroring voters' preferences—as opposed to one that also provides a setting for persuasion and bargaining. For example, if the Electoral College in the United States were hypothetically set up with each state's electoral votes proportional to its number of voters, and these electoral votes were split in proportion to the state's popular vote instead of being cast as a bloc for the plurality winner, there would surely be less complaint about weighted voting, because it would mirror the popular vote and thus be transmitting views perfectly. But for a legislature wherein bargaining and vote trading play a heavy role, arguments against weighted voting are telling, including those highlighting the nonlinear relationship between voting weight and voting power.

7. The adverse effects of weighted voting will depend, to some extent, on the values of the weights. These effects may be less if the values can vary over time, enabling shifts in legislative voting strength to take place.

8. Weighted legislative voting should not, by itself, disqualify a PR system from being acceptable. Other criteria, such as degree of nonmanipulability and degree of representativeness, which are topics I take up next, are also important in comparing different PR systems.

6.12. NONMANIPULABILITY

My main focus in this chapter has been on (i) showing the usefulness of integer programming in determining winners under various PR voting systems, and (ii) developing a framework, based on integer programming, that subsumes different voting systems. In this and the next section, I offer some brief comments on two important properties, nonmanipulability and representativeness, of voting systems.

With respect to nonmanipulability, Monroe (1995) made several observations. He suggested that his method of minimizing misrepresentaton would be relatively resistant to manipulation if used with AV. This appears to be a reasonable conjecture in view of the relative nonmanipulability of AV (Brams and Fishburn, 1983, 2007, ch. 2).

Monroe also contended that his method, if used with rankings, would be easy to manipulate and therefore impractical. He reasoned that because the Borda count is highly manipulable, his system, when used with rankings, would be likewise vulnerable because, for the special case of $m = 1$, it is the same as the Borda count.

I have already noted the variety of systems that are amenable to integer programming. A system that selects the m winner(s) to be the m candidate(s) with the highest Borda score(s), whether $m = 1$ or $m > 1$, is a special case in which the μ_{ij}'s are based on rankings, with $k = m$ and $q = 1$.

But another special case, wherein the μ_{ij}'s are based on rankings, is a system that can be rendered arbitrarily close to Tullock's. It occurs if one chooses $k = 1$, sets m slightly below M and q at or near 0, and uses weighted voting. Because the Borda count with $m = 1$ (highly manipulable) and the Tullock system (essentially nonmanipulable) both fall within the framework developed herein, it is evident that PR voting systems based on rankings can run the gamut with respect to nonmanipulability.

6.13. REPRESENTATIVENESS

Representativeness may have different meanings. Consider the simple case in which all issues before a legislature have only two sides. (For example, a legislature may or may not decide to declare war against an adversary.) Suppose that, on every issue, the majority side is the same as the majority side in the electorate.

Although a legislature satisfying this property would properly reflect the desires of the electorate, it would hardly be considered representative if, on every issue, it voted unanimously (or nearly unanimously) on the majority side. This could occur, even with a closely divided electorate, under very different election systems, including ones in which members are elected either by district (if the districts are similar enough in their breakdowns between majority and minority) or at-large. The result would be that the minority side would not have representation in either deliberation or decision.

At a more stringent level of representativeness, an appropriate ideal, or "gold standard" to strive for, would be to require the percentage split in the legislature to be the same as the percentage split in the electorate on every issue. At least in theory, Tullock's system satisfies this criterion, because any voters who do not find a candidate in agreement with their views can vote for themselves, thereby effecting perfect representation.

Whether judged by either criterion—the majority in the legislature on each issue is (i) the majority in the electorate, or (ii) a perfect reflection of the majority proportion in the electorate—any of Monroe's systems will be unrepresentative to the extent that the candidate who is assigned to a voter is not that voter's first choice. (Of course, Monroe's system seeks to minimize such misrepresentation.) In some cases, however, an improvement in representativeness can result from nothing more than a change in the field of candidates.

For example, if $2n/m$ voters each rank candidate A first and candidate B second but have little enthusiasm for B, then Monroe's system will assign no more than half of these voters to A and assign the remainder to a candidate unappealing to them. The situation of the A supporters would be improved, of course, if the ballot included not only A but also a clone of A.

The representativeness of some PR systems so far considered can run amok of certain anomalies, two of which I illustrate next. First, suppose that $m = 5$ winners are to be chosen from $M = 6$ candidates, three (A1, A2, and A3) from party A and three (B1, B2, and B3) from party B, with the μ_{ij}'s based on AV and with $q = 1$. Also suppose that three-fifths of the n voters approve of just A1, A2, and A3, and the other two-fifths approve of only B1, B2, and B3.

There are essentially two possible outcomes that can occur, depending on whether one of the A-candidates or one of the B-candidates is the loser. One would expect a PR system to choose three A-candidates and two B-candidates.

This is indeed what happens when $k = 1, 2,$ and 5 candidates per voter. But if $k = 4$, the two outcomes are tied, because total misrepresentation is $(8/5)n$, regardless of whether or not the winning set excludes an A-candidate or a B-candidate. Worse, any winning set for $k = 3$ has only two of the A-candidates, because total misrepresentation is $(3/5)n$ if an A-candidate is excluded, and $(4/5)n$ if a B-candidate is excluded.[14]

As a second example, consider an election in which there are two issues and $n = 60$ voters. Suppose that the voter positions on the issues are (Yes, No), or (Y, N), for 27 voters; (N, Y) for another 27; and (Y, Y) for the remaining 6. There are $M = 4$ candidates, A, B, C, and D, with respective positions (Y, Y), (Y, N), (N, Y), and (N, N).

With the μ_{ij}'s based on rankings, let Monroe's system ($q = 1, k = 1$) be used to choose $m = 3$ winners. Suppose that the voters vote their sincere preferences, and that these are (B, D, A, C) for the first 27 voters, (C, D, A, B) for the second 27, (A, B, C, D) for 3 of the last 6, and (A, C, B, D) for the remaining 3. Total misrepresentation is then 26 for the set {B, C, D} (for which candidate D is assigned to 10 of the first 27 voters and 10 of the second 27), and 28 for the set {A, B, C}.

Thus, {B, C, D} wins. Whereas the legislature opposes both issues by 2 to 1, the electorate favors both by 33 to 27. Moreover, candidate A, who would be the choice of the "voice of reason" (Monroe, 1995, p. 935) in a single-winner election, is now the sole loser. Because neither candidate A nor any of the

---

[14] To verify informally this disconcerting result when $k = 3$, assume $n = 5$. Construct two 5×5 grids in which the rows of each grid represent the five voters (three A-voters and two B-voters), the columns of the first grid represent the candidate set {A1, A2, B1, B2, B3}, and the columns of the second grid represent the set {A1, A2, A3, B1, B2}. One then places in the 25 squares of each grid 15 X's (where each X represents the assignment of a voter to a candidate) so that there are three X's in each row and in each column, and the total misrepresentation is as low as possible. It is easy to show that the minimal misrepresentation is 3 for the first grid and 4 for the second. The result for $k = 4$ can be demonstrated similarly.

other three candidates is a Condorcet winner, these perplexing results are perhaps not so surprising.

6.14. HIERARCHICAL PR

So far I have considered several types of PR systems that minimize the overall misrepresentation of voters. Integer programming, which is applicable both to ranking systems like the Borda count and to nonranking systems like AV, provides a systematic way of finding the set of winning candidates.

In the United States, there are three levels of federal office, with varying numbers filling each level: one president and one vice president; 100 senators; and 435 representatives. These three levels, written into the federal constitution, are mirrored in all states except Nebraska, which has, like the other states, one governor but a unicameral rather than a bicameral legislature.

I think it impractical as well as unwise to attempt to change the U.S. Constitution, or state constitutions, to elect senators and representatives nationwide, or state legislators statewide, using the kind of PR system discussed here. It seems more appropriate for electing city councils or other representative bodies of, say, five to twenty-five members. Such a PR system could also be adapted, as I next show, to allow for the election of not only regular members of a council but also a council chair or a mayor.

If separate elections are held for a mayor and a city council, as is usually the case today, the more ambitious candidates, who prefer to be mayor but think it safer to run for city council, are forced to make a difficult choice between offices for which to seek election. By contrast, under a system I call *hierarchical PR*, different levels of representation can be achieved in just one election.

To illustrate how hierarchical PR would work, consider a municipality in which one mayor and five members of a city council are to be elected. Under hierarchical PR, the mayor would be the one candidate, among all those running for the city council, who minimizes misrepresentation across all the voters. If voters can rank the candidates, this person would be the Borda count winner; if the voters can approve of as many candidates as they like, this person would be the AV winner.

Once chosen, the mayor is eliminated from further consideration. The 5-member council would then be elected from the remaining candidates to minimize total misrepresentation.[15]

---

[15] If the μ_{ij}'s (the misrepresentation values) are based on rankings of the candidates, there are two ways to define these values after elimination of the mayor: they can be left as they are, or they can be redone based on rerankings without the mayor. In the latter case, μ_{ij} stays the same, decreases by 1, or decreases by 1/2 according to whether, respectively, voter j ranks candidate i above, below, or tied with the mayor. It is not clear which of the two ways is preferable.

Suppose, for example, that AV is used with $k = 1$ (i.e., one candidate is assigned to each voter) and $q = 1$ (i.e., each candidate is assigned to the same number of voters), and there are twenty voters who approve of the candidates as follows:

- voters 1–3, 5–7, 9–11, 13–15, and 17–19 (75 percent) approve of candidate A
- voters 1–4 (20 percent) approve of candidate B
- voters 5–8 (20 percent) approve of candidate C
- voters 9–12 (20 percent) approve of candidate D
- voters 13–16 (20 percent) approve of candidate E
- voters 17–20 (20 percent) approve of candidate F

In the race for mayor, A would be elected. In the race for council, {B, C, D, E, F} would be elected (which would also be true if there were other candidates who were approved of by fewer voters). In the case of the council, observe that every one of the twenty voters is represented by exactly one of the five winners in such a way that total misrepresentation is zero.

Now consider a situation in which there is no election for mayor but only an election for the 5-member city council (assume the council chooses one of its elected members as mayor). Then the same council as before, {B, C, D, E, F}, would be chosen. This is fine as far as it goes: all the voters, as before, have exactly one representative of whom they approve. The rub is that the most popular candidate, A, loses, because any 5-member council that includes A would misrepresent one voter.[16]

Admittedly, this example is an artificial one designed to illustrate a certain type of possible outcome under hierarchical PR for two offices—one with one winner and the other with five winners in this case—when $k = 1$. Less artificial is Monroe's example, in which the *Independent* (I) would have been chosen for the first "office" because it is the overall most popular newspaper (recall that misrepresentation in his example was based on ranks).

If there were two winners for the next "office" (instead of the five assumed in the city council example), they would not, of course, include the *Independent,* because it was eliminated. As shown in Table 6.1, the *Independent* and the *Times* (T) are the two newspapers that would have been chosen if there were two winners for a *single* office. This illustrates that (i) hierarchical PR need not give the same results for a second-level office as does a nonhierarchical election with the same number of winners, and (ii) the winners in the nonhierarchical election need not exclude the winner of the top office under hierarchical PR.

If there are two or more levels of representation, hierarchical PR has the advantage of not requiring different elections for the different levels (offices),

---

[16] It is worth mentioning that whoever is chosen as mayor would be supported by only four voters and can, therefore, hardly be considered representative of the entire electorate.

relieving candidates of the necessity of making the kind of difficult choice alluded to earlier (to run for mayor or for city council). Instead, the system itself selects the most popular candidate for the highest office, and the most representative choices for lower offices, making endogenous who is most appropriate at each level.

I think this is an important advantage of hierarchical PR. Popular candidates, as I suggested earlier, often face the agonizing choice of keeping a relatively safe seat (e.g., on a city council) or running for higher office (e.g., mayor), which is almost always riskier. Some of these candidates will choose not to take this risk. This is especially unfortunate if they are well qualified for the higher office.[17]

Under hierarchical PR, such candidates can, as the old saw goes, have their cake and eat it too. They can campaign to gain widespread approval and become mayor; but that failing, they will probably stand a good chance of winning a council seat, the earlier example notwithstanding. Democracy is well served, I believe, when the electorate, *through the voting system,* makes the choice of the office for which a candidate is most qualified—rather than force the candidate to make the choice, or leave it in the hands of officeholders rather than the voters.

It is worth noting that a higher office might well have several winners. For example, this office might be a 5-member executive committee, which has administrative functions that would be unwieldy for, say, a 25-member council, of which the executive committee is a part, to carry out. Although traditionally it is the council or the council chair that selects such a committee, under hierarchical PR it would be the voters themselves who choose this committee.

The remaining council members would be the twenty other candidates who, after the five executive-committee members are chosen, minimize—at a second level—the total misrepresentation of the voters. Integer programming would be applied twice, first to determine the composition of the executive committee and then to select the rest of the council, even though only one election to the 25-member council would have to be conducted.

I assume that the voters and the candidates would know that the executive committee would be chosen first. In a campaign, therefore, those "running" for the executive committee would presumably make their appeals sufficiently broad so that, together with a group of four other candidates, they might "cover" the electorate. By comparison, those running for the regular council would presumably make narrower, more ideological appeals, seeking only enough support so that, together with nineteen other council candidates, they also "cover" the electorate.

---

[17] To be sure, hierarchical PR would not be appropriate if the different offices required different skills, such as a legal background to be district attorney that a mayor need not have.

Other than at-large elections in which the top vote getter assumes a top position in a governing body, I know of no other voting system in which hierarchical levels can be derived from just the ballot information in a single election. I hesitate, however, to offer prescriptions about the optimal number of levels—except to observe that most political systems have a single top officeholder (executive committees notwithstanding). Below the president, governor, or mayor, there is usually a bicameral legislature or a unicameral council.

Practically speaking, I think hierarchical PR is most relevant to the election of single executives or small executive committees, on the one hand, together with larger councils or boards of which they are a part, on the other. Municipal and county elections in the public sphere, and corporate boards and their executive committees in the private sphere, come quickly to mind as appropriate venues for hierarchical PR. In academia, of course, there is a hierarchy in both departments and faculties, for which I think hierarchical PR would also be an attractive choice.

Hierarchical PR can be used with most of the PR voting systems considered in this chapter. One exception is Tullock's system, which would not be suited to electing a specified number of persons (one or more) to the top office. Additionally, in cases where there is to be only one winner at the top level, the use of a system based on rankings would be open to question, because this would be equivalent to the use of the highly manipulable Borda count to elect one candidate.

6.15. SUMMARY AND CONCLUSIONS

In this chapter I have shown how integer programming provides a means for determining election winners under different PR voting systems. Extending the research of Monroe (1995), I have also developed a framework, based on integer programming, to minimize misrepresentation. It encompasses a large class of PR systems.

Within this framework, the variables that a designer of a PR system would need to specify include the nature of the misrepresentation values (à la Monroe), the number of candidates assigned to a voter (k), a variable (q) that controls the extent to which the number of voters assigned to a candidate can vary among winners, and whether or not weighted legislative voting is to be used. Other variables are the number of candidates (M), the number of winners (m), and the number of voters (n).

Which PR system is best? Does Tullock's system deserve resuscitation after years of neglect? Is the Chamberlin-Courant system viable in some form? Or are both these systems fatally flawed because of their use of weighted voting?

If so, should one turn to one of the Monroe systems, perhaps using AV, or to some hybrid system within the framework? Or is there a method outside of the integer-programming framework, such as STV, that is better than anything within the framework in certain situations?

These are thorny questions that involve a number of issues. To give some structure to them, I offer the following thoughts and recommendations on the PR systems studied in this chapter:

1. The best PR system to use will depend on such conditions as the nature of the institution, the size of the electorate, its diversity, the number of parties (if any), and, of course, what is acceptable to the electorate.
2. Most PR systems I have considered have pitfalls not obvious on first examination. There may be some as yet undetected flaws in these systems, which will require further scrutiny.
3. It is reasonable to ground PR systems either on rankings or on AV. Rankings demand more of voters, whereas AV is more appropriate when information on all the candidates is relatively scarce.
4. It is not evident what the value of k, the number of winning candidates assigned to each voter, should be. The answer may depend not only on "objective" circumstances but also on taste. On the one hand, the choice of $k = 1$ offers the advantages of simplicity and interpretability, encourages voters to support fewer candidates in the case of AV, and gives each voter a single champion of his or her views. On the other hand, I showed cases wherein $k = 1$ gives less representative results than some larger k.
5. Fractional assignments have several advantages over integer assignments and no apparent disadvantages. They are, however, less intuitive.
6. At least for a two-party system, the PR method that elects legislators in three-member districts using AV and $k = 3/2$ seems promising. The main barrier to adoption may be the difficulty of explaining the use of $3/2$ for k.
7. The use of weighted voting presents a mixed picture. On the positive side, some of the arguments against weighted voting do not apply to a Tullock-like system, which scores high on nonmanipulability and representativeness. On the negative side, there are institutional problems, including (i) the apparent nonegalitarian nature of weighted voting, which makes some representatives more equal than others; and (ii) that weightier representatives have the same finite and unexpandable time as less weighty ones to take care of the burden of representation, even if they are given larger staffs to serve their more numerous supporters.
8. Most PR systems can be extended to provide hierarchical levels of representation, whereby the most representative candidates for *different* offices can all be chosen in one election rather than being required to run in different elections. Hierarchical PR seems not only an efficient and effective

way to build representation at different levels—because it requires only one election—but also is more democratic than having, say, an elected body choose a leader, as occurs today in most parliamentary systems.

To expand on this last point, I believe it is the voters themselves who should have the opportunity to choose, for highest office, the most representative candidate or candidates (e.g., a prime minister and also possibly a cabinet).[18] This process can then be repeated at lower levels, with voters in each case choosing representatives who minimize misrepresentation at those levels. Thereby voters would better be able to express themselves at all levels, which seems to me the cornerstone of democratic representation, especially in a federal system.

APPENDIXES

A. LARGE ELECTORATES

Integer programming is computationally demanding. At some point a PR integer program will become so large that computing time will be prohibitive. In order to render the method more suitable for larger electorates, it is useful to examine possible ways to reduce computer time. Of course, the computational burden is likely to become less of an issue as rapid improvements in computer technology continue.[19]

The following modification does not affect the solution for the winners, but it reduces computer time by decreasing both the number of x-variables and the number of constraints. In a large electorate, there will be cases wherein two or more voters mark their ballots alike and thus have identical misrepresentation values. The modification consists of consolidating into one group each set of voters who mark their ballots in the same way.

Define the index j so that it applies to these groups rather than to individual voters. Although j runs, as before, from 1 to n^*, n^* is redefined to be the number of groups of voters who all vote alike (i.e., cast the same ballot) and is no longer equal to n, the number of voters.

Let v_j denote the number of voters in group j ($j = 1, 2, \ldots, n^*$). Continue to use n to denote the number of voters, where now $n = \sum_j v_j$. Although x_i has the same meaning as before, x_{ij} is now the number of voters, out of the v_j voters in group j, to whom candidate i is assigned.

---

[18] This is also possible under "coalition voting" (Brams and Fishburn, 1992b), one aspect of which is "yes-no voting" (Brams and Fishburn, 1993) that I briefly discuss in chapter 7.

[19] For an in-depth analysis of this problem with respect to the voting systems analyzed in this chapter, see Procaccia, Rosenschein, and Zohar (2007).

The objective function, z, has the same formula as before, but two of the constraints change. Constraint (6.8), which is the generalization of (6.2), changes to

$$\sum_i x_{ij} = kv_j \quad \text{for each } j, \tag{A6.1}$$

which specifies that the number of candidate assignments to the v_j voters in group j is kv_j. Constraint (6.6) becomes

$$x_{ij} \text{ is an integer less than or equal to } v_j$$
$$\text{for each } (i, j) \text{ combination.} \tag{A6.2}$$

Thus, the constraints are now (6.1), (A6.1), (6.3)–(6.4) or (6.7), (6.5), and (6.2). L, U, and B are as before, as described in section 6.4.[20]

It is instructive to apply this modification to Monroe's data. Voters 1 and 30 both ranked the candidates (F, T, I), with the other three candidates unranked, and voters 3 and 25 ranked the candidates (I, T, G). There were no other cases in which voters voted alike. Thus, $v_j = 2$ for the two j-values corresponding to these two groups, $v_j = 1$ for the remaining 29 j-values, $n = 33$ as before, and $n^* = 31$.

To illustrate the solution for $m = 4$ and $k = 1$, {F, G, I, T} is the winning set of candidates with total misrepresentation of 13, just as before. The new solution also yields $x_{2j} = 2$ for the first group, $x_{5j} = x_{6j} = 1$ for the second group, and 0 for the remaining x_{ij}'s in those two groups. Hence, F is assigned to both voter 1 and voter 30. Whether T is assigned to voter 3 and I to voter 25, or vice versa, does not matter, because the effect on total misrepresentation is the same.

The type of voting system that is used affects n^*—which, in turn, affects computing requirements, because a smaller n^* will reduce computer time. For AV, n^* cannot exceed $2^M - 2$ (disregarding ballots that rate all the candidates the same). For voting based on rankings, the maximum n^* is much larger, except when M is very small, because it is equal to $M!$ even if each voter ranks all M candidates. If one allows voters to leave candidates unranked (but only at the bottom of the rankings), the maximum n^* is[21]

---

[20] To prove in general that replacing constraint sets (6.8) and (6.6) with (A6.1) and (A6.2) does not affect the solution for the winners, one has to show that, for any set of x_{ij}'s satisfying (A6.1) and (A6.2) for a given group j, it is possible to distribute the kv_j assignments so that k candidates are assigned to each of the v_j voters such that (for each i) candidate i is assigned to x_{ij} of these voters. To show this, one can simply cycle the assignments through the v_j voters, one winning candidate at a time. For example, if $v_j = 3$, $k = 2$, and $m = 4$, and if winning candidates A, B, C, and D have respective x_{ij}'s of 2, 0, 3, and 1, then one can assign A to voter 1, A to voter 2, C to voter 3, C to voter 1, C to voter 2, and D to voter 3. With some simple emendations, this technique works even if fractional assignments (discussed later in this appendix as well as the text) are allowed.

[21] The first term in (A6.3) is the number of ways of ranking all M candidates. The gth term ($g = 2, 3, \ldots, M - 1$) in (A6.3) is the number of ways of ranking the candidates if g candidates are unranked at the bottom. (There is, of course, no term for a single unranked candidate at the bottom.)

$$M! + M!/2! + M!/3! + \cdots + M!/(M-2)! + M!/(M-1)!. \qquad \text{(A6.3)}$$

The value (A6.3) is bounded above by $(e-1)M!$, or about $1.72M!$.

For voting based on rankings, the maximum n^* is $M!$, instead of about $1.72M!$, if a ballot with g unranked candidates at the bottom ($g = 2, 3, \ldots,$ $M-1$) is tallied as $(1/g!)$th of a ballot for each of the $g!$ ways of completing the ballot. Even though such a change would decrease the maximum n^*, the actual n^* could be either higher or lower, because the splits into $g!$ parts affect n^* both positively and negatively. It is possible for the change to lead to a difference in the solution for the winners,[22] except that for $k = m$ there is no difference because the winners are based on Borda-count scores, which will not change.

Fractional assignments, analyzed in section 6.6, can be used together with the aforementioned technique for grouping like ballots. Let constraint (A6.2*) denote the relation (A6.2) with the words "an integer" deleted. The applicable set of constraints with fractional assignments is then (6.1), (A6.1), (6.7) with $B = kn/m$ (which need not be an integer), (6.5), and (A6.2*).

Because fractional and integer assignments are more likely to produce different sets of winners in smaller electorates, the earlier discussion is pertinent mainly to such electorates. For larger electorates, fractional assignments also deserve serious consideration, but for a different reason: they could bring about reductions in computer time by sharply reducing the number of x-variables constrained to be integers. (If k is not an integer, a provision for fractional assignments would be mandatory rather than optional.)

B. TIES

The minimum value of total misrepresentation may be attainable at more than one set of values of the x-variables. Frequently, in fact, the minimum is attainable at differing sets of values of the x_{ij}'s. This type of multiple solution (or tie) causes no problem under Monroe's system, however, because the x_{ij}'s reflect assignments for computational purposes only. Indeed, voters will generally not even be aware of which candidates were assigned to them by the solution.

---

[22] The following example proves that splitting an incomplete ballot into $g!$ equal parts can lead to a solution whose set of winners differs from that obtained when unranked candidates simply receive the average of the unused ranks. Suppose that $n = 60$, $k = 1$, $m = 2$, and $M = 3$, with the candidates labeled A, B, and C. Let there be 18 voters who vote (C, B, A), 20 who vote (B, C, A), and 22 who place A first but do not rank the other two candidates. With the ballot-splitting method, $(\mu_{1j}, \mu_{2j}, \mu_{3j})$ will be (2, 1, 0) for 18 ballots, (2, 0, 1) for 20 ballots, (0, 1, 2) for 11 ballots, and (0, 2, 1) for 11 ballots. Total misrepresentation, when minimized, is 26 for {A, B}, 28 for {A, C}, and 23 for {B, C}, so {B, C} is the winning set. With the average-rank method, misrepresentation values are as before, except that they are (0, 1.5, 1.5) for each of the last 22 voters. Although total misrepresentation is again 26 for {A, B} and 28 for {A, C}, it is now 33 for {B, C}, so {A, B} wins.

A second type of tie, though less frequent (especially in large electorates), is more significant when it does occur. If minimum total misrepresentation is attainable at more than one set of values of the x_i's, then there will be a tie among winners. Although integer-programming software will find a solution that minimizes total misrepresentation, it will not then go on automatically to find other solutions that also attain that minimum, but with a different set of winners.

One can check for ties by running additional integer programs, however. In fact, a search for ties in all thirty elections shown in Tables 6.1 and 6.2 found there to be no other ties than those indicated in the tables. One checking method consists of running an integer program that is the same as the one run initially, except that there is an added restriction that rules out the set of winners that was already found.

To illustrate in the case of $m = 4$ and $k = 3$ in either Table 6.1 or Table 6.2, the initial solution showed {F, G, I, T} to be the winning set. To prevent a rerun from yielding this set again, add the restriction $x_1 + x_4 \geq 1$, which forces either D or H into the set of winners. Because the new solution has greater total misrepresentation than the original one, there is no tie. But if a tie had occurred, then one would have to run another integer program to check for another tie.

To break a tie, one can choose randomly, with equal probability, from among the tied sets. Other tie-breaking conventions can also be used.

C. FILLING VACANCIES

If a vacancy arises before the end of a legislator's term, there needs to be a method for filling that vacancy. That method should not affect the seats held by other legislators.

With any of the PR integer-programming methods, it is possible to fill a vacancy without making an earlier winner a loser. For example, consider the case of $m = 3$ and $k = 1$ in Table 6.1 or 6.2, in which F, I, and T are the winners.

Now suppose that T's seat is vacated, and that D, G, or H is to be chosen as a successor, based on the original ballots. After redoing the ballots with T removed, one simply runs the integer program, except that two constraints, $x_2 = 1$ and $x_5 = 1$, are added.[23] Without such added constraints, an earlier winner could be unseated.[24]

---

[23] If the μ_{ij}'s are based on rankings, there are two methods of revising them with one candidate removed, as explained in note 15 for an analogous situation. The example in note 24 is constructed so that the outcome is the same regardless of which method is used.

[24] That this is so is shown by an election under Monroe's system with $M = 4$, $m = 2$, and $k = 1$, where the rankings (D, A, B, C), (A, B, D, C), (C, B, D, A), and (C, D, B, A) each appear on 1/4 of the ballots. The winners are A and C. Suppose that C vacates. A recalculation with C removed yields B and D as the new pair of winners if no constraint is added.

Under the Chamberlin-Courant system, nonvacaters will also be able to retain their seats when a vacancy is filled, though all weights will be determined
anew. A change in weight can be either up or down.[25]

---

[25] Although an ad hoc method could be devised to prevent any weights from decreasing, this
restriction could give rise to new problems, thereby replacing one imperfection with others.

7

Selecting Winners in Multiple Elections

7.1. INTRODUCTION

In this chapter, I take a tack that is different from previous chapters, wherein I described and analyzed voting procedures that seem well suited for different kinds of elections. Here, by contrast, I begin by analyzing so-called aggregation paradoxes, which arise when voters are asked to make choices in multiple elections, and ask how these paradoxes might best be resolved.

The elections, for example, might involve voting on a bill and its amendments in a legislature, or voting on several propositions in a referendum. I show not only that different aggregation procedures may produce radically different outcomes but also that new procedures may work better than extant procedures at finding consensus choices. Some of these procedures, like AV, are already familiar, but others, like "yes-no voting," are not.

Aggregation paradoxes abound in the literature on statistics and social choice. Roughly speaking, they describe situations in which the sum of the parts is, in some sense, not equal to the whole, such as an election outcome that fails to mirror the "will" of the electorate. They include the Condorcet paradox (cyclical majorities) and Arrow's impossibility theorem (Arrow, 1951, 1963; Dietrich and List, 2007), Anscombe's paradox (Anscombe, 1976; Wagner, 1983, 1984), Ostrogorski's paradox (Rae and Daudt, 1976; Deb and Kelsey, 1987; Kelly, 1989; Laffond and Laine, 2006), and Simpson's paradox (Simpson, 1951; Gardner, 1976; Wagner, 1982). An interesting attempt to "understand, classify, and find new properties" of such paradoxes is given in Saari (1995); more generally, see Saari (1994, 2001c), Lagerspetz (1996), and Nurmi (1999).

I start by analyzing an aggregation paradox of direct relevance to elections: that of multiple elections, in which voters may not know the results of one election before they vote in another. Such voting is commonplace, as when a voter votes, at the same time, for president, senator, or representative in a presidential election year, or for or against several propositions in a referendum. The paradox is also applicable to multiple votes on a bill and its amendments in a legislature, whereby voting is sequential so the voter acquires some

Note: This chapter is adapted from Brams, Kilgour, and Zwicker (1998) with permission of Springer Science and Business Media; see also Brams and Fishburn (1993) and Brams, Kilgour, and Zwicker (1997).

information, as votes are taken, but does not know the results of future votes.

I call the set of winners in each of the individual elections the *winning combination*. Surprisingly, as I will show, few if any voters may actually have voted for this combination. Indeed, if there are as few as three propositions in a referendum—on which voters can vote either yes (Y) or no (N)—the outcome of the election may, for example, be YYY (i.e., all propositions pass), even though not a single voter voted for this winning combination.

I call this phenomenon *the paradox of multiple elections* and give an example, and several variants of this example, in section 7.2.[1] In section 7.3 I offer a theoretical analysis of the paradox, showing the conditions necessary and sufficient for a combination to win and for the paradox to occur. I also show how the occurrence of the paradox depends on the "incoherence" of support for the winning combination.

In section 7.4 I give a real-life example of the paradox, based on the choices made by 1.8 million Los Angeles county voters choosing among the twenty-eight propositions in the 1990 California general election. In addition, I discuss other empirical examples that are not full-blown paradoxes but, nonetheless, indicate a discrepancy between the most popular parts of a combination and the less popular whole.

Divided government, whereby the president is of one party and one or both houses of Congress are controlled by the other party, may be interpreted as a manifestation of this discrepancy.[2] In 1980, for example, the Republicans won control of the presidency and the Senate, while the Democrats retained control of the House, which I indicate by RRD for control of the presidency, Senate, and House, respectively. Based on the outcomes in congressional districts, however, this winning combination came in only fourth out of the eight possible combinations. This outcome was decidedly more paradoxical than the DDD outcome of the previous presidential election (1976), which was also the most popular combination.

---

[1] Some readers will not view this result as paradoxical because, once illustrated and explained, the apparent contradiction disappears. Nonetheless, the idea that a winning combination can receive zero votes seems surprising and counterintuitive to most people on first hearing, which conforms with the informal sense in which "paradox" is used in political science (Brams, 1976; Maoz, 1990).

[2] Sometimes "divided government" is used to mean that only the House of Representatives is controlled by a different party from that of the president, but the exact definition is not important. For a discussion of the merits and demerits of divided government, see "Symposium: Divided Government and the Politics of Constitutional Reform" (1991), Brady (1993), and McKay (1994). Of course, because control of the House and Senate depends on which party wins a majority of seats, which does not necessarily mean having a majority of votes across all House and Senate elections, there is some ambiguity about any purported discrepancy between the winning combination and the number who supported it.

In section 7.5 I consider two legislative examples from the House of Representatives in which there was an apparent Condorcet paradox (the Wilmot Proviso of 1846 and the Revenue Act of 1932). I explicate the sense in which this paradox is a special case of the more general multiple-election paradox.

In section 7.6 I consider certain normative and social-choice consequences of the paradox. For example, should voters be presented with the opportunity to choose combinations on ballots? If so, should they be allowed to vote for more than one combination under AV, or to rank the combinations under the Borda count? Answers to these questions, and their implications for making coherent social choices, are explored in the context of democratic political theory.

In section 7.7 I briefly explore "yes-no voting," whereby voters can indicate those alternatives they unconditionally support ("yes") and those they do not ("no"), leaving open those that are neither yes nor no. This obviates the need for voters to designate each and every subset they approve of while still allowing one to determine the most approved subset—not by aggregating votes for alternatives but instead by aggregating votes for subsets. In section 7.8 I offer some concluding remarks.

7.2. REFERENDUM VOTING: AN ILLUSTRATION OF THE PARADOX OF MULTIPLE ELECTIONS

Consider a referendum in which voters can vote either Y or N on each proposition on the ballot. The paradox of multiple elections occurs when the combination of propositions that wins receives the fewest votes, or is tied for the fewest. I first illustrate the *basic paradox*, without ties for fewest, after which I illustrate both stronger and weaker versions.

Example 7.1. Basic Paradox, without Ties for Fewest: Three Propositions

If there are three propositions, there are $2^3 = 8$ combinations, because each voter can make one of two choices (i.e., Y or N) on each proposition. (The possibility of abstention will be considered later.) Suppose 13 voters cast the following numbers of votes for each of the eight combinations:

YYY: 1   YYN: 1   YNY: 1   NYY: 1   YNN: 3   NYN: 3   NNY: 3   NNN: 0.

On each proposition i, the total number of voters voting for Y or N is indicated by $Y(i)$ and $N(i)$, where $i = 1, 2$, or 3. The election results for each proposition are

$$N(1) > Y(1), \quad N(2) > Y(2), \quad N(3) > Y(3),$$

each by 7 votes to 6.

Thus, when votes are aggregated separately for each proposition, which I call *proposition aggregation*, the winning combination is NNN. However, when votes are aggregated by combination, which I call *combination aggregation*, this combination (i.e., NNN) comes in last, because it is the only one that receives 0 votes.

This example is *minimal*, like those that follow, in the sense that no example with fewer voters can meet the stated conditions of the paradox. The construction depends on assigning the fewest votes (i.e., 0) to the paradoxical winner NNN, the next-fewest (i.e., 1) to combinations that agree on one proposition, and then finding the smallest number for the combinations that agree on two propositions (i.e., 3) so as to create the paradox—that is, so that NNN (barely) wins under proposition aggregation.

The paradox vividly illustrates the difference that may arise from aggregating votes by proposition and by combination. It also illustrates how proposition aggregation may leave no voter completely satisfied with the outcome: NNN is not supported by any of the 13 voters.

More generally, a *paradox of multiple elections* occurs when the winning combination under proposition aggregation receives the fewest, but not necessarily zero, votes (as in Example 7.1).[3] Of course, if the winning combination receives some support, there will be some voters completely satisfied with the outcome.

Note in Example 7.1 that YNN, NYN, and NNY are tied for first place under combination aggregation. However, this is not to say that most voters would prefer one of these combinations to the proposition-aggregation winner, NNN—only that NNN is not the first choice of any voters if they are sincere in their voting.

The paradox, which describes a conflict between two different aggregation procedures, does not depend on either sincere or strategic voting. Voters may be perfectly sincere in voting for their preferred position on every proposition, or they may be strategic (in some sense). The paradox says only that majority choices by proposition aggregation may receive the fewest votes when votes are aggregated by combination.

Example 7.1 is not the minimal example of the basic paradox if there may be ties for the fewest votes. Example 7.2 shows that the paradox can occur with only 3 voters:

Example 7.2. Basic Paradox, with Ties for Fewest: Three Propositions

Suppose 3 voters cast the following numbers of votes for the three combinations:

$$\text{YYN: 1} \quad \text{YNY: 1} \quad \text{NYY: 1.}$$

---

[3] Scarsini (1998) shows that not only the winning combination, but all combinations sufficiently close to it, may receive the fewest votes, which he calls a "strong paradox of multiple elections."

YYY is the winning combination, according to proposition aggregation, by 2 votes to 1 for each proposition; yet it receives 0 votes as a combination. But so do the other four combinations (YNN, NYN, NNY, and NNN). In this example, therefore, the winning combination ties for the fewest votes, unlike Example 7.1, in which it is the *only* combination with the fewest (0) votes.

A more pathological form of the paradox can occur if there are four propositions, which gives $2^4 = 16$ possible voting combinations.

Example 7.3. Complete-Reversal Paradox, without Ties for Fewest: Four Propositions

Suppose 31 voters cast the following numbers of votes for the sixteen different combinations:

YYYY: 0   YYYN: 4   YYNY: 4   YNYY: 4   NYYY: 4
YYNN: 1   YNYN: 1   YNNY: 1   NYYN: 1   NYNY: 1   NNYY: 1
YNNN: 1   NYNN: 1   NNYN: 1   NNNY: 1   NNNN: 5.

Then it is easy to show that Y beats N for each of the four propositions by 16 votes to 15. However, YYYY is the only combination to receive 0 votes. The new wrinkle here is that NNNN, the "opposite" of the proposition-aggregation winner, YYYY, has the most votes (i.e., 5) and therefore wins under combination aggregation.

The complete-reversal paradox can occur with fewer votes if there may be a tie for combinations with the fewest votes.

Example 7.4. Complete-Reversal Paradox, with Ties for Fewest: Four Propositions

Suppose 11 voters cast the following numbers of votes for five combinations:

YYYN: 2   YYNY: 2   YNYY: 2   NYYY: 2   NNNN: 3.

There is a complete-reversal paradox, because YYYY ties ten other combinations for the fewest votes (0) but, nevertheless, wins according to proposition aggregation by 6 votes to 5 against its opposite, NNNN, the combination winner with 3 votes.

It is not difficult to show that the basic paradox cannot occur if there are only two propositions, but a milder version of this paradox can arise— namely, that the winner by proposition aggregation can come in as low as third (out of four) by combination aggregation. I call this the *two-proposition paradox*.

Example 7.5. Two-Proposition Paradox, without Ties for Fewest

Suppose 5 voters cast the following numbers of votes for the four combinations:

YY: 1   YN: 2   NY: 2   NN: 0.

While YY (1 vote) wins under proposition voting by 3 votes to 2 on each of the two propositions, it is behind both YN and NY (2 votes each) under combination aggregation; it is ahead only of NN (0 votes), putting it in next-to-last place.

This relatively low rank for the proposition-aggregation winner in the two-proposition case may still describe a possibly serious discrepancy between the two aggregation procedures. Indeed, there is no theoretical limit on *how far* behind the proposition-aggregation winner can be from the first-place and second-place combination winners (though it is easy to demonstrate that it will always place above the fourth-place combination).

If there are only two propositions, but there is a third option of abstaining (A), then the basic paradox can occur even when there are only two propositions.

Example 7.6. Basic Paradox, with Abstention and with Ties for Fewest: Two Propositions

Suppose 15 voters cast the following numbers of votes for the $3^2 = 9$ different combinations:

YY: 0   YN: 3   NY: 3   YA: 3   AY: 3   NA: 1   AN: 1   NN: 1   AA: 0.

There is a paradox, because YY wins under proposition voting by 6 votes to 5 but, as a combination, it ties for the fewest votes (i.e., 0) with AA. The notion that AA receives 0 votes—or any other number—is somewhat misleading, however, because it is often impossible to ascertain the numbers who "chose" AA if these voters did not go to the polls (later I will count, as abstainers, voters who cast ballots but abstain on the propositions being considered).

It is worth noting in this example that YY wins in large part because 6 voters (the YA and AY voters) abstained on one office and supported Y for the other. An A, however, is not necessarily to be interpreted as a vote against YY. A's are qualitatively different from other votes, in part because they are never a component of a winning combination.

Example 7.7. Complete-Reversal Paradox, with Abstention and without Ties for Fewest: Three Propositions

Suppose 52 voters cast the following numbers of votes for the 27 (3^3) combinations:

YYY: 0	YYN: 4	YNY: 4	NYY: 4	YYA: 4	YAY: 4	AYY: 4
YNN: 1	NYN: 1	NNY: 1	YAA: 1	AYA: 1	AAY: 1	
NAA: 1	NAN: 1	NNA: 1	NYA: 1	ANY: 1	YAN: 1	AAA: 5
ANN: 1	ANA: 1	AAN: 1	AYN: 1	NAY: 1	YNA: 1	NNN: 5.

There is a complete-reversal paradox, because YYY wins under proposition voting by 20 votes to 16 on each proposition but, as a combination, it has the fewest (i.e., 0) votes. On the other hand, the other two "pure" combinations, AAA and NNN (the latter might be considered the opposite of YYY),[4] have the most votes (i.e., 5).

The foregoing examples illustrate a range of discrepancies between aggregating votes by proposition and aggregating them by combination. Next I analyze the general conditions that give rise to these discrepancies, focusing on the basic paradox in the two-option, three-proposition case and the "coherence" of voter support.

7.3. THE COHERENCE OF SUPPORT FOR WINNING COMBINATIONS

Having demonstrated the existence of a multiple-election paradox and some variants of it, I turn in this section to the analysis of conditions that give rise to it. In particular, I distinguish between voting directly for a combination and voting indirectly for it by supporting some of its parts.

This distinction is illustrated by Example 7.2. The three voters who vote for YYN, YNY, and NYY give Y a 2-to-1 margin of victory for each proposition, resulting in the choice of YYY by proposition aggregation. But this indirect support by the three voters for YYY is indistinguishable under proposition aggregation from the direct support that one hypothetical voter, voting for YYY, could give to this combination.

In effect, this one voter could contribute three times as much support to YYY as do any of the three voters who "tilt" toward YYY by agreeing with it on two of the three propositions. Not only is the support of this one voter more potent, but it is also more "coherent" because there is no question that if YYY prevails, the YYY voter supported it.

---

[4] "Opposite" is ambiguous, of course, when there are more than two options. In the case of the three options postulated in Example 7.7, for instance, the opposite of YYY might be not only NNN but also the seven other combinations that do not include any Y's. It turns out that not even a basic paradox can be constructed when each of the eight non-Y combinations must have more votes than each of the nineteen Y combinations (with, say, zero votes each)—and one of the latter combinations must also be the proposition-aggregation winner. Necessary and sufficient conditions for the paradox in the three-proposition case, but without abstention, are given in section 7.3, but these conditions can be generalized to nonbinary elections in which, for example, A is a third option.

To make these ideas more precise, define a quantitative measure Q of the support for some combination, UVW, comprising three propositions. Q possesses two properties:

1. It is the sum of the coherent (C) and incoherent (I) contributions of voters:

$$Q(UVW) = C(UVW) + I(UVW). \tag{7.1}$$

   where the C and I components will be defined shortly.

2. The winning combination according to proposition aggregation is that which maximizes Q (to be proved in Proposition 7.1).

To construct the C and I terms, let $n(UVW)$ denote the number of votes cast for combination UVW. Define four differences between "opposites," using the original binary distinction between Y votes and N votes on three propositions:

$$n_0 = n(YYY) - n(NNN)$$

$$n_1 = n(YYN) - n(NNY)$$

$$n_2 = n(YNY) - n(NYN)$$

$$n_3 = n(NYY) - n(YNN).$$

These differences are set up to favor YYY, with the positive term in each difference agreeing with YYY in more than half the propositions and the negative term agreeing with YYY in fewer than half the propositions.

Given these differences, we define

$$C(YYY) = 3n_0 \quad \text{and} \quad I(YYY) = n_1 + n_2 + n_3,$$

based on the intuition, in the example just discussed, that a direct vote has three times the effect of indirect votes that tilt in favor of YYY.[5] Substituting into (7.1),

$$Q(YYY) = 3n_0 + n_1 + n_2 + n_3.$$

Q values for combinations other than YYY are similarly defined, but they require the insertion of some minus signs to compensate for the arbitrary choices of signs in the definitions of $n_0, \ldots, n_3$. For example,

$$Q(NNY) = -3n_1 - n_0 + n_2 + n_3,$$

because negative values of n_1 and n_0 indicate agreement with NNY in more than half the propositions.

Proposition 7.1. *The winning combination according to proposition aggregation maximizes Q.*

---

[5] As a measure of the tilt toward YYY, I(YYY) is analogous to the "spin," or the cyclic component, of the total vote (Zwicker, 1991).

Proof. Define the difference (d) for proposition 1 as

$$d_1(Y > N) = \# \text{ of voters voting Y} - \# \text{ of voters voting N}$$

on this proposition. Note that $d_1(Y > N) = -d_1(N > Y)$. Similarly, define d_2 and d_3 to be the differences for propositions 2 and 3.

Given any combination, such as NNY, define the following sum (S):

$$S(NNY) = d_1(N > Y) + d_2(N > Y) + d_3(Y > N).$$

Note that the N or Y for each proposition in NNY matches the N or Y that is assumed greater in each d_i term on the right-hand side of the equation. It is apparent that NNY will win the election if and only if each of the d_i's is positive (I ignore here the possibility of ties and how they might be broken to determine a winner).

Assume NNY is the winning combination according to proposition aggregation. The S for any nonwinning combination sums the same three numbers as for YNN, but with one or more sign changes. Necessarily, at least one of the numbers for the nonwinning combinations is negative. Not only is the winning combination the only one for which each of the three d_i's is positive, but this combination is also the one that maximizes S.

To complete the proof, it remains only to show that $S(UVW) = Q(UVW)$ for any combination UVW. I do this for YYY and leave the other combinations for the reader to check:

$$d_1(Y > N) = n_0 + n_1 + n_2 - n_3$$
$$d_2(Y > N) = n_0 + n_1 - n_2 + n_3$$
$$d_3(Y > N) = n_0 - n_1 + n_2 + n_3.$$

Summing the three d_i's given above gives

$$S(YYY) = 3n_0 + n_1 + n_2 + n_3 = Q(YYY),$$

as desired. ∎

Proposition 7.1 shows that the winning combination according to proposition aggregation is the one with the largest Q value. The fact that this value has both a C and an I component enables one to judge the extent to which a victorious combination owes its triumph to *coherent*, or direct, support rather than to incoherent, or indirect ("tilt"), support.

The paradox of multiple elections, as illustrated in all the examples in section 7.2 except Example 7.5 (the two-proposition paradox), describes the extreme case wherein *all* the support for each winning combination is incoherent— no voter votes for this combination. To be sure, the paradox may occur when the winning combination receives some, but fewer, votes than any other combination.

It is worth noting that the Q value bears some resemblance to the Borda count. Imagine that a person voting for the combination UVW actually awards some points to each of the eight combinations, with the rule being that +1 point is awarded for each proposition on which UVW agrees with the combination in question, and −1 point for each proposition on which UVW disagrees. For example, a vote for YYY awards +1 + 1 + 1 = 3 points to YYY itself, whereas it awards −1 + 1 + 1 = 1 to NYY.

There are four possible levels of agreement and disagreement. Specifically, a vote for YYY awards

+3 points to YYY
+1 point to YYN, YNY, NYY
−1 point to YNN, NYN, NNY
−3 points to NNN.

It is easy to show that if adding the total number of points awarded by all voters to a particular combination (e.g., YYY), the resulting sum equals $Q(YYY)$.[6] Hence, the winning combination according to proposition aggregation—that is, the combination that maximizes Q—is the one with the highest Borda score (as interpreted here).

The voting system defined by the preceding system of awarding points is fully equivalent to the system of proposition aggregation currently in use. This correspondence shows how the present system of voting on propositions presumes an underlying cardinal evaluation of combinations of propositions. Thus, a ballot cast for YYY gives, effectively, a ranking of the eight combinations—but only an *incomplete* ranking, truncated to four levels of agreement and disagreement separated by equal intervals of 2 points, as I just showed. Not only does a YYY ballot contribute more to, say, YYN than YNN, but it does so by the same amount that other ballots that agree in two versus one proposition do to other combinations.

Of course, if the standard version of the Borda count were applied directly to the combinations, a voter could give a complete ranking of all eight combinations. I consider this possibility later.

Next consider under what conditions $Q(YYY)$ is maximal and, therefore, the combination YYY wins. Then I will discuss the more stringent conditions that render this winning combination paradoxical.

Proposition 7.2. *A necessary and sufficient condition for YYY to win is that* $Q(YYY) > 2n_0 + 2[max\{n_1, n_2, n_3\}]$. *(Other combinations are governed by similar inequalities.)*

---

[6] The Borda score for YYY can be seen as identical to $Q(YYY)$ and $S(YYY)$, with points grouped differently. Generalizations of this scoring system to elections with any number of offices, and with abstention allowed, are given in Brams, Kilgour, and Zwicker (1997).

Proof. What we need to show is that YYY maximizes the Q value, and is therefore the winning combination, if and only if the above inequality is satisfied. To do this, note that YYY is winning precisely when each term of

$$d_1(Y > N) + d_2(Y > N) + d_3(Y > N)$$

is positive, or

$$n_0 + n_1 + n_2 > n_3$$
$$n_0 + n_1 + n_3 > n_2 \qquad (7.2)$$
$$n_0 + n_2 + n_3 > n_1.$$

I begin by showing that the three inequalities given by (7.2) are all necessary. Assume that $n_3 \geq \max\{n_1, n_2\}$, which I refer to as case 1. Then the second and third inequalities are true if the first is true. Interchanging n_3 and n_2, and thereby obtaining the second inequality from the first, preserves the truth of the inequality, as does interchanging n_3 and n_1. But the first inequality must be true for all three to be satisfied. Therefore, all three inequalities are true if and only if the first is true.

Now the first inequality is equivalent to

$$n_0 + n_1 + n_2 + n_3 > 2n_3,$$

which is equivalent to

$$3n_0 + n_1 + n_2 + n_3 > 2n_0 + 2n_3.$$

This is the same as

$$Q(YYY) > 2n_0 + 2n_3,$$

which implies

$$Q(YYY) > 2n_0 + 2[\max\{n_1, n_2, n_3\}]. \qquad (7.3)$$

Inequality (7.3) is the condition of Proposition 7.2. Because it is symmetric in n_1, n_2, and n_3, it is similarly implied by either of the other two cases, corresponding to the possibility that n_2 or n_1 is largest, or tied for largest, among n_1, n_2, and n_3.

To show sufficiency, note that if inequality (7.3) holds, regardless of which case prevails, the steps of the earlier proof are reversible, as they are with the interchange of n_2 and n_3, or n_1 and n_3. Thereby, the three inequalities of (7.2) all hold, ensuring that YYY is the winning combination. ■

To obtain further insight into the conditions of the paradox, observe that Proposition 7.2 says that YYY is the winning combination if and only if

$$n_0 + n_1 + n_2 + n_3 > 2[\max\{n_1, n_2, n_3\}]. \qquad (7.4)$$

Inequality (7.4) says that the total margin by which voters favor combinations with more Y's than N's over their opposites must be greater than twice the maximum of the margins corresponding to indirect support.

If YYY receives the fewest votes—or ties for the fewest—then $n_0 \leq 0$ in inequality (7.4), which makes this inequality more difficult to satisfy than were $n_0 > 0$. Momentarily drop the n_0 term from (7.4) (e.g., assume $n_0 = 0$ because YYY and NNN receive equal numbers of votes). Then inequality (7.4) becomes

$$n_1 + n_2 + n_3 > 2[\max\{n_1, n_2, n_3\}]. \tag{7.5}$$

Roughly speaking, inequality (7.5) is satisfied when n_1, n_2, and n_3 are all positive (i.e., all the tilt terms favor YYY over NNN) and, in addition, n_1, n_2, and n_3 are close to one another in size (i.e., the tilt is evenly spread).

Returning to inequality (7.4), how does the presence of n_0 affect this observation? The more direct support that YYY receives over its opposite, NNN (i.e., the larger n_0 is), the less uniform the tilt must be in order for YYY to prevail.

On the other hand, if YYY receives no votes—making a win paradoxical—the tilt must be spread fairly evenly for YYY to win. At the same time, the larger the vote for NNN (i.e., the more negative n_0 is), the more evenly spread as well as larger the tilt must be to produce the paradox.

These considerations suggest two conditions sufficient to guarantee that the paradox does *not* occur for YYY. First, if n_0 is negative with absolute value either equal to or greater than the largest of n_1, n_2, or n_3, then inequality (7.4) cannot be satisfied and, hence, there can be no paradox. Second, the paradox is also precluded if even one of the tilt terms, n_1, n_2, or n_3, is less than or equal to 0; this happens when one or more of the combinations—YYN, YNY, or NYY (i.e., those that tilt toward YYY)—receive no (net) votes.

In summary, there must be more-or-less-equal positive differences between the mixed combinations that favor Y (YYN, YNY, and NYY) and their opposites (NNY, NYN, and YNN) for the paradox to occur. These positive differences overwhelm the greater direct support that NNN enjoys over YYY, enabling YYY to win even though it receives fewer votes than any other combination.

I turn next to three empirical cases. The first involves voting on multiple propositions, the second has a divided-government interpretation, and the third raises questions about the coherence of legislative choices. A genuine multiple-election paradox occurred in the first case. Although there was no full-fledged paradox in either of the latter two cases, which involved voting for different offices in an election and different bills in Congress, they illustrate situations in which there was a discrepancy between the two kinds of aggregation I have discussed in a referendum.

7.4. EMPIRICAL CASES

Case 1. Voting on Propositions

On November 7, 1990, California voters were confronted with a dizzying array of choices on the general election ballot: 21 state, county, and municipal races, several local initiatives and referendums, and 28 statewide propositions. I analyze here only voting results on the 28 propositions, which concerned such issues as alcohol and drugs, child care, education, the environment, health care, law enforcement, transportation, and limitations on terms of office.

The data are images from actual ballots cast by approximately 1.8 million voters in Los Angeles county (Dubin and Gerber, 1992). Voters could vote yes (Y), no (N), or abstain (A)—abstention being the residual category of voting neither Y nor N—with a proposition passing if the number of its Y's exceeded the number of its N's, and failing otherwise. In Los Angeles County, 11 of the 28 propositions passed, but several of these were defeated statewide, and some of the defeated propositions in Los Angeles County passed statewide.

For the purposes of this analysis, I consider only the results for Los Angeles County and ask how many voters voted for the winning combination,

$$\text{NNNYNNYNYNNNNNNYNYYYNYYNNYNY,}$$

on propositions 124–151. The answer is that nobody did, so there was a multiple-election paradox.

Because there are $2^{28} \approx 268.4$ million possible Y-N combinations, however, this is no great surprise. With fewer than 2 million voters, more than 99 percent of the combinations *must* have received 0 votes, even if each of the voters voted for a different combination.

In fact, however, the latter was not the case. "All abstain" received the most votes (1.75 percent), and "all no" was a close second (1.72 percent). Ranking fifth (0.29 percent) among the combinations were the recommended votes of the *Los Angeles Times*, and ranking ninth was "all yes" (0.20 percent).[7] Thus, the effect of the *Times* recommendations, at least for the complete list of propositions, was marginal. Nonetheless, it was greater than what Mueller (1969) found in the 1964 California general election, in which absolutely nobody in his sample of approximately 1,300 voters backed either all, or all except one, of the *Times* recommendations on 19 propositions.

Although a paradox occurred in voting on all 28 propositions, it was not a complete-reversal paradox, because the opposite of the winning combination

[7] The total for all abstain, all no, and all yes is only 3.7 percent. Thus, the vast majority of voters (96.7 percent) were not "pure" types but discriminated among propositions by choosing mixed combinations that included both Y's and N's.

did not garner the most votes (it, too, received zero votes). Moreover, because voters evidently considered abstain (A)—in addition to Y and N—as a voting option, it seems proper to use the $3^{28} \approx 22.9$ trillion combinations, which include A as well as Y and N as choices, in asking whether anybody voted for the winning combination. While A was "selected" by between 7.1 percent and 16.3 percent of voters on each of the propositions, its choice over Y or N could never elect A but could influence whether Y or N won.

What a voter's choice of A on any proposition did preclude was that voter's voting for the winning combination, thereby decreasing the already small likelihood that the winning combination received any votes. One could, of course, count A as a vote for both Y and N, thereby increasing the number of combinations that a voter supports; alternatively, one could give each voter one vote, splitting it among all combinations with which he or she agrees on each proposition by choosing Y, N, or A. These different ways of counting abstentions can produce different winners (Brams, Kilgour, and Zwicker, 1997).

Because the 28 propositions on the California ballot dealt with a bewildering variety of different issues, it seemed reasonable to ask whether combination voting might make more sense if it were applied to a much smaller subset of related issues on which voters could, presumably, make coherent choices (Brams, Kilgour, and Zwicker, 1997). In fact, three propositions involved the issuance of bonds in varying amounts to support the environment in different areas:

P130. Forest Acquisition ($742 million)
P148. Water Resources ($340 million)
P149. Park, Recreation, and Wildlife Enhancement ($437 million)

Pro-environment voters would vote for all three, whereas anti-spending voters would vote for none. But many voters will want to support just one or two of the propositions, whereas they would have a much harder time in deciding on combinations of propositions that involved wildly disparate issues.

Although there was not a multiple-election paradox in voting on these three propositions, the winning combination according to proposition aggregation was YNY, but it was supported by fewer than 6 percent of the voters, placing it fifth out of the eight possible combinations. While not a full-fledged paradox, the poor showing of YNY illustrates how an unpopular compromise may defeat more popular "pure" combinations (YYY was supported by 26 percent of the voters, NNN by 25 percent). The winning combination in this case was not only incoherent in the mathematical sense used earlier but also in a more substantive sense: It was pro-environment on two bond issues, anti-spending on the third, rendering policy choices by the voters somewhat of a hodgepodge that had little direct support. In fact, two other mixed combinations, YNN and NYY, received more support (about 8 percent each).[8]

Case 2. Voting in Federal Elections

The multiple-election paradox may occur not only in voting on multiple propositions but also in voting for multiple offices in an election. For example, in a presidential election year, a voter might vote for the Republican candidate for president, the Democratic candidate for Senate, and the Republican candidate for House—that is, the combination RDR. Just as votes are aggregated by proposition in a referendum, votes can be aggregated by office in an election (which I call *office aggregation*), so in theory RDR could receive, as a combination, the fewest votes.

Insofar as the federal government is conceived as a single entity, normative arguments can be made that the most popular combination *should* win. But most voters, it seems, do not think in terms of electing a combination, at least not at a conscious level (Fiorina, 1992, pp. 65). Nevertheless, in voting for their favorite candidates for each office, they may worry about the consequences of divided (or unified) government—and possibly act on this concern in choosing a combination.

In so doing, voters seem to apply different criteria in selecting presidents and legislators. For example, voters until 1994 tended to favor Democrats for the benefits, protections, and services they provided at the district and state level, but Republicans for the discipline and responsibility, especially on economic matters, that they exercised at the national level (Zuppan, 1991; Jacobson, 1991, pp. 71–75). Thereby they hedged their bets, especially if they were "sophisticated," and opted for "balance" in the government (Alesina and Rosenthal, 1995). In the view of some (e.g., Conlan, 1991; Mayhew, 1991; Fiorina, 1992), divided government, which has been the norm since 1968 (it occurred in twenty-eight of the forty years, or 70 percent, in the period 1968–2008), did not impede the passage of major legislation.

---

[8] Only if all voters have separable preferences will the winning combination assuredly be a Condorcet winner that can defeat all other combinations in pairwise comparisons (Lacy and Niou, 2000). (A voter's preferences are *separable* if the way he or she votes on one proposition is not affected by outcomes on other propositions, so the propositions may be thought of as independent of each other.) This will not be the case for the voter V if he or she is concerned about the state's overspending on bonds, like those that support the environment. Whether V supports one bond is likely to depend on whether he or she thinks other bonds will pass and, by doing so, "break the budget." (Thereby, V will not have separable preferences on these propositions.) Of course, it is impossible for V to know what will pass in a referendum, which is why Lacy and Niou argue that sequential voting on such issues—whereby the outcomes of prior votes are known—is desirable. But this may not always be feasible, which is why I suggest later that voting for combinations, in some manner, may be desirable. For a variety of examples, both hypothetical and real-life, in which voters do not have separable preferences, see Brams, Kilgour, and Zwicker (1997) and Lacy and Niou (2000).

To most voters, there is nothing incoherent in voting for a combination like RDR. Moreover, if this combination wins according to office aggregation, there is nothing paradoxical about the fact that the voters, collectively, chose divided government.

It is instructive to contrast two cases. In 1976, a Democratic president, Senate, and House were elected, giving DDD. In the absence of reliable combination voting data for the three offices (either from actual ballots cast by individual voters for the three offices or from voter surveys), I treat the 435 congressional districts *as if* they were voters, which raises difficulties I will discuss shortly. I then classified the subset of districts with senatorial races (about two-thirds) according to the eight combinations, depending on which party (D or R) won each of the three offices in a district.

I caution that, unlike the hypothetical examples for propositions given in section 7.2, I consider the winners in the Senate and House to be the party that wins a majority of seats in each house, not the party with the greatest number of Senate or House votes nationwide. Also, because many voters base their choices less on party than on the individual candidates running, interpreting the combination that wins as indicating a preference for either divided or unified government is somewhat questionable.

Bearing these caveats in mind, the results for 1976 are that the DDD combination was the most popular, being the choice of 40.8 percent of the 316 districts with senatorial races, whereas the next-most-popular combination, RRR, won in only 20.6 percent of the districts (see Table 7.1). In short, unified government garnered more than three-fifths of the vote that year, at least

TABLE 7.1

Combination Returns for 1976 and 1980 Presidential Elections, by Congressional Districts, in States with Senatorial Contests

Combination	1976 (Percentages)	1980 (Percentages)
1. DDD	40.8[a]	22.2
2. DDR	5.7	2.5
3. DRD	2.6	2.2
4. DRR	1.9	0.6
5. RDD	11.7	16.8
6. RDR	8.2	12.7
7. RRD	8.5	14.3[a]
8. RRR	20.6	28.6
Total	100.0 ($n = 316$)	100.0 ($n = 315$)

[a] Winner by office aggregation.

as indicated by the congressional district results with senatorial races. The most popular of these combinations, DDD, concurred with the office-aggregation winner.

By contrast, RRD was the winning combination in 1980, but it was only the fourth most popular voting combination (again, by congressional districts with senatorial races, of which there were 315), as shown in Table 7.1. As in 1976, the straight-ticket voting combinations won in the most districts (28.6 percent for RRR and 22.2 percent for DDD). Although RRD was not the least popular combination, its fourth-place finish with 14.3 percent seems at least semiparadoxical.

To be sure, the contrast between 1976 and 1980, with unified government coinciding with the winning combination in 1976 and divided government coinciding with the fourth-place combination in 1980, could be happenstance. Unfortunately, combination-voting data for the three federal offices seem to have been collected only for 1976 and 1980 (Gottron, 1983), so I cannot test for the paradox in other presidential election years.[9]

I have, however, analyzed combination-voting data for the two-office elections of president/senator and president/representative for the five presidential elections between 1976 and 1992. In such elections, it will be recalled from Example 7.5 in section 7.2, the winning combination by office aggregation can rank as low as third out of four. Treating the fifty states (actually, only the thirty-three or thirty-four states that had senatorial contests in each year) as if they were voters in the president/senator comparisons, and the 435 House districts as if they were voters in the president/representative comparisons, in only two of the ten comparisons—the president/representative comparisons in 1980 and 1988—did the winning combination by office aggregation (i.e., RD) come in even as low as second (RR in each of these years won according to combination aggregation).[10]

[9] Inexplicably, voting returns reported in Congressional Quarterly's *Congressional Districts in the 1990s: A Portrait of America* (CQ Press, 1993) for the 1988 presidential election are for the 1992 congressional districts, based on the 1990 census, so they do not accurately reflect the results of the 1988 election. While Congressional Quarterly's annual *Politics in America* and National Journal's annual *Almanac of American Politics* give presidential-election returns by congressional district and state, senatorial returns for each congressional district are not available (except in 1976 and 1980). Although senatorial returns are broken down by county in Congressional Quarterly's *America Votes* series, congressional districts often divide counties, requiring that one use precinct-level data to determine senatorial results by district. While such data for approximately 190,000 precincts have been collected for the period 1984–1990 by a now-defunct group called "Fairness for the 90s," an officially nonpartisan and nonprofit organization, most of the data are not currently in a form amenable to computer analysis (King, 1996).

[10] In the president/senator comparison in 1988, there was a tie for first, according to combination aggregation, between RD and RR (RD was the office-aggregation winner that year). Although the Republican presidential candidate (George Bush) prevailed in both combinations, this fact does not ensure that such a candidate, who may win in a majority of states, would win a majority of either popular or electoral votes should these states be mostly small.

The absence of even a mildly paradoxical third-place finish of the combination winner (the two-office paradox) may well be attributable to aggregating voters by district and state and treating these large units as if they were individual voters. It seems likely that this aggregation wipes out numerous mixed combinations of individual ballots, one of which might win with relatively few votes according to office aggregation.[11] Furthermore, the fact that the second-place finishes occur only in the president/representative comparisons and not in the president/senator comparisons is *prima facie* evidence that more aggregative units (i.e., states rather than congressional districts) have this wipe-out effect, decreasing the probability of observing a paradox.

Case 3. Voting on Bills in Congress

There was a two-office paradox in the case of what are generally acknowledged to be the two most important votes to come before the House of Representatives during the first two years of Bill Clinton's administration—that on the budget on August 5, 1993, and that on the North American Free Trade Agreement (NAFTA) on November 17, 1993. On these two bills, NY got 36 percent, YN got 32 percent, YY got 18 percent, and NN got 14 percent; the winning combination was YY. The explanation for why only 18 percent of the House—all Democrats—supported President Clinton on *both* bills lies in the fact that the party split was very different on the two bills: 84 percent of Democrats and no Republicans voted Y on the budget bill, whereas 40 percent of Democrats and 75 percent of Republicans voted Y on the NAFTA bill.

It is appropriate to ask whether there is anything problematic about the winning combination of YY receiving so few votes, given that it can be viewed as a compromise between the two more popular alternatives, NY and YN. Indeed, 86 percent of House members saw their preferred position enacted on *at least one* of the two bills. By this measure of satisfaction, YY is better than YN (64 percent satisfied), NY (68 percent satisfied), and NN (82 percent satisfied).

I will return to a consideration of the normative implications of the paradox in the concluding section. But first I take up the connection between the multiple-election paradox and the most venerable of all aggregation paradoxes—the Condorcet paradox.

7.5. RELATIONSHIP TO THE CONDORCET PARADOX

To illustrate the linkage between these two paradoxes, I present two examples from voting in Congress. Because the Condorcet paradox assumes that voters

---

[11] In particular, a district that roughly reflects the nation might appear to vote for the national winning combination, even though that vote in fact represents a paradoxical combination that individual ballot data would have revealed.

have certain preferences over a set of alternatives, the previous analysis, based solely on a numerical comparison of winning combinations under two different aggregation methods, must be extended. The preferences I assume enable one to create an isomorphism (i.e., a one-to-one correspondence, which I will illustrate shortly) that renders the multiple-election paradox a natural generalization of the Condorcet paradox.

The first congressional example concerns the Wilmot Proviso, which prohibited slavery in land acquired from Mexico in the Mexican war. On August 8, 1846, there were several votes in the House of Representatives for attaching this proviso to a $2 million appropriation to facilitate President James K. Polk's negotiation of a territorial settlement with Mexico. The three possible outcomes were (i) appropriation without the proviso, (ii) appropriation with the proviso, (iii) no action. Riker (1982, pp. 223–227) reconstructs the preferences of eight different groups of House members for these outcomes, where *xyz* indicates a group prefers x to y, y to z, and x to z (the groups are assumed to have transitive preferences).[12] The preferences of these groups, which comprise a total of 172 House members, are shown in Table 7.2.

To simplify the subsequent analysis, assume that the 8 border Democrats split 4–4 for each of their two possible preference scales shown in Table 7.2, and the 3 border Whigs split $1\frac{1}{2} - 1\frac{1}{2}$ for their two possible preference scales (not actually possible, of course, but the subsequent results do not depend on how one splits the votes of either group). Then it is easy to show that majorities are cyclical: b beats a (as happened) by 93 to 79 votes, a beats c by $129\frac{1}{2}$ to $42\frac{1}{2}$ votes, and c beats b by 107 to 65 votes. Consequently, there is no Condorecet winner that defeats each of the other outcomes in pairwise comparisons, which makes the social choice an artifact of the order of voting.

To establish an isomorphism between the paradox of voting and the multiple-election paradox, assume that the eight votes actually taken on the proviso in the House on August 8, 1846, can be reduced to three hypothetical pairwise contests between (i) a and b, (ii) b and c, and (iii) c and a. Assume further that, given its preferences, each group can answer yes (Y) or no (N) about whether it prefers the first member of each pair to the second. (I assume, as before, that the border Democrats and border Whigs split 50–50 on their preferences for second and third choices.)

Answers to these three questions give an *answer sequence*. For example, an answer sequence of YYN indicates that the group prefers a to b, b to c, but not c to a, so its preference scale is *abc*. (In the remainder of this section, I assume for simplicity that preferences are strict.) Likewise, five other mixed answer sequences of Y's and N's can be associated with the preference scales shown below each:

---

[12] I follow Riker's interpretation of events here but note that it has been criticized by Mackie (2003, ch. 11), who offers a different interpretation. Whether there was cycling (Riker) or not (Mackie), however, the main purpose of this section is to relate the multiple-election paradox to the Condorcet paradox.

TABLE 7.2
Preferences of Groups of Members of the House of Representatives in Voting on the
Wilmot Proviso (1846)

Group	No. of Members	Preferences
Northern Administration Democrats	7	*abc*
Northern Free Soil Democrats	51	*bac*
Border Democrats	8	*abc* or *acb*
Southern Democrats	46	*acb*
Northern prowar Whigs	2	*cab*
Northern antiwar Whigs	39	*cba*
Border Whigs	3	*bac* or *bca*
Southern and border Whigs	16	*acb*
Total	172	

Source: Riker 1982, p. 227.

Preference:	*abc*	*cab*	*bca*	*acb*	*bac*	*cba*	?	?
Sequence:	YYN	YNY	NYY	YNN	NYN	NNY	YYY	NNN

The question marks indicate intransitive or cyclic preferences. Thus, for a group to answer Y to all three questions indicates a preference cycle *abca*; to answer N to these questions reverses the direction of the cycle, giving *cbac*. Although I assume that groups of like-minded House members have, like individuals, transitive preferences, I will return to this matter later.

In voting on the Wilmot Proviso, observe that the winner by combination aggregation is *acb* (YNN) with 66 votes, comprising 4 border Democrats, 46 Southern Democrats, and 16 Southern and border Whigs:

Sequence:	YYN	YNY	NYY	YNN	NYN	NNY	YYY	NNN
Votes:	11	2	$1\frac{1}{2}$	66	$52\frac{1}{2}$	39	0	0

By contrast, the winner by what I will call *bill aggregation*, which is analogous to proposition aggregation and office aggregation, in pairwise contests (1), (2), and (3) is NNN. Specifically, N beats Y in contest (i) by 93 to 79 votes, in contest (ii) by 107 to 65 votes, and in contest (iii) by $129\frac{1}{2}$ to $42\frac{1}{2}$ votes. Thus, there is a basic multiple-election paradox: the winner by bill aggregation (NNN) ties for the fewest votes with the other intransitive sequence (YYY).[13]

This coincidence of the Condorcet paradox and the multiple-election paradox is no accident. If there is a Condorcet paradox, the outcome is cyclical

---

[13] It is perhaps more accurate to call this the "multiple-vote paradox," because the multiple votes in Congress are not really multiple elections, as in the case of voting on propositions in a referendum or for different offices in an election. For simplicity, I will stick with "multiple-election paradox," but it is worth noting two distinct features of voting in Congress: (i) the non-election quality of votes on bills and amendments, and (ii) the sequential nature of voting, which gives voters information about the results of previous votes that simultaneous voting on multiple propositions, or for multiple offices, does not provide.

majorities, which in the isomorphism translates into either YYY or NNN. But since no group with transitive preferences has these sequences, they must, according to combination aggregation, receive 0 votes. Consequently, the winning combination according to bill aggregation (either YYY or NNN) when there is a Condorcet paradox must tie for the fewest votes (with the other intransitive sequence), giving

Proposition 7.3. *If the preferences of individual voters (or like-minded groups) are transitive with respect to pairwise contests among three or more alternatives, then a Condorcet paradox, based on the pairwise contests, implies a multiple-election paradox.*

Whether the reverse implication holds turns on the number of alternatives being ranked. For three alternatives, it turns out that none of the six mixed combinations can win, according to bill aggregation, and also receive 0 votes when YYY and NNN do, too. To show that there is no such example for three pairwise contests, associate the following numbers of voters with the six mixed-answer sequences:

Sequence:	YYN	YNY	NYY	YNN	NYN	NNY
Number:	0	v	w	x	y	z

Without loss of generality, assume that YYN receives the fewest votes, and that this number of votes is 0. The other numbers are all nonnegative. In order for Y to win the first and second offices by bill aggregation, one requires

$$v + x > w + y + z$$

$$w + y > v + x + z.$$

Adding these inequalities gives $0 > 2z$, which is impossible since $z \geq 0$. This contradiction shows that YYN cannot win by bill aggregation and receive the fewest votes.

I next consider an example with four alternatives.

Example 7.8. Basic Paradox, with Ties for Fewest: Four Outcomes and Six Pairwise Contests

Suppose there are 3 voters, whose preferences among the alternatives a, b, c, and d are as follows:

$$bacd: 1 \quad cabd: 1 \quad dabc: 1.$$

Then their votes on the six questions of whether their first alternative is preferred to their second for the six possible pairwise contests—a and b, b and c, c and d, a and c, a and d, and b and d—will be

$$\text{NYYYYY: 1} \quad \text{YNYNYY: 1} \quad \text{YYNYNN: 1.}$$

Now YYYYYY is the winning combination according to bill aggregation, corresponding to the transitive ordering *abcd* for which none of the voters voted. Thus, this example of a multiple-election paradox does not arise from a Condorcet paradox, given the aforementioned enumeration of pairwise contests. Note that the Condorcet winner, *a*, is not ranked first by any of the voters.

More generally, this example, together with the earlier argument that a transitive combination cannot win according to bill aggregation when there are only three pairwise contests, yield the following strengthening of Proposition 7.3.

Proposition 7.4. *Assume there are three or more alternatives over which voters have transitive preferences. Then every paradox of voting corresponds to a multiple-election paradox. The reverse correspondence holds for three alternatives but fails for more than three.*

Proposition 7.4 shows that, given the isomorphism, the multiple-election paradox is a generalization of the paradox of voting, because whenever the latter occurs so does the former, but not vice versa if there are more than three alternatives.

I caution that Proposition 7.4 should not be construed as an empirical law in situations in which voters may, for a variety of reasons, not express transitive preferences and, therefore, not meet the condition of Proposition 7.4. For example, it may not be clear at the outset that they will vote in a particular sequence in three pairwise contests, so the question of being consistent is not a primary consideration.

Even if it is, voters may decide to vote YYY or NNN if such ostensibly inconsistent behavior on the part of enough voters leads to a contradiction, which in turn triggers a default option that these voters prefer. For example, assume that an NNN sequence indicates that a voter votes "no" on three pairwise contests between three levels or types of regulation; if none wins, the status quo of no regulation prevails, which the voter prefers. Then the apparent contradiction of preferring none of the three levels—when matched in pairs against each other—is really no contradiction, given a preference for the default option.

That some voters are, at least on the surface, inconsistent in this sense is observable in actual legislative contests. Blydenburgh (1971) studied the voting behavior of members of the House on Representatives in voting on three provisions of the Revenue Act of 1932: the first to delete a sales tax, the second to add an income tax, and the third to add an excise tax.

Let *a* be the status quo (SQ) without a sales tax, *b* be the SQ with an income tax, and *c* be the SQ with an excise tax. Based on his reconstruction of voter preferences, Blydenburgh (1971) argued that there was a Condorcet paradox *abca*, so majorities would answer YYY in each of the three pairwise contests.

In fact, however, there were no such contests, because the voting was sequential under the amendment procedure. The first contest was *a* versus SQ; when *a* passed, the second contest was *a* versus *a* plus *b* (i.e., SQ with both a sales and an income tax); when the latter failed, the third contest was *a* versus *a* plus *c* (i.e., SQ with both a sales and an excise tax), which passed. Thus, the winner by bill aggregation in these three pairwise contests was YNY. This combination was chosen by 38 voters, ranking fifth of the eight combinations according to combination aggregation.[14]

The fact that there was no multiple-election paradox shows there is obviously some slippage between the theoretical results and their empirical reality.[15] I take the fifth-place finish of the winning combination, nonetheless, as partial confirmation of a discrepancy between—if not a paradoxical aspect of—the two different ways of aggregating votes.

The significance of this discrepancy is underscored by the linkage of the multiple-election paradox to the Condorcet paradox. The Condorcet paradox has produced an enormous literature since the pioneering work of Black (1958) and Arrow (1951, 1963) that extended and generalized the original paradox discovered by Condorcet in the late eighteenth century (see Black, 1958; McLean and Urken, 1995). The multiple-election paradox casts the paradox of voting in a new light that illuminates, especially, its implications for making coherent social choices using different aggregation procedures.

7.6. NORMATIVE QUESTIONS AND DEMOCRATIC POLITICAL THEORY

I return to the case of voting on propositions, with which I introduced the analysis. Given that the winner under proposition aggregation can receive the fewest votes under combination aggregation—and even that the two methods of aggregation can produce diametrically opposed social choices (when there is a complete-reversal paradox)—it is legitimate to ask which choice, if either, is the proper one. In defining "proper," one might apply such social-choice criteria as the election of Condorcet winners (if they exist), the selection of Pareto-efficient outcomes, the existence of incentives to vote sincerely, and so on.[16]

---

[14] The winning combination was YYY with 85 votes. Because the pairwise contests were not among *a*, *b*, and *c* but partially overlapping sets of alternatives (see text), it is not inconsistent for individual voters to have a preference order associated with YYY in this case.

[15] Mackie (2003, ch.15) also questions the presence of a cycle.

[16] Benoit and Kornhauser (1994) focus on the Pareto-inefficiency of assemblies elected by office aggregation, which they show may occur even when voters have separable preferences over all possible combinations of candidates for the assembly. (A *Pareto-inefficient* assembly is one in which the candidates elected by office aggregation are worse for all voters than some other assembly—possibly one elected by combination aggregation—and so might receive zero votes

I will not pursue this line of inquiry here, however, but instead ask an explicitly normative question: Is a conflict between the proposition and combination winners necessarily bad?

In addressing this question, consider first whether this conflict comes as any great surprise. If there is one lesson that social-choice theory over the last several decades teaches us, it is that strange things may happen when we try to aggregate individual choices into some meaningful whole. Thus, the whole may lose important properties that the parts had, such as transitivity of preferences when there is a Condorcet paradox.

Whether the intransitivity of social preferences caused by the Condorcet paradox is a serious social problem has been much debated in the literature (see Riker, 1982, Miller, 1983, and Regenwetter, Grofman, Marley, and Tsetlin, 2006, for different views). The multiple-election paradox shows up a different aspect of this problem by drawing attention to the discrepancy between aggregating votes by proposition and by combination. From a theoretical viewpoint, what is interesting about the multiple-election paradox is that it is a more general phenomenon than the Condorcet paradox—at least under the isomorphism—but those situations that give the multiple-election paradox, and not the Condorcet paradox, have not been analyzed to see precisely where the differences between the two paradoxes lie.

From a practical viewpoint, one is led to ask whether, given the multiple-election paradox, it would be advisable for voters to vote directly for combinations rather than for individual propositions, offices, and bills. I have doubts in the case of different offices, in part because it is not clear how combination aggregation would work in the election of bodies like the Senate or House. In the case of the president, one could prescribe that if the winning combination includes, say, D for president, the Democrat would be elected. But if the winning combination turns out to be DDD, as occurred in 1992, what does it mean to elect a Democratic Senate and a Democratic House, and in what proportions in what states?[17]

---

when pitted against it.) A crucial difference between the present model and Benoit and Kornhauser's is that there is no restriction on the number of propositions that can pass in a referendum, or bills in a legislature, whereas in their model the number of representatives to be elected to the legislature is predetermined. Although the multiple-election paradox is based purely on numerical comparisons, it may be explicitly linked to preference-based models like that of Benoit and Kornhauser, as illustrated in the case of the Condorcet paradox through the answer-sequence isomorphism. See also Brams, Kilgour, and Zwicker (1997), Lacy and Niou (2000), and Hodge and Schwaller (2006), who analyze referendums in which voters have nonseparable preferences. If voters have separable preferences, Barberà, Sonnenschein, and Zhou (1991) show that voting for multiple candidates or alternatives—by indicating the subset a voter considers acceptable, as under AV—is nonmanipulable or strategy-proof.

[17] Fiorina (1992, p. 120) argues that the eight combinations the voter can choose for the three federal offices in the United States are more numerous than voters have in many multiparty systems; furthermore, unlike in multiparty systems, voters can "vote directly for the

In voting in other arenas, such as on a referendum or in a legislature, the choices that legislators and voters can currently make substantially restrict their ability to express their preferences. Thus, legislators cannot express support for different packages of amendments, such as the amendments sequentially voted on in the 1932 Revenue Act (section 7.5). If they vote YYY, for example, this contributes nothing to NNN, even though the latter package might be their second choice. Likewise, there is no way under the present system that legislators can support exactly the six mixed combinations.

A possible solution to this problem is to use AV, whereby voters could, in the present instance, vote for as many combinations as they wish. Thus, a proponent of all the amendments or of none—assuming he or she regards these as the only acceptable packages—could indeed vote for both YYY and NNN, just as a proponent of some but not all of the amendments could vote for from one to six of the mixed combinations.

But using AV for combinations is not the only way of expanding voter choices. Other means for producing more coherent social choices, in light of the paradox, include allowing voters to rank the combinations under a system like the Borda count. This would enable voters to make more fine-grained choices than does the crude variation of the Borda count, discussed in section 7.3, that corresponds to the present system.

To be sure, if there are more than eight or so combinations to rank, the voter's task becomes burdensome. How to package combinations (e.g., of different propositions on a referendum, different amendments to a bill) so as not to swamp the voter with inordinately many choices—some perhaps inconsistent— is a practical problem that will not be easy to solve.

7.7. YES-NO VOTING

Yes-no (Y-N) voting (Fishburn and Brams, 1993), whereby a voter can indicate multiple *subsets* of alternatives that he or she supports, would render practicable voting for many alternatives. Here is how it would work in a referendum with multiple propositions. Each voter would mark every proposition as

- Y—must be included in every subset of propositions he or she approves
- N—must be excluded from every subset of propositions he or she approves
- neither (no mark)—may or may not be included in every subset of propositions he or she approves

coalition they most prefer." As the multiple-election paradox dramatically demonstrates, however, this *expression* of preference for a combination means little, because the combination with the least support can actually win, vitiating the vaunted "popular will" from being expressed.

As an example, suppose that there are five propositions, and voter V's preference for the passage of each is as follows: $1 > 2 > 3 > 4 > 5$. If V strongly approves of the passage of proposition 1 and strongly disapproves of the passage of proposition 5, he or she might well mark proposition 1 as Y and proposition 5 as N. Because every subset of propositions that V approves of must include 1 and must exclude 5, V would approve of the following eight subsets:

$$\{1\}, \{1, 2\}, \{1, 3\}, \{1, 4\}, \{1, 2, 3\}; \{1, 2, 4\}; \{1, 3, 4\}; \{1, 2, 3, 4\}.$$

Among the three 2-member sets, it is reasonable to suppose that V would most prefer passage of $\{1, 2\}$; and among the three 3-member sets, V would most prefer passage of $\{1, 2, 3\}$.[18] But one cannot infer what V's preference would be among passage of $\{1\}$, any 2-member set, any 3-member set, or $\{1, 2, 3, 4\}$, because Y-N voting does not ask V to indicate a preference among these different-size subsets.

But now assume that V's ideal is the passage of $\{1, 2, 3\}$. Then V can express this preference exactly by marking these three propositions Y and $\{4, 5\}$ N. The problem with such a specific strategy is that if $\{1, 2, 3\}$ has little chance of passage but V's second choice—say, $\{1, 2, 4\}$—does, V's strategy provides $\{1, 2, 4\}$ with no support, whereas V's less restrictive strategy of marking proposition 1 Y and proposition 5 N does—and supports, as well, seven other subsets of propositions.

Fishburn and Brams (1993) analyze such trade-offs between restrictive and nonrestrictive Y-N voting strategies, but in the context of choosing a governing coalition in a parliament rather than in voting for multiple propositions in a referendum.[19] In chapter 8 I analyze two other procedures for selecting a governing coalition, but Y-N voting may also be well suited for this task, especially if there are a relatively large number of political parties that are candidates for membership in a governing coalition.

Although Y-N voting mimics AV insofar as it enables voters to approve of multiple options, it does not allow them to approve of all subsets they might like to support. Thus in the example above, it does not permit them to vote for *exactly* $\{1, 2, 3\}$ and $\{1, 2, 4\}$. If V marks propositions 1 and 2 as Y and proposition 5 as N, he or she will approve of these two subsets, $\{1, 2\}$, and $\{1, 2, 3, 4\}$. But V may not desire the latter subset if he or she thinks the passage of both propositions 3 and 4 would be too costly.

---

[18] This inference depends on whether V's preferences are *additive*. If not, the utility of V for, say, $\{1, 3\}$ might be greater than that for $\{1, 2\}$ if the former are complements and the latter are not. But with additivity (and separability, on which it depends), V will value the passage of $\{1, 2\}$ more than $\{1, 3\}$, because the utilities for each subset are the sum of the utilities of their members.

[19] They combine Y-N voting with choosing a governing coalition in a voting system called "coalition voting" (Brams and Fishburn, 1992b).

It is probably not feasible to ask voters to indicate their approval or nonapproval of the $2^5 = 32$ subsets that are possible if there are five propositions, not to mention the 268.4 million subsets that were possible with 28 propositions in the 1990 California election. In my opinion, Y-N offers a pragmatic solution for dealing with multiple propositions, and the much larger number of possible subsets, even though it will usually not enable many voters to express their preferences perfectly.

7.8. SUMMARY AND CONCLUSIONS

The paradox of multiple propositions demonstrates how two different ways of aggregating votes in multiple elections can lead to diametrically opposed results, whereby the winner under one method may receive the fewest votes under another method. I showed, via an isomorphism, that this paradox is a generalization of the well-known Condorcet paradox.

One real-life example of the paradox involved voting on 28 propositions in California, in which not a single voter voted on the winning side on all propositions. This paradox manifested itself in milder form in voting on a subset of three environmental propositions in this referendum.

Several empirical variants of the paradox occurred in federal elections—one of which led to divided government—and legislative votes in the U.S. House of Representatives. As an illustration of the latter, I showed that in votes on NAFTA and the budget in 1993, the subset of members on the winning sides on both bills was less than 20 percent, reflecting very different party alignments. But this outcome was not particularly problematic, because (i) these bills were not strongly related, and (ii) the winning combination at least partially satisfied more members (86 percent) than any other.

The multiple-election paradox tends to create the greatest problem when the issues being voted on in a referendum are linked because voter preferences are nonseparable, but voters must make simultaneous choices. As a case in point, suppose 1/3 of the electorate favors proposition A = do a alone, 1/3 B = do a and b, and 1/3 C = do a, b, and c. (These measures might represent different levels of environmental cleanup, which I will call low, medium, and high.) Now if a voter votes only for his or her first choice, then 1/3 of the electorate will vote YNN (low), 1/3 YNY (medium), and 1/3 NNY (high), yielding NNN under proposition aggregation—so none passes.

But the choice that best reflects the will of the electorate is B (medium), a moderate level of cleanup, which completely satisfies one-third of the voters and partially satisfies the remaining two-thirds. It seems likely that B would have won had AV been applied to the eight combinations. On the other hand, Y-N voting, a set-theoretic voting procedure, might be the most feasible election method if there are several propositions and many more combinations.

At a minimum, a heightened awareness of the multiple-election paradox alerts one to unintended and often deleterious consequences that may attend the tallying of votes by proposition or office aggregation. The paradox does not just highlight problems of aggregation and packaging, however, but strikes at the core of social choice—both what it means and how to uncover it. In my view, the paradox shows there may be a clash between two different meanings of social choice, leaving unsettled the best way to clarify what this elusive quantity is.

PART 2. *Fair-Division Procedures*

8

Selecting a Governing Coalition in a Parliament

8.1. INTRODUCTION

In most parliamentary systems, it is rare for a single party to win a majority of seats and thereby be able to govern by itself. Typically, two or more parties that together hold a majority of seats will form a governing coalition, which cannot be overthrown unless some of its members defect.

This coalition will usually be led by the largest party in parliament, whose leader becomes prime minister. But this may not be desirable if the parliament is ideologically fractured and this party is relatively extreme (on the left or the right).

In this chapter, I propose a procedure for choosing a governing coalition that harks back to fallback voting (FV) in chapter 3 and the minimax procedure in chapter 5. Such a procedure may well diminish unseemly or unscrupulous bargaining, put a government in place sooner, and please more parliamentary members. The coalition chosen by such a procedure, because its formation is based on the choices of parliamentary members, is likely to be viewed as more legitimate and, consequently, less vulnerable to defections.

In game theory, coalitions are collections of players. In this chapter, I treat the individual members of parliament, not the political parties to which they belong, as players. But there is nothing inherent in the procedures I analyze that prevents the parties from being the players. In fact, this is likely to be the norm.

A key question in the formation of coalitions is how stable they are likely to be: Will their members have an incentive to stay in the coalition, or will they be tempted to defect to an opposing coalition? I show that the process by which players come together and form coalitions may critically affect how enduring a coalition will be.

To determine which coalitions are likely to form and be stable, I assume that each player ranks all other players as coalition partners. At the outset, I suppose that players report their rankings truthfully, which is an assumption I reconsider later. A coalition of k players, or k-coalition, is *stable* if no member would prefer to be in another k-coalition.

Note: This chapter is adapted from Brams, Jones, and Kilgour (2005) with permission of Springer Science and Business Media.

It is apparent that there is always at least one stable coalition—the grand coalition, or n-coalition, that comprises all n players—because there is no other n-coalition. But below the grand coalition, what coalitions will form, and how stable they will be, is unclear. The two coalition-formation processes I postulate, and which I will interpret as procedures by which to choose a governing coalition, clarify this question and also enable one to distinguish two levels of stability.

To rule out strategic issues that arise because of differences in player size or capability,[1] I assume that (i) all players are of equal weight (as in a parliament in which each member has one vote), and (ii) winning coalitions are those with at least a simple majority, m, of members. While I focus on nonstrategic processes of coalition formation, later I consider the manipulability of these processes as well as the possibility that the players—if they are political parties instead of their members—may be of different size or weight.

I assume coalitions form according to two processes:

- *Fallback (FB)*: Players seek coalition partners by descending lower and lower in their preference rankings until some majority coalition, all of whose members consider each other mutually acceptable, forms.
- *Build-up (BU)*: Same descent as FB, except only majorities whose members rank each other highest form coalitions.

Both these processes are driven by players' preferences for each other, not their preferences for coalitions.[2]

I begin with notation and definitions in section 8.2. In section 8.3 I show that several FB majority coalitions with $k \geq m$ members may form simultaneously. Call the set of FB coalitions that form first (i.e., at the lowest value of k) FB_1. The analogous set, BU_1, comprises a unique coalition, which is stable. If the preferences of the players are single-peaked (to be defined), FB_1 coalitions may be disconnected, but a BU_1 coalition is always connected (in a sense to be made precise later).

BU_1 contains all coalitions in FB_1, which may have fewer members than BU_1. This raises the question of which majority coalition is most likely to form—smaller FB_1 majority coalitions, in which some members may prefer players outside the coalition to players inside, or the BU_1 coalition, in which

---

[1] For example, two ideologically distant players might join together if it would enable them to win, but neither would join a smaller more centrally located player if the resulting coalition were not winning. Such considerations may come into play if parties control the voting of their members, which I comment on at the end of this section.

[2] In Cechlárová and Romero-Medina (2000), each player uses its preference rankings of all other players to evaluate coalitions according to two criteria—the most-preferred, and the least-preferred, members they contain. Other criteria are postulated in an agent-based simulation model in a neural-network framework, wherein political parties seek to attract a majority of players in a spatial voting game (Iizuka, Yamamoto, Suzuki, and Ohuchi, 2002). Related work on coalition-formation models is discussed in section 8.3.

this cannot happen. I call smaller FB_1 majority coalitions *semi-stable* if at least some of their members are attracted to outside players, whereas BU_1 coalitions are stable.

In section 8.4 I show that semi-stable FB_1 coalitions are *manipulable* in that a player, by announcing a false preference ranking, can induce a majority coalition that it prefers. By contrast, BU_1 stable coalitions are not manipulable. A manipulator may be able to induce a smaller majority coalition with a false announcement, but this coalition will not necessarily be preferred. The reason is that the larger BU_1 coalition, which forms when the manipulator is truthful, must contain at least one member that the manipulator prefers to some player in the smaller majority coalition—so the manipulator will not assuredly prefer the smaller coalition.

In section 8.5 I investigate the properties of stable coalitions. BU_1 may be a simple-majority coalition, the grand coalition, or any size in between. More generally, stable coalitions of any size between m and $n-1$ may or may not exist. Two stable coalitions (of any size) are either disjoint, or one contains the other. A *bandwagon strategy* may enable a player to be a member of a winning coalition sooner than it would be otherwise, but it will not necessarily be a winning coalition that it prefers.

In section 8.6 I show that if all player preferences are equally likely, the probability that a randomly chosen majority coalition is stable first decreases to some minimum between m and $n-1$, then increases to 1 when the grand coalition forms, yielding a bimodal distribution, with peaks at minimal majority and unanimity. This finding also holds for the distribution of first-forming majority coalitions when preferences, whether single-peaked or not, are randomly chosen.

Empirical data on the size of U.S. Supreme Court majorities, which I present in section 8.7, show the distribution to be bimodal, with most being either minimal (5-person) or maximal (9-person) majorities. I illustrate the formation of majorities on the Court with an 8–0 decision (one justice recused himself) and a 5–4 decision. Data on the size of majorities in the U.S. House of Representatives also show the distribution to be bimodal.

I conclude that FB and BU mirror different real-life coalition-formation processes. BU yields larger and nonmanipulable majority coalitions, compared to the more wieldy but vulnerable majority coalitions of FB. Together these models show how the stability of outcomes is inextricably linked to the processes that generate them.

But FB and BU also define procedures that can be used to select a governing coalition in a parliamentary democracy.[3] In a parliament of, say, one hundred

---

[3] In this sense, they are both explanatory and prescriptive. They explain coalition formation in real-world voting bodies, and they suggest procedures for finding governing coalitions in parliaments that, because they mirror the preferences of their members, are likely to persist.

or more members, I would not expect each member to strictly rank all other members. Rather, parties are likely to instruct their members to rank members of other parties in the same way (e.g., by giving them the average ranking of that party's members). Effectively, this would mean that parliamentary members would rank parties, which will generally be of different sizes and cast different numbers of votes, rather than their individual members, who all have one vote.

I would not preclude parliamentary members from ranking individuals if they so desired. In my view, they should be able to speak for themselves instead of being forced to follow the dictates of their party, though I suspect such deviance would be rare in a disciplined party system.

8.2. NOTATION AND DEFINITIONS

Assume that all players, $1, 2, \ldots, n$, strictly rank each other as coalition partners, as illustrated in Example A, where $n = 5$.

Example A.   **1:** 2 3 4 5   **2:** 1 3 4 5   **3:** 4 5 2 1   **4:** 3 2 1 5   **5:** 4 3 2 1.

Further assume that each player ranks itself first—that is, it most desires to be included in any majority coalition that forms. In Example A, player 1, after itself, most prefers player 2 as a coalition partner, followed by players 3, 4, and 5 in that order. A complete listing of all players' preferences, as illustrated in Example A, is a *preference profile*.

It is clear that if there are n players, there are $[(n - 1)!]^n$ possible preference profiles. In the model of a random society discussed later, all preference profiles are assumed to be equiprobable.

Sometimes I assume that the players can be placed along a line—in order 1, $2, 3, \ldots, n$, from left to right—so that the preference profile is single-peaked. That is, each player's preference for coalition partners declines monotonically to the left and right of its position in this ordering. A preference profile that satisfies this condition is called *ordinally single-peaked* (Brams, Jones, and Kilgour, 2002). Such profiles are commonly assumed in spatial models of candidate and party competition.

To express single-peakedness in another way, consider the set of players in a coalition; call the left-most player l and the right-most player r. The set is *connected* if it is of the form $\{l, l + 1, \ldots, r\}$: it contains exactly the players from l to r, inclusive. Then a preference profile is single-peaked if and only if, for each $k = 1, 2, \ldots, n$, every player's k most-preferred coalition partners, including itself, form a connected set—that is, no party is "skipped over" in the left-right ordering of its members.

Thus in Example A, when $k = 3$, the most-preferred 3-coalitions of players 1 (123), 2 (213), 3 (345), 4 (432), and 5 (543) are all connected sets. For all other

k between 1 and 5, it is easy to see that all most-preferred k-coalitions are connected, so the preference profile of Example A is ordinally single-peaked.

In fact, such a preference profile may or may not be geometrically realizable in the following sense. If n points can be positioned along a line such that a player's preference decreases as distance from its position increases, then the preference profile is called *cardinally single-peaked*. To see that this condition is not satisfied in Example A, assume that player i is located at position p_i on the line. Define the distance between two positions, p_i and p_j, to be $d_{ij} = |p_i - p_j|$. From player 3's preference ordering, $d_{54} < d_{53} < d_{32}$, whereas from player 4's ordering, $d_{32} < d_{42} < d_{54}$. This contradiction shows that the preference profile of Example A is ordinally but not cardinally single-peaked.[4]

8.3. THE FALLBACK (FB) AND BUILD-UP (BU) PROCESSES

The *fallback (FB)* process of coalition formation unfolds as follows (Brams, Jones, and Kilgour, 2002; Brams and Kilgour, 2001b):

1. The most preferred coalition partner of each player is considered. If two players mutually prefer each other, and this is a majority of players, then this is the majority coalition that forms. The process stops, yielding a level-1 majority coalition because only first-choice partners are considered.
2. If there is no level-1 majority coalition, then the next-most preferred coalition partners of all players are also considered. If there is a majority of players that mutually prefer each other at this level, then this is the majority coalition (or coalitions) that forms. The process stops, yielding a level-2 majority coalition.
3. The players successively descend to lower and lower levels in their reported rankings until a majority coalition (or coalitions), all of whose members mutually prefer each other, forms *for the first time*. The process stops, with the set of *largest* majority coalition(s)—not contained in any others at this level—designated FB_1.

This process mirrors fallback voting (FV) in section 3.4, except FV may not find a majority winner if voters truncate their preferences (e.g., by ranking only one candidate), whereas I assume players using FB (and later BU) are required to give complete rankings.

What does FB yield in Example A? At level 1, observe that player 1 prefers player 2, and player 2 prefers 1, so designate 12 as a level 1 coalition, as is also coalition 34.[5] Descending one level, player 3 likes player 5 and player 5 likes

---

[4] In other words, the players' ordinal rankings are inconsistent with every possible cardinal representation of their positions.

[5] These preferences are truthful; I consider later the possibility that the players strategically misrepresent their preferences.

player 3, yielding 35 as a coalition at level 2. Descending one more level, majority coalitions 124 and 234 form for the first time: each player in these coalitions finds the other two players acceptable at level 3. In summary, the following coalitions form at each level:

Level 1: 12, 34   *Level 2*: 35   *Level 3*: 124, 234.

Notice that coalitions are listed at the level at which they form, except that subcoalitions are never listed. Thus at level 3, pairs 14, 23, and 24 form but do not appear in the listing, because they are proper subsets of coalitions 124 or 234.

Since coalitions 124 and 234 are the first majority coalitions to form, the process stops, rendering $FB_1 = \{124, 234\}$. Observe that players 2 and 4 are common to both coalitions; player 2 prefers coalition 124, and player 4 prefers coalition 234. Obviously, players 1 and 3 prefer the coalition of which each is a member.

Despite the players' preferences being single-peaked, one of the two FB_1 coalitions (124) is *disconnected*: there is a "hole" due to the absence of player 3. The reason that player 3 is excluded from coalition 124 is that whereas players 1 and 2 necessarily rank player 3 higher than player 4 (because of single-peakedness), player 3 ranks players 2 and 1 at the bottom of its preference order. In particular, player 3 does not consider player 1 acceptable at level 3, which excludes player 3 from coalition 124.

While FB is grounded in preferences of players for each other, it could as well be based on their preferences for different features that a policy might include. Thus in Example A, assume players rank five features, $\{a, b, c, d, e\}$, in the same way that they rank each other. Then at level 1, player 1 would find feature *a* acceptable, and at level 2 feature *b*; likewise, player 2 would find both *a* and *b* acceptable at level 2. Consequently, at level 2 (rather than level 1) the coalition 12 would form because of the two players' concurrence on both *a* and *b*. In this example, the level at which coalitions form changes, but not their membership, as players switch from ranking each other to ranking policy features.[6]

The *build-up* (BU) process of coalition formation is the same as FB, with one major difference. As players descend to lower and lower levels, coalitions form if and only if two or more players consider each other mutually desirable *and consider players not in the coalition less desirable*. In other words, all players in a BU coalition rank each other—and no players outside the coalition—highest. In Example A, this yields the following coalitions at each level:[7]

Level 1: 12, 34   *Level 4*: 12345.

---

[6] The number of policy features need not match the number of players. If there are more features than players, coalitions will form later than if there are fewer features than players. For examples, see Brams and Kilgour (2001b).

[7] In Brams, Jones, and Kilgour (2002), a different BU model is proposed in a cardinal-utility context. Coalition members fuse into a single player whose position is the average of its members when preferences are defined by points on the real line.

At levels 2 and 3, no new BU coalitions form after coalitions 12 and 34 form at level 1. Only at level 4 does the first majority coalition appear; it is the grand coalition, so $BU_1 = \{12345\}$, or just 12345. Note that no member would prefer to be in another 5-coalition—there is none!—proving that this majority coalition is not only *stable* but uniquely so.

Compare this outcome with that produced by FB, which gave FB_1 coalitions 124 and 234 at level 3. These coalitions are *semi-stable*: Even though all their members consider each other acceptable at level 3, some members of each coalition consider some excluded players more desirable as coalition partners. For coalition 124, players 1 and 2 prefer excluded player 3 to included player 4; for coalition 234, player 2 prefers excluded player 1 to included players 3 and 4, and player 3 prefers excluded player 5 to included player 2.[8]

Proposition 8.1. *BU_1 contains a unique stable coalition. If FB_1 forms at the same level as BU_1, $FB_1 = BU_1$. Otherwise, FB_1 forms earlier (i.e., at a lower level), in which case all FB_1 coalitions are semi-stable and proper subsets of the BU_1 coalition.*

Proof. Because the grand coalition is a BU coalition, BU_1 is well defined and never empty. Suppose it contains two majority coalitions. Because both are the same size, say k, they must contain at least one common member i.[9] Since the other members of both coalitions must be exactly i's k most-preferred coalition partners, the two coalitions in BU_1 must be identical. Hence, the set BU_1 contains a single member (coalition), which I henceforth call BU_1. Moreover, because all members of BU_1 rank each other highest, BU_1 is stable.

Every BU coalition is an FB coalition since the process of descent is the same. If the level of FB_1 is the same as the level of BU_1, then BU_1 belongs to FB_1. Because there cannot be any other coalition in FB_1, then $FB_1 = BU_1$.

Now suppose that the level at which BU_1 forms is k, and the level at which FB_1 forms is $j < k$. Consider any coalition C in FB_1. While the members of C consider each other acceptable at some level, there is at least one player in C that prefers some player not in C. (If this were not the case, then C would be BU_1, and j would equal k.) This makes C semi-stable. Moreover, because both C and BU_1 are majority coalitions, they must have a member in common, say i. But BU_1 contains i and i's k most-preferred coalition partners, whereas C contains i and a proper subset of i's most-preferred coalition partners, which has j members. Therefore, C is properly contained in BU_1. ∎

---

[8] The exclusion of preferred players from a coalition, and its manipulability (section 8.4), are two indicators of its instability. While "there is only a relatively small number of results that guarantee the existence of a 'stable' coalition structure" (Greenberg and Weber, 1993, p. 60), even fewer models offer insight into the step-by-step processes of coalition formation that may (or may not) contribute to stability.

[9] If preferences are single-peaked and this common member is unique, it must be the median player.

Example A illustrates Proposition 8.1. Semi-stable FB_1 coalitions 124 and 234 are contained in stable BU_1 coalition 12345. There are no stable majority coalitions smaller than this grand coalition.

The next example illustrates that BU_1 need not be the grand coalition.

Example B.   **1**: 2 3 4 5   **2**: 3 4 1 5   **3**: 4 2 1 5   **4**: 1 2 3 5   **5**: 4 3 2 1.

The FB coalitions at each level are:

Level 2: 23, 24   *Level 3*: 1234.

Whereas no two players consider each other mutually acceptable at level 1, at level 2 two pairs do. At level 3, the first majority coalition forms, so $FB_1 = \{1234\}$. But this 4-player coalition is also BU_1, because all its members consider each other, and no others, acceptable. Thus in Example B, the FB and BU processes produce exactly the same majority coalition, which is neither minimal nor grand. To be sure, the grand coalition is also stable, but it seems unlikely to form since players 1–4 are united in their opposition to player 5, which they all rank last.

If $FB_1 \neq BU_1$, smaller FB_1 coalitions, which are semi-stable, form earlier in the descent, only later to be subsumed by a larger BU_1 coalition that is stable. Thus in Example A, semi-stable FB_1 coalitions 124 and 234 (level 3) are proper subsets of stable BU_1 coalition 12345 (level 4).

Proposition 8.2. *If preferences are single-peaked, at least one FB coalition of two players must form at level 1.*

Proof. Single-peakedness requires that every player rank an adjacent player highest. Let C be the subset of players whose top-ranked coalition partners are players to their right—that is, all players i for which $i + 1$ is first choice. Note that $1 \in C$ (because there is no player to the left of 1) and that $n \notin C$ (because there is no player on n's right). Let r be the right-most (highest-numbered) player in C and note that $r < n$. Then $r + 1$ must be r's top choice, and r must be $(r + 1)$'s top choice, so the coalition $\{r, r + 1\}$ must form at level 1. ∎

In Example A, two coalitions, 12 and 34, form at level 1, whereas in Example B no coalitions form at level 1 because its preference profile is not single-peaked.

Proposition 8.3. *If preferences are single-peaked, then (i) FB_1 coalitions may be disconnected, but (ii) BU_1 is connected.*

Proof. Example A, with disconnected FB_1 coalition 124, proves (i). To prove (ii), assume that the left-most (lowest-numbered) member of BU_1 is player l, and the right-most (highest-numbered) player is r, where $l < r$. I next show that BU_1 must also contain any i satisfying $l < i < r$. If the level of BU_1 is k, then BU_1 comprises l and l's k most-preferred coalition partners. By single-peakedness, these must be players $l+1$, $l+2$, ..., $l+k$. It follows that $l+k=r$, and $i \in BU_1$, rendering BU_1 connected. ∎

It is worth mentioning linkages to other work on coalition-formation processes. Grofman (1982) and Straffin and Grofman (1984) show, in a dynamic model of coalition formation that somewhat resembles the BU model, that coalitions will always be connected in one dimension but not necessarily in two or more dimensions. But under FB, as I illustrated in Example A, coalitions need not be connected, even in one dimension, if preferences are ordinally single-peaked.[10]

I next turn to the question of whether players can manipulate either the FB or the BU processes to their advantage. FB, it turns out, is vulnerable to manipulation, but BU is not.

8.4. THE MANIPULABILITY OF FB AND BU

Call a process *manipulable* if one player, by reporting a preference ranking different from its true preference ranking, can induce a majority coalition that it prefers.

Proposition 8.4. *FB is manipulable.*

Proof. Consider the following example:

Example C. **1**: 2 3 4 5   **2**: 3 4 1 5   **3**: 2 4 1 5   **4**: 3 5 2 1   **5**: 4 3 2 1.

The FB coalitions at each level are:

Level 1: 23   *Level 2*: 34, 45   *Level 3*: 123, 234   *Level 4*: 12345.

Now assume player 4 misrepresents its preferences as follows:

   4: 3 2 5 1.

Then FB gives the following:

Level 1: 23   *Level 2*: 234   *Level 3*: 123   *Level 4*: 12345.

When player 4 is truthful, $FB_1 = \{123, 234\}$, whereas when player 4 misrepresents its preferences, $FB_1 = \{234\}$. Because player 4 prefers coalition 234 to coalition 123, misrepresentation, which precludes the possibility of coalition 123, is rational, rendering FB manipulable.[11] ∎

---

[10] For references to more recent models in this vein, and tests of these models in party-coalition formation in parliamentary systems, see Brams, Jones, and Kilgour (2002).

[11] Thus, truthful reporting is not a Nash equilibrium under FB, given the strategies of players are to either be truthful or not truthful in a noncooperative game (player 4 would have an incentive not to be truthful in Example C). As I will show next, however, a player cannot assuredly do better by misrepresenting its preferences under BU. Consequently, *when* the process of coalition

By comparison, misrepresentation will not be rational for player 4 if the comparison is between the (apparent) BU_1 coalition that forms with misrepresentation and one that forms without misrepresentation. With misrepresentation, $BU_1 = 234$; without misrepresentation, $BU_1 = 12345$. Because the larger coalition, 12345, includes both a preferred player (5) and a nonpreferred player (1) compared to player 2 in the smaller coalition, 234, one cannot say that player 4 prefers 234 to 12345 or vice versa. Thus, by reporting a preference ranking different from its true preference ranking, player 4 cannot induce a majority coalition that it *assuredly* prefers, illustrating the nonmanipulability of BU.[12]

Proposition 8.5. *BU is not manipulable.*

Proof. Assume BU_1 has k members. Then a majority coalition that any member i of BU_1 prefers cannot have more than k members, because it would contain members that i ranks lower than those in BU_1 and no members that i ranks higher.

Suppose that i prefers a majority coalition with fewer members—specifically, with j members such that $m \leq j < k$, where m is a simple majority. To induce this smaller majority coalition through misrepresentation, i would have to reduce its ranking of some player P not in the j-coalition, and raise some player Q into the j-coalition.

The resulting j-coalition, though an apparent BU coalition, does not contain i's $j - 1$ most-preferred coalition partners since it contains player Q. Moreover, i will not necessarily prefer the j-coalition to BU_1, because although it is smaller, it does not contain player P, which i prefers to Q. Thus, i is not able to induce through misrepresentation a smaller coalition that it definitely prefers. ∎

In Example C, as I showed earlier, player 4 can induce through misrepresentation coalition 234—by raising player 2 (Q) and lowering player 5 (P) in its reported ordering—making it an (apparent) BU_1 coalition at level 2. But player 4 will not necessarily prefer coalition 234, which is an FB_1 coalition, to the grand coalition, 12345.

8.5. PROPERTIES OF STABLE COALITIONS

After the appearance of BU_1, larger and larger BU majority coalitions may— or may not—appear at subsequent levels of descent. Each larger BU majority

---

formation terminates affects the stability of outcomes generated under it, underscoring the fact that "the process matters." Put another way, outcomes are not independent of the process that produces them.

[12] To be sure, if there were more information about preferences—in particular, cardinal valuations of different coalitions by each player—it would be possible to say whether player 4 prefers 234 to 12345 or vice versa. In the absence of such information, however, player 4 does *not* have an incentive to depart from reporting its true preference that yields 12345.

coalition contains all smaller BU majority coalitions, as illustrated next with a cardinally single-peaked example.

Example D.   **1**: 2 3 4 5 6 7   **2**: 1 3 4 5 6 7   **3**: 2 1 4 5 6 7   **4**: 3 2 1 5 6 7
                5: 6 4 3 2 1 7   **6**: 5 4 3 2 1 7   **7**: 6 5 4 3 2 1.

Geometrically, the preferences of these players can be represented by placing them at points along the real line:

```
1 2  3     4            5 6                                        7
```

Thus, for example, the members of pairs 12 and 56 are each other's most-preferred coalition partners, for they are closer to each other in distance than to any other players. Because player 3 prefers players 2 and 1 to player 4, player 3 is farther from player 4 than from player 1. Likewise, players 5 and 6 are farther from player 7 than from player 1, because they rank player 7 last.

I list below all the FB coalitions, not contained in any others at each level, distinguishing those that are also BU coalitions:

Level 1: 12 (BU), 56 (BU)   *Level 2*: 13, 23   *Level 3*: 1234 (BU)
Level 4: 2345   *Level 5*: 123456 (BU)   *Level 6*: 1234567 (BU).

Observe that the first FB majority coalition to appear, 1234 at level 3, is also a BU majority coalition, so $FB_1 = BU_1 = 1234$. As the descent continues, there is no BU coalition at level 4, but at level 5 a 6-member BU coalition forms. Finally, the grand coalition, which is always a BU coalition, appears at level 6.

Given a cardinally single-peaked preference profile, define the *spread* of a coalition to be the distance between its extreme players. Thus, the spread of coalition 1234 is the distance between player 1 on the left and player 4 on the right, or d_{14}. That this distance is less than d_{45} ensures that coalition 1234 forms before player 5 is brought into the fold. But because player 5 ranks player 6 above all other players, player 5 does not find player 1 acceptable at level 4—only players 2, 3, 4, and 6 are acceptable at this level.

Coalition 12345, therefore, is not a BU coalition. On the other hand, because the spread of coalition 123456 is less than the distance between player 6 and player 7, coalition 123456 is a BU coalition at level 5, as is the grand coalition, 1234567, at level 6.

If players' preferences are cardinally single-peaked, it is easy to discern the stable coalitions that form from the players' positions and distances between them.

Proposition 8.6. *If preferences are cardinally single-peaked, then a subset of players is a BU coalition if and only if it is connected and its spread is less than the distances from each extreme member (other than 1 and n) to the nearest player not in the subset.*

Proof. Suppose that the connected subset $C = \{l, l + 1, \ldots, r\}$ has the properties that either $l = 1$ or $d_{l(l-1)} > d_{rl}$, and either $r = n$ or $d_{(r+1)r} > d_{rl}$. Clearly, the remaining members of C are l's top choices as coalition partners, and similarly for r. Also, if $i \in C$, $j \in C$, and $k \notin C$, then $d_{ij} < d_{rl} < \min\{d_{ik}, d_{rk}\}$, which shows that the remaining members of C are also i's top choices as coalition partners, making C a BU coalition. The converse is obvious. ∎

Put more informally, a coalition that is disconnected cannot be a BU coalition, because members would rank the left-out member higher than some members of the coalition. Now assume a coalition is connected but that the distance of an extreme member to an adjacent nonmember—either on the left or on the right—is less than the spread. Then the adjacent nonmember will be ranked higher by the extreme member than some player in the coalition, so the coalition cannot be a BU coalition.

Proposition 8.6 provides a characterization of BU coalitions if the players have cardinally single-peaked preferences, thereby enabling one to "read" the BU coalitions from the geometric representation. In general, members of a BU coalition must be sufficiently isolated from players outside it to rank only each other as tops.

Whether players' preferences are cardinally single-peaked or not, it is always possible to ensure the existence—or nonexistence—of BU majority coalitions at any level from $m - 1$ (simple majority coalition) to $n - 1$ (grand coalition).

Proposition 8.7. *BU majority coalitions may appear—or not appear—at any level, up to the appearance of the grand coalition.*

Proof. See appendix.

In the proof of Proposition 8.7, I show that, with one exception, it is possible to construct a cardinally single-peaked preference profile whose majority BU coalitions are of any size or combination of sizes. The exception occurs when $n = 3$; in this case, ordinally and cardinally single-peaked preferences are identical and produce a BU_1 coalition of size 2. When preferences are not single-peaked, BU_1 is of size 3. I will describe this case in detail in section 8.6.

The algorithm used to prove Proposition 8.7 yields, in the case of Example D, the following positions p_i of players i:

i:	1	2	3	4	5	6	7
p_i:	0	1	3	7	16	16.5	34

These positions are approximated by the representation given on the real line earlier.

The construction in the proof of Proposition 8.7 is a quantitative one that yields three stable majority coalitions (1234, 123456, and 1234567) in this example. But I emphasize that it is the ordinal rankings that determine the stability of coalitions. Thus, $BU_1 = 1234$ in Example D, because players 1–4 all rank each other highest. There is no 5-member BU coalition, because player 5 does not rank players 1–4 in its top four places (it ranks player 1 lower, and player 6 higher, than fourth place). There is next a 6-member BU coalition, 123456, because players 1–6 rank each other highest. The grand coalition, 1234567, is, as always, a BU coalition.

Notice that less-than-majority FB coalitions, but not BU coalitions, form at level 2 in Example D. These two 2-member coalitions at level 2 become part of BU majority coalition 1234 at level 3, and the 4-member FB majority coalition, 2345, at level 4 becomes part of BU majority coalition 123456 at level 5.

The level 1 BU coalition 56 remains apart until level 5. Although player 5 is acceptable to players 1–4 at level 4, player 5 does not find player 1 acceptable at this level. Consequently, player 5 does not get absorbed into a majority coalition until the descent reaches level 5, when player 6—player 5's most-preferred coalition partner—also gets absorbed.

Example D illustrates that a less-than-majority BU coalition (56) and a majority BU coalition (1234) can coexist. However, two different BU *majority* coalitions, which of necessity overlap, cannot coexist, as is possible under FB (see Example A for an illustration).

Proposition 8.8. *If two BU coalitions intersect, then one contains the other.*

Proof. Suppose that two BU coalitions—one with j members and the other with $k \geq j$ members—have a member i in common. The members of the j-coalition must be player i and the first $j - 1$ players in i's preference ranking. The members of the k-coalition must be player i and the first $k - 1$ players in i's preference ranking. Clearly, the k-coalition contains the j-coalition. ■

A consequence of Proposition 8.8 is that any BU majority coalition of a specific size is unique. In particular, BU_1 contains only a single coalition, as already noted in Proposition 8.1. In Example D, the BU coalitions that form at levels 1 and 3 are contained in the level 5 BU coalition, which in turn is contained in the level 6 BU coalition. But BU coalition 56, which forms at level 1, is disjoint from BU majority coalition 1234 that forms at level 3. In general, if BU coalitions coexist, then at most one is of majority size.

In section 8.4, I showed that members of BU_1 cannot, in general, induce a preferred majority coalition. However, they might be able to speed up the formation of an apparent (smaller) BU_1 coalition. But what if a nonmember of BU_1 desires to be part of a BU coalition? I next show that such a player can conceivably benefit from a *bandwagon strategy*, which enables it, by misrepresenting its preferences, to be part of a larger BU majority coalition sooner than it would be if it were truthful.

To illustrate, suppose that player 5 in Example D reports its preference ranking to be

5: 4 3 2 1 6 7.

At level 5, BU majority coalition 12345 will form, which includes player 5. By comparison, if player 5 were truthful, the next BU majority coalition to form—after $BU_1 = 1234$—would be 123456. Because player 6 is player 5's most-preferred coalition partner, player 5 does not necessarily benefit from a bandwagon strategy, even though this strategy puts it into a smaller BU majority coalition at level 4 rather than level 5.

If there is a benefit, it would come by misrepresenting one's preferences in order to join the winning coalition early (i.e., "jumping on the bandwagon"). Indeed, there is evidence from U.S. national party conventions of delegates' shifting to the expected winner—allegedly to demonstrate party unity—as soon as the handwriting of his or her victory is on the wall. Such proclamations of support may well be motivated by cold-blooded calculations of the direct benefit (e.g., a government appointment) that sometimes accrues to former opponents (Brams, 1978, ch. 2).

8.6. THE PROBABILITY OF STABLE COALITIONS

Because BU coalitions may or may not exist at every level from simple majority to the grand coalition, it is useful to ask when they are most likely to form. To illustrate in a simple case, assume there are three players. Then each player can rank the two others in two ways. For example, player 1 can rank players 2 and 3 as follows:

(i) **1**: 2 3   (ii) **1**: 3 2.

Suppose (i) obtains. If player 2 has the following ranking,

2: 1 3,

$BU_1 = 12$, whatever the ranking of player 3 (two cases). Suppose (ii) obtains. If player 3 has the following ranking,

3: 1 2,

$BU_1 = 13$, whatever the ranking of player 2 (two cases). Whether the preferences of player 1 are (i) or (ii) (two cases), $BU_1 = 23$ if

2: 3 1
3: 2 1.

Altogether, there are six cases in which a 2-member BU coalition forms.

By comparison, there are two cases in which $BU_1 = 123$:[13]

1: 2 3	**1:** 3 2
2: 3 1	**2:** 1 3
3: 1 2	**3:** 2 1.

If the $2^3 = 8$ cases are equally likely, the probability that there is a 2-member stable coalition is $6/8 = 3/4$. On the other hand, because the grand coalition is always stable, the probability that there is a 3-member stable coalition is 1.

I next generalize this result by finding a formula, $P(n, k)$, for the probability that a k-coalition ($k \geq m$) is stable if all strict preference rankings of n players are equally likely, which I call a *random society*. The following proposition describes the behavior of this probability as the size of a majority coalition increases from m to n.

Proposition 8.9. *The probability of a BU coalition, starting at $k = m$, decreases to a minimum at some intermediate value of k before increasing to 1 at $k = n$. More precisely, for each $n \geq 3$, there exists an integer $k_0(n) = k_0$, satisfying $m \leq k_0 < n$, such that $P(n, k + 1) < P(n, k)$ if $m \leq k < k_0$, $P(n, k_0 + 1) \geq P(n, k_0)$, and $P(n, k + 1) > P(n, k)$ if $k > k_0$. Moreover, $k_0(n) > m$ whenever $n \geq 5$.*

Proof. See appendix.

For small values of n and k, I have calculated not only $P(n, k)$ but also $Q(n, k)$, the probability that a k-coalition ($k \geq m$) is stable when all preference rankings of the n players are ordinally single-peaked and equally likely to occur. In addition, using the method of inclusion-exclusion (Brualdi, 1999, pp. 159–168), I have made analogous calculations of the probabilities, $P_1(n, k)$ and $Q_1(n, k)$, that stable majority coalitions form *for the first time*—that is, form at size k but not earlier. All these probabilities are given in Table 8.1 for values of n between 3 and 9, and all values of k between m and n.

In the 3-player case just described, $P(3, 2) = 0.75$ and $P(3, 3) = 1$, as I showed. When preferences are restricted to those that are ordinally single-peaked, $Q(3, 2) = Q(3, 3) = 1$, because there are no instances in which BU_1 coalitions do not form at level 1.

The probabilities of BU_1 coalitions appearing for the first time are $P_1(3, 2) = 0.75$ and $P_1(3, 3) = 0.25$, because two of the eight preference profiles yield BU_1 coalitions when $k = 3$. But when preferences are ordinally single-peaked, $Q_1(3, 2) = 1$ and $Q_1(3, 3) = 0$, because all six preference profiles yield BU_1 coalitions when $k = 2$.

---

[13] The preferences of players in these two cases lead to a Condorcet paradox.

TABLE 8.1
Probabilities of Stable (BU) Coalitions (P, Q) and First-Forming Stable (BU_1) Coalitions (P_1, Q_1)

A. All preference profiles equiprobable

	$k=2$	$k=3$	$k=4$	$k=5$	$k=6$	$k=7$	$k=8$	$k=9$
$P(3,k)$	0.75	1						
$P_1(3,k)$	0.75	0.25						
$P(4,k)$		0.1481	1					
$P_1(4,k)$		0.1481	0.8519					
$P(5,k)$		0.0463	0.0195	1				
$P_1(5,k)$		0.0463	0.0166	0.9371				
$P(7,k)$			2.19×10^{-4}	2.77×10^{-5}	1.50×10^{-4}	1		
$P_1(7,k)$			2.19×10^{-4}	2.76×10^{-5}	1.50×10^{-4}	0.9996		
$P(9,k)$				7.50×10^{-8}	2.72×10^{-9}	2.67×10^{-9}	5.36×10^{-7}	1
$P_1(9,k)$				7.50×10^{-8}	2.72×10^{-9}	2.67×10^{-9}	5.36×10^{-7}	0.999999

B. All ordinally single-peaked preference profiles equiprobable

	$k=2$	$k=3$	$k=4$	$k=5$	$k=6$	$k=7$	$k=8$	$k=9$
$Q(3,k)$	1	1						
$Q_1(3,k)$	1	0						
$Q(4,k)$		0.4444	1					
$Q_1(4,k)$		0.4444	0.5556					
$Q(5,k)$		0.3333	0.1875	1				
$Q_1(5,k)$		0.3333	0.1042	0.5625				
$Q(7,k)$			0.0640	0.0640	0.0308	1		
$Q_1(7,k)$			0.0640	0.0591	0.0272	0.8497		
$Q(9,k)$				8.55×10^{-3}	2.75×10^{-3}	2.12×10^{-3}	4.81×10^{-3}	1
$Q_1(9,k)$				8.55×10^{-3}	2.54×10^{-3}	2.01×10^{-3}	4.65×10^{-3}	0.9822

Now consider the P values in Table 8.1A. For fixed n, these probabilities are virtually identical when $n=7$ and $n=9$. They first decrease going from $k=m$ to some intermediate value of k, and then increase to almost 1 in the case of $P_1(n,k)$, and to 1 in the case of $P(n,k)$.

What Proposition 8.9 does not indicate, though the numerical values of both $P(n,k)$ and $P_1(n,k)$ do, is that even when $k=m$, these probabilities are very small compared with their values when $k=n$. In other words, almost all BU_1 coalitions in a random society form—in fact, form for the first time—only when the grand coalition appears.

It is evident that the probability that *any* BU majority coalition (except the grand coalition) forms in a random society becomes vanishingly small as n increases. This reflects the fact that there is at most one BU coalition at each majority size, and that stability is a certainty only for the grand coalition.

While the probability values in Table 8.1 may not be empirically accurate, the *distributions* may be qualitatively correct in many situations. As I will illustrate later, majority coalitions in real-life voting bodies often do cluster around simple majority and grand—that is, their distribution is V-shaped between $k = m$ and $k = n$, as the BU model predicts.

To be sure, the bimodal distribution of the probability values for general preferences concentrates almost all the support on the grand coalition. This support is dampened somewhat if preferences are restricted to profiles that are ordinally single-peaked (see the Q values in Table 8.1B). When $n = 5$, for example, $Q_1(5, 3) = 0.333$ and $Q_1(5, 4) = 0.104$, compared with $P_1(5, 3) = 0.046$ and $P_1(5, 4) = 0.017$. Thus, in the former case there is a 44 percent chance that BU_1 will not be the grand coalition, whereas in the latter case there is only a 6 percent chance.

Of course, coalition formation does not generally occur in a random society. Subsets of players, such as political parties in a national legislature, will have members with similar preferences. In such situations, we would expect less-than-grand coalitions to form more frequently and be stable.

I conjecture that the distribution of FB_1 semi-stable majority coalitions in a random society, for which I have not made detailed calculations, is also V-shaped, whether preferences are general or ordinally single-peaked. But instead of the V's being so heavily weighted on the side of the grand coalition—such that the V looks more like a J—preliminary calculations indicate that the FB_1 distribution will be considerably flattened. Thereby, there will be more weight in the middle and around a simple majority.

In the next section, I present empirical data on the distribution of majorities in the U.S. Supreme Court and illustrate coalition formation on the Court with two cases. In addition, I present data on the distribution of majorities in the U.S. House of Representatives, showing that, like the Court, the distribution is bimodal.

8.7. THE FORMATION OF MAJORITIES IN THE U.S. SUPREME COURT

In the 9-person U.S. Supreme Court, majority coalitions fit the bimodal probability distribution found for BU, with majorities tending to be either minimal winning or unanimous. Between 1962 and 1997, the distribution, with the minimum occurring at majority size 7, is as follows:[14]

---

[14] These data are drawn from Edelman and Sherry (2000), who also note the bimodal character of the Supreme Court majority decisions. They explain it in terms of a Markov process of coalition formation, using the Supreme Court voting data to calculate the probability of different absorbing states. By contrast, I suggest a V-shaped distribution on theoretical grounds, independent of any data.

Majority size:	5	6	7	8	9
Percent of cases:	24	21	13	14	27

These statistics, however, do not elucidate the process by which justices actually coalesce, either in divided 5–4 majorities or in consensual 9–0 decisions. For this purpose I consider two examples, one in which the Court was unanimous and the other in which it was sharply divided.

Consider the unanimous decision in *United States v. Nixon* (1974), the infamous White House tapes case that was actually decided by an 8–0 vote.[15] Before this case reached the Court, it looked like it would be contentious, based on an unofficial poll by Justice William Brennan. The four Nixon appointees favored the president's claim of executive privilege on withholding the White House tapes, and four took the other side, with Justice Byron White, who usually kept his counsel, inscrutable (Woodward and Armstrong, 1979, p. 289). Because the decision in this case triggered an unprecedented event— the resignation of a president—Schwartz (1996, p. 145) views it as "the most spectacular case decided by the Burger Court."

In the Court's deliberations, Chief Justice Warren Burger, a Nixon appointee, initially sided with the president on executive privilege but was opposed by the rest of the Court. Acting as a kind of rump committee, the other justices redrafted Burger's original opinion, with different twosomes and threesomes rewriting its seven parts (Schwartz, 1996, p. 147). While *Nixon* was not the only such case of "decision by committee" during the Burger reign—another exception was *Buckley v. Valeo* (1976), an important campaign-finance case (Schwartz, 1996, pp. 143–144)—*Nixon* is particularly insightful on how the buildup toward a final consensus was achieved.

The analysis of *Nixon* that follows does not do justice to Woodward and Armstrong's (1979) 63-page blow-by-blow account, or even Schwartz's (1996) 4-page insider account that includes a page from the personal files of one justice. Although the justices agreed that executive privilege was neither absolute nor unreviewable (especially *in camera*), they differed on how much confidentiality should be accorded presidential conversations and papers.

Because of the paramount importance and extreme public interest in the case, most justices believed that the decision should be the strongest possible—in particular, one delivered as a joint opinion, not written by a particular justice.[16] But

[15] Justice William Rehnquist recused himself because of his earlier service in the Nixon administration.

[16] It was not just a matter of writing a strong opinion; the justices were also extremely concerned that President Nixon would not abide by their decision unless it was "definitive," a term used but never defined by Nixon that was widely interpreted to mean unanimous. Brams (1978, ch. 5) argues that Nixon's implicit threat set up a game between Nixon and his two appointees, Burger and Harry Blackmun, who had to decide whether to make the decision unanimous by siding with the 6-person majority; Nixon, in turn, had to decide whether to comply with the decision or not. Rational strategies in the game are for Burger and Blackmun

Chief Justice Burger refused to go along, saying, "The responsibility is on my shoulders." Schwartz (1996, pp. 145–146) summarizes the situation that then developed: "He [Burger] would prepare the opinion and would circulate its different parts as he finished them. But the drafts he sent around took a more expansive view of presidential power than the others were willing to accept. The Justices refused to go along and virtually wrested the opinion-drafting process from the Chief Justice in order to secure an opinion that they could join."

In granting *certiorari* under a provision that allowed for expedited review of cases "of imperative public importance," five—rather than the usual four—justices needed to agree to review the case. Justice William Brennan took the lead in putting together the votes, making the following calculation (Woodward and Armstrong, p. 290): "He could count on [William] Douglas and [Thurgood] Marshall. Douglas was eager to come to grips with his long-time antagonist. . . . They might well be joined by [Potter] Stewart. . . . [Byron] White could be within reach. Burger was beyond hope. . . . Blackmun was a possible cert vote. . . . It was difficult to tell where [Lewis] Powell stood." In the end, the expedited *certiorari* decision received six votes.

Brennan and Douglas, the core of the coalition against Nixon, worried that the other justices might resent their doing most of the writing if Burger did not go along. Marshall signed on next, and Stewart seemed receptive. But even at the start of deliberations, Brennan found Douglas's draft opinion "rais[ed] issues that were likely to derail consensus" that "did not need to be addressed" (Woodward and Armstrong, 1979, p. 297).

Subsequently, these drafts were modified, but Powell and Stewart still had misgivings, not wanting to put the functionings of government "into a goldfish bowl," exposed to all and undermining the principle of confidentiality. Blackmun had another misgiving, fretting that if the Court reached a consensus against the president, Burger might assign the case to himself. This would raise serious questions about his impartiality if the president were impeached and, subsequently, if there were a trial in the Senate, over which Burger as chief justice would preside (Woodward and Armstrong, 1979, pp. 298, 301).

In the end, a liberal coalition comprising Brennan, Douglas, and Marshall, and a moderate coalition comprising Blackmun, Powell, and White, formed, with Stewart the linchpin that brought the two sides together.[17] He was assisted by Brennan, the greatest consensus builder then sitting on the Court.

---

to side with the majority, and for Nixon to comply, which is, of course, what happened and forced Nixon's resignation seventeen days later. But treating Burger and Blackmun as a single player belies newer evidence, cited here, indicating that Burger alone was the only significant holdout from the majority until the end. For a more informal treatment of strategizing on the Supreme Court that includes a statistical analysis of cases, see Maltzman, Spriggs, and Wahlbeck (2000).

[17] Stearns (2000, p. 236) classifies Blackmun as liberal, though he almost always voted with Burger, whom he classifies as conservative. In fact, Burger and Blackmun had the highest

It is reasonable to suppose that each of the three-member coalition members ranked Stewart fourth, followed by members of the other coalition, with Burger, who held out until the end, ranked last by all the other justices. Under FB, the first majority coalition of five or more members to form would comprise Stewart and either the liberal coalition and one or more conservatives, or the conservative coalition and one or more liberals. The first BU coalition to form would then include all justices except Burger.

Once the other seven justices had reached agreement, however, the pressure was on Burger, who felt that the others had been "merciless" and that he had been "sandbagged" (Woodward and Armstrong, 1979, pp. 340–341). Ultimately he acquiesced, not wanting to be a minority of one—but not without claiming authorship of the "committee" decision, which he made his own.[18]

I next turn to the 5–4 case (*Miller v. California*, 1973). The Supreme Court has considered many obscenity cases over the past forty years, and almost all its decisions have been divided and contentious. Instead of examining the buildup of coalitions on any single case, I jump to the final stage of this decision. With the Court deadlocked 4–4, it turned to Blackmun to cast the fifth and decisive vote. Between the two protagonists, "he [Blackmun] could make his new friend Brennan or his old friend the Chief [Burger] author of the majority opinion" (Woodward and Armstrong, 1979, p. 252).

Blackmun worried that Burger's broad definition of obscenity could lead to the banning of much worthwhile literature. When Blackmun threatened not to support Burger, Burger reluctantly agreed to incorporate a more limited definition of obscenity into his opinion. Blackmun then became Burger's fifth vote; subsequently, Brennan revised his opinion as a dissent (Woodward and Armstrong, 1979, p. 252).

In effect, Blackmun ranked Burger coalition members above Brennan coalition members, and they him, making this 5-member coalition both an FB and a BU majority coalition. But, of course, it did not become a winning coalition until Burger made a concession, illustrating that the rankings presumed in the FB and BU models may not be set in stone.

In fact, players change their minds, sometimes because of a change of heart and sometimes for strategic reasons (as Burger did in the two cases considered here). As an example of a more sincere switch, Brennan renounced all definitions of obscenity after the 1972 case, allying himself with Douglas's more

---

agreement level of any pair of justices in the 1970–1974 period (83.5 percent), which is why they were called the "Minnesota twins."

[18] And how might this decision be described? "It was now virtually impossible to trace the turns and twists the opinion had taken: ideas articulated by Douglas and Powell, modified by Brennan, quickly sketched by the Chief [Burger]; a section substituted by White; a footnote dropped for Marshall; Blackmun's facts embroidered over the Chief's; Stewart's constant tinkering and his ultimatum" (Woodward and Armstrong, 1979, p. 344).

liberal view and ending their sixteen years of disagreement on this issue (Woodward and Armstrong, 1979, p. 253).

There are other voting bodies to which the models seem applicable, including the U.S. House of Representatives. Indeed, as in the Supreme Court, there is a bimodal distribution of majority sizes, based on the 12,688 roll call votes between 1955 and 1990:[19]

Percent majority:	50–60	60–70	70–80	80–90	90–100
Percent of roll calls:	26	19	14	11	30

Although the minimum occurs in the 80–90 percent range, not the middle 70–80 percent range, the two modes are the near-majority and near-unanimity ranges, consistent with the BU model.

In the final section I assess the stability of the grand and smaller majority coalitions in light of both the FB and BU models. I also suggest some empirical observations and data that might be useful in further testing the models and indicate how they might be used to find governing coalitions acceptable to a majority of parliamentary members.

8.8. SUMMARY AND CONCLUSIONS

BU seems most applicable to studying coalition formation in multimember courts and legislatures, in which small subsets of members coalesce and build up to a majority, all of whose members rank each other highest and are therefore stable. FB probably better describes the formation of a governing coalition in parliamentary democracies, wherein disconnected coalitions sometimes form. Because parties in such coalitions rank some parties outside the coalition higher than parties in it, these coalitions are at best semi-stable.[20]

Insofar as voters' preferences are single-peaked, the coalition governments that form are usually connected. Indeed, they are often described by such terms as "left-center" or "center-right." On occasion, however, the left and right do get together and form national-unity governments—sometimes in response to a crisis, like the threat of war—in which many members may be far from each other's favorite coalition partners. In fact, BU coalitions tend to be either grand or minimal-majority coalitions, so the occurrence, on occasion, of national-unity governments is not a surprise.

[19] These data were compiled by David W. Rohde for the Inter-University Consortium for Political and Social Research in January 1995.

[20] Because the significant players in parliamentary democracies are different-size parties, strategic considerations come into play that the FB and BU do not take into account of (see note 1). Data on coalition governments in Western Europe can be found in Müller and Strøm (2000).

Oversized semi-stable coalitions, which may be disconnected, tend not to last. According to Riker's (1962) size principle, some of their members grow disaffected and leave if there are insufficient resources to reward them in an oversized coalition.

Through manipulation, players can disrupt semi-stable coalitions by announcing false preferences. Not all these changes, however, may be purely opportunistic. For example, Jim Jeffords, a U.S. senator from Vermont, switched from the Republican Party to become an independent in 2001, turning the Democratic Party into the majority party in the Senate. He seems to have been motivated by a genuine belief that he could better serve his state of Vermont and his country by changing his party affiliation. By contrast, the preference changes that create bandwagons may not be so sincere.

Patently, coalition-formation processes affect the size and stability of the coalitions they generate. If stability can be measured by durability, then the models may provide insight into why parliamentary coalitions in a country like Italy are less durable than those in the Scandinavian countries, where government coalitions sometimes do not include even a simple majority of members.

The models may also enhance one's understanding of the stability of coalitions in other arenas, including international relations. Some international alliances, like NATO, have been long-lasting, whereas others have been ephemeral. Is the process that led to the former more BU-like, the latter more FB-like? The models, I believe, provide tools for investigating such questions.

Finally, FB and BU can be used as procedures to find parliamentary coalitions acceptable to a majority of members. While parliamentary members may, in principle, be able to rank other individual members, it is likely that political parties would propose that their members rank all members of each party the same. Thus, parties would, in effect, be ranking each other.[21]

FB and BU could then be used to find either semi-stable (FB) or stable (BU) majority coalitions, with the former tending to be smaller than the latter. Such coalitions, while they might reflect prior agreements among parties, would formalize the process of putting together a governing coalition. This process would be less subject to the whims of party leaders and more democratic—at least if parliamentary members are not required to follow party directions in their rankings. In addition, the use of FB or BU would circumvent much of the haggling and delay that often accompany the formation of governments today, which in some countries can stretch into weeks or even months.

---

[21] In city councils or county legislatures with many fewer members than state legislatures or a national parliament—wherein there are not parties if elections are nonpartisan—FB or BU could be used to choose committees with only a few members. For example, the descent process might stop when at least three members have been chosen. In such a case, members will generally not have to rank all other members but only some fraction that is likely to yield a committee of the desired size.

APPENDIX

Proposition 8.7. *BU majority coalitions may appear—or not appear—at any level, up to the appearance of the grand coalition.*

Proof. Assume that there are $n > 3$ players and that $m = \lceil (n+1)/2 \rceil$. The following algorithm positions player i at x_i on the real line. Players' preferences are then defined by Euclidean distance, whereby player i prefers player j over player k if the distance between x_i and x_j is less than the distance between x_i and x_k. The following algorithm constructs cardinally single-peaked preferences in which the only stable majority coalitions are of size k_h, $h = 1, 2, \ldots, t$, where $m \le k_1 < k_2 \cdots < k_t = n$. Note that $t \ge 1$.

Assign $x_1 = 0$. For $1 < i \le m$, let player i's position be defined recursively by $x_i = 2x_{i-1} + 1$. By construction, for $1 < i < m$, player i's single-peaked preferences are

$$i: i-1 \quad i-2 \ldots 2 \quad 1 \quad i+1 \quad i+2 \ldots m-1 \quad m \quad m+1 \ldots n.$$

Notice that player i most prefers player $i - 1$ as a coalition partner.

The construction considers two separate cases. First suppose $t = 1$ (i.e., only the grand coalition is stable and $k_1 = n$.) Let $x_{m+1} = x_m + 1/2$, and define x_i for $i > m + 1$ by $x_{m+k+1} = x_{m+k} + x_{m-k+1} - x_{m-k}$. Since the distance between x_{m+k+1} and x_{m+k} is the same as the distance between x_{m-k+1} and x_{m-k}, it follows that, for $m + 1 < j < n$, player j most prefers player $j + 1$.

Player m has the following preferences, depending on whether n is odd or even, respectively,

$$\begin{array}{llllllll} \textbf{\textit{m} odd:} & m+1 & m-1 & m+2 & m-2 \ldots m+k & m-k \ldots n & 1; \text{ or} \\ \textbf{\textit{m} even:} & m+1 & m-1 & m+2 & m-2 \ldots m+k & m-k \ldots n & 2\ 1. \end{array}$$

Because preferences are single-peaked, any k-stable coalition for $k \ge m$ must contain the median voter(s). Hence, we can focus on player m's preferences.

Any k-stable majority coalition must comprise the first $k - 1$ players in player m's ranking. If $k - 1 = 2l$, then player m ranks player $m - l$ in position $k - 1$ of its ranking. So a k-stable coalition must contain players m and $m - l$. But player $m - l$ most prefers player $m - l - 1$. Since player m does not rank player $m - l - 1$ among its $k - 1$ most-preferred coalition partners (because player $m - l$ is the least-preferred player in the coalition), the coalition is not k-stable. Similarly, if $k - 1 = 2l - 1$, then player m ranks player $m + l$ in position $k - 1$ of its ranking.

A k-stable coalition must contain players m and $m + l$. But player $m + l$ most prefers player $m + l + 1$ as a coalition partner, which is not among player m's top $k - 1$ players. Hence, no majority coalition of size $k < n$ is stable.

Now suppose $t > 1$, so $k_1 < n$. For $m \leq i < k_1 < n$, let j satisfy $i = m + j$. Then player i's position is given by $x_i = x_m$ for $j = 0$ and $x_i = x_{i-1} + \dfrac{1}{2^j}$ for $j \geq 1$. Similarly, for $k_1 < i \leq k_2$, let j satisfy $i = k_1 + j$. Define $x_i = 2x_m + 2$ for $j = 1$ and $x_i = x_{i-1} + \dfrac{1}{2^{j-1}}$ for $j > 1$. Finally, for $k_{s-1} < i \leq k_s \leq n$, let j satisfy $i = k_{s-1} + j$. Player i's position is given by $x_i = 2x_{s-1} + 2$ for $j = 1$ and $x_i = x_{i-1} + \dfrac{1}{2^{j-1}}$ for $j > 1$.

It is possible to describe player i's preferences for $i \geq m$, too. If $k_{s-1} < i \leq k_s$ where j satisfies $i = k_{s-1} + j$, then player i's ranking of coalition partners is

$$i: \quad i+1 \quad i+2 \ldots k_s \quad i-1 \quad i-2 \ldots 2 \quad 1 \quad k_s+1 \quad k_s+2 \ldots n.$$

Any majority stable coalition that contains player i must contain a player j where $1 < j < m$. But, by construction, player j most prefers player $j - 1$. This implies that all of the players less than i must be in the coalition. Since player i's most-preferred coalition partners are players $i + 1$, $i + 2$, $i + 3$, ..., and k_s, only coalitions of the form $\{1, 2, \ldots, k_s\}$ are stable. The same argument holds for i where $m \leq i \leq k_h$ as well. ■

Proposition 8.9. *The probability of a BU coalition, starting at $k = m$, decreases to a minimum at some intermediate value of k before increasing to 1 at $k = n$. More precisely, for each $n \geq 3$, there exists an integer $k_0(n) = k_0$, satisfying $m \leq k_0 < n$, such that $P(n, k + 1) < P(n, k)$ if $m \leq k < k_0$, $P(n, k_0 + 1) \geq P(n, k_0)$, and $P(n, k + 1) > P(n, k)$ if $k > k_0$. Moreover, $k_0(n) > m$ whenever $n \geq 5$.*

Proof. Because each player can rank all other players in $(n - 1)!$ ways, there are a total of $[(n - 1)!]^n$ preference profiles. Suppose a k-coalition is stable, where $2 \leq k \leq n$. Then for each member of the coalition, the first $k - 1$ players in its preference ranking must be the other members of the coalition. It follows that the number of preference rankings admitting a stable coalition of k members is

$$\binom{n}{k} [(k - 1)!(n - k)!]^k [(n - 1)!]^{n-k}. \tag{A1}$$

To justify this formula, note that there are $\binom{n}{k}$ ways to choose the members of the stable k-coalition. For each of the k members of this coalition, the other members (which come highest in its preference ranking) can be arranged in $(k - 1)!$ ways; each of the nonmembers of the coalition, which come lower in its preference ranking, can be arranged in $(n - k)!$ ways. Finally, there are $(n - 1)!$ ways to choose the preference rankings of each of the $n - k$ nonmembers of the coalition.

Formula (A1) is not very useful for small values of k, because it double-counts instances when there are two or more disjoint stable coalitions with k members. But for $m \leq k \leq n$, formula (A1) gives the number of preference rankings, out of $[(n-1)!]^n$, in which a (unique) stable majority coalition exists (see Proposition 8.8).

It follows from the preceding argument that in a random society with n players, the probability that there is a BU coalition with k members, where $m \leq k \leq n$, is

$$
P(n,k) = \frac{\binom{n}{k}[(k-1)!(n-k)!]^k[(n-1)!)]^{n-k}}{[(n-1)!]^n}
$$

$$
= \binom{n}{k}\left[\frac{(k-1)!(n-k)!}{(n-1)!}\right]^k = \frac{\binom{n}{k}}{\binom{n-1}{k-1}^k}. \tag{A2}
$$

By construction, $0 < P(n, k) \leq 1$. Also, it is easy to check that $P(n, n) = 1$.

If $k < n$, it follows from (A2) that

$$
\frac{P(n,k+1)}{P(n,k)} = \frac{\binom{n}{k+1}\binom{n-1}{k}^{-(k+1)}}{\binom{n}{k}\binom{n-1}{k-1}^{-k}}
$$

$$
= \frac{\binom{n}{k+1}}{\binom{n}{k}}\left[\frac{\binom{n-1}{k-1}}{\binom{n-1}{k}}\right]^k \frac{1}{\binom{n-1}{k}}
$$

$$
= \frac{n-k}{k+1}\left(\frac{k}{n-k}\right)^k \frac{1}{\binom{n-1}{k}}. \tag{A3}
$$

From (A3) it follows that $P(n, k + 1) \geq P(n, k)$ if and only if

$$
\binom{n-1}{k} \leq \frac{n-k}{k+1}\left(\frac{k}{n-k}\right)^k. \tag{A4}
$$

Moreover, $P(n, k + 1) > P(n, k)$ if and only if strict inequality holds in (A4).

Now suppose that k satisfies $m \leq k < n$ and that $P(n, k + 1) \geq P(n, k)$. Therefore, (A4) holds. One can show that $P(n, k + 2) > P(n, k + 1)$ by showing that strict inequality holds on the right side of (A4) when $k + 1$ is substituted for k. Because $k + 1 > k > n/2$, it follows that

$$\binom{n-1}{k+1} < \binom{n-1}{k} \leq \frac{n-k}{k+1}\left(\frac{k}{n-k}\right)^k < \frac{n-k-1}{k+2}\left(\frac{k+1}{n-k-1}\right)^{k+1}. \quad \text{(A5)}$$

The final equality of (A5) is true because

$$\frac{k}{n-k} < \frac{k+1}{n-k-1}$$

and

$$\frac{n-k}{k+1} < \frac{n-k-1}{k+2}\left(\frac{k+1}{n-k-1}\right) = \frac{k+1}{k+2}. \quad \text{(A6)}$$

To verify the inequality of (A6), note that it is equivalent to

$$2k^2 - (n-4)k - (2n-1) > 0,$$

which is easily shown to be true because $k > n/2$.

   It is not difficult to show that when $n \geq 5$ and $k = m$, the right side of (A4) is approximately equal to $e^{3/2} \approx 4.48$. Therefore, k_0, the minimum value of k for which (A4) holds, exceeds m if and only if $n \geq 5$. ∎

9

Allocating Cabinet Ministries in a Parliament

9.1. INTRODUCTION

How coalition governments in parliamentary democracies form and allocate cabinet ministries to political parties is the subject of a substantial theoretical and empirical literature. By and large, a rule of proportionality, whereby parties are given more ministries or more prestigious ministries (e.g., finance, foreign affairs, or defense) in proportion to their size, is followed. However, small centrist parties that are pivotal in coalitions (e.g., the Free Democrats in Germany) have successfully bargained for larger-than-proportional allocations (Browne and Dreijmanis, 1982; Budge and Keman, 1990; Warwick, 2001; Warwick and Druckman, 2001).

The degree to which party leaders are satisfied with their allocations will depend on their ability to use the ministries as levers of power to advance their programs or personal ambitions (Laver and Shepsle, 1994, 1996; Warwick, 2001). Their success renders coalition defections less likely and governments more durable—as measured, for example, by their tenure in office (Warwick, 1994; Grofman and van Roozendal, 1997).[1]

Today the governments of most parliamentary democracies tend to be either center-left or center-right. Even when there is ideological coherence, however, the task of putting together a coalition that will (i) agree on major policy decisions, and (ii) meet with public favor can be a formidable task in the absence of a formal allocation mechanism (see chapter 8).

This task is complicated when less-than-compatible parties, like the Christian Democrats and the Greens, join the same coalition. While fiscal conservatism and protecting the environment are often at odds, these parties may still be accommodated if, for instance, the Christian Democrats are given the

---

Note: This chapter is adapted from Brams and Kaplan (2004) with permission of Sage Publications Ltd.

[1] The tenure of a government, however, is not the only gauge of satisfied party leaders and stable government. Italy has had more than sixty coalition governments since World War II, but some parties (e.g., the Christian Democrats) have been regular members of almost every government. Their leaders have been little affected by the rise and fall of different governments and, therefore, probably remain quite satisfied.

finance ministry and the Greens the environmental-protection ministry, and each has major influence over policies in its area.

To facilitate the allocation of cabinet ministries to political parties, I propose procedures that take into account both party interests and party size. They rely on a mechanism that shifts the burden of making cabinet choices from the prime minister designate, or *formateur*, who is usually the leader of the largest party in a coalition government, to party leaders that join the government.[2] Thereby these procedures give party leaders primary responsibility for the makeup of the coalition government.

I assess the fairness of this procedure, based on different criteria of fairness. This analysis is inspired by O'Leary, Grofman and Elklit (2005), who analyzed an apportionment method that has been used in Northern Ireland since 1999, and in four Danish cities that commenced with Copenhagen's use of it in 1938. This method is also used to select committee chairs in the European Parliament.

This method determines the *sequence* in which parties make ministry choices. It works such that the largest party in a coalition gets first choice; presumably, it would choose the position of prime minister, mayor, or other top executive position. After that, the apportionment method determines the order of choice.

For example, suppose there are three parties, ordered by size A > B > C, and there are six ministries to be allocated. If the sequence is ABACBA, A will receive three ministries, B two ministries, and C one ministry. But beyond these numbers, the sequence says that A is entitled to a second choice before C gets a first choice, and C gets a first choice before B gets a second choice.

If parties have complete information about one another's preferences, I show that it may not be rational for them to choose *sincerely*—that is, to select their most-preferred ministry from those not yet chosen. Rather, a party (e.g., A) may do better postponing a sincere choice and, instead, selecting a less-preferred ministry if (i) that ministry might be the next choice of a party that follows it in the sequence (e.g., B or C), and (ii) A's sincere choice is not in danger of being selected by B or C before A's turn comes up again. Such *sophisticated choices*, which take into account what other parties desire, can lead to allocations very different from sincere ones.

---

[2] In some models, the choice of the *formateur* (e.g., a president or monarch) is assumed to be strategic; conditions under which such an agenda setter can be decisive in selecting a preferred minimal winning coalition are analyzed in Bloch and Rottier (2002) and articles cited therein; see also Tsebelis (2002, ch. 4). For a theory and empirical tests of coalition formation in multiparty systems, see Schofield and Sened (2006); for an empirical analysis of the effects of voting and voting rules on political behavior in more than thirty countries, see Norris (2004). Issues in the design of electoral systems to mitigate interethnic conflict and to create cross-cutting coalitions, especially in new democracies, are analyzed in, among other places, Reilly (2001), and debated in Diamond and Plattner (2006).

If there are only two parties, sophisticated choices and sincere choices both yield efficient or Pareto-optimal allocations: no parties, by trading ministries, can do better, based on their ordinal rankings of ministries. However, this is not true if there are three or more parties that make sophisticated choices, which was first demonstrated for sequential choices made in professional sports drafts (Brams and Straffin, 1979).

Another problem that crops up is that of *nonmonotonicity*. A political party may do worse by choosing earlier in a sequence, independent of the efficiency or Pareto-optimality of the sophisticated choices. Hence, the apparent advantage that a party's size gives it by placing it early in a sequence can, paradoxically, work to its disadvantage—it may actually get preferred choices by going later.

Like Pareto-nonoptimal allocations, nonmonotonicity *cannot* occur if parties are sincere. Thus, it is fair to ask how sincere choices might be recovered—or induced in the first place if the parties know they cannot "get away with" insincere choices. While there is an allocation mechanism that makes sincerity optimal for two parties, there are difficulties in extending it to more than two parties.

By putting the choice of ministries in the hands of party leaders, these leaders are made responsible for their actions. Ultimately, I believe, party leaders will be more satisfied making their own choices rather than having to bargain for them. Moreover, this greater satisfaction should translate into more stable coalition governments, which is a subject that has been extensively studied by many scholars, including Taagepera and Shugart (1989), Laver and Shepsle (1994, 1996), Blondel and Cotta (1996), van Deemen (1997), Laver and Schofield (1998), and Müller and Strøm (2000).

The allocation procedures analyzed here could go a long way toward minimizing the horse trading that typically ensues when a *formateur* bargains with party leaders over the ministries they will be offered. By cutting down on the rents extracted in the bargaining process, a coalition government is likely to form more expeditiously and be less costly to maintain. In addition, a government that so forms is not subject to a vote of confidence before it is installed in office.

This is not to say that the procedures I discuss solve all problems. Because ministries are indivisible, there will not generally be a perfect match of the claims of each party in a governing coalition and its allocation. Furthermore, there are certain problems that are ineradicable, whatever allocation procedure is used. For example, it may not be possible to make an equitable assignment among two equally entitled parties—one may be allotted ministries it ranks higher than the other. Nevertheless, the procedures I describe offer a promising start to attenuating conflicts that have plagued the formation of coalition governments and, not infrequently, led to their downfall.

9.2. APPORTIONMENT METHODS AND SEQUENCING

Most coalition governments comprise political parties holding a majority of seats in a parliament, though several parliamentary democracies since World War II have had one or more minority governments (Strøm, 1990). The leader of the largest party is usually offered the opportunity to put together a coalition of parties with a majority of seats. This may take days, weeks, or even months to complete (Müller and Strøm, 2000; Carmignani, 2001).

Typically, the bargaining to form a coalition begins soon after an election, which fixes the number of seats each party has in a parliament. Of course, even after a coalition has formed, there can be problems in keeping it together, as frequent government failures in Israel and Italy have demonstrated.

The number of cabinet ministries is usually a small fraction of the number of parliament members in the government coalition. Although most parliament members in the coalition cannot, therefore, be ministers, each party in the coalition is usually awarded at least one ministry.

Under the procedures to be analyzed, there is no bargaining over either the number of ministries to which each party is entitled or which ministries it will obtain.[3] To illustrate how ministries are allocated, suppose a parliament has 400 members, so 201 members constitute a simple majority. Suppose five parties, with a total of 201 members, form a coalition government. If 15 ministries are to be allocated, there will be no problem in giving every party at least one ministry if the smallest party in the coalition has 15 members. For even though this party constitutes only 7.5 percent (15/201) of the coalition, this percentage would entitle it to 1.12 of the 15 ministries.

But now assume that smallest party has only 10 members, entitling it to almost 5 percent (10/201) of the 15 ministries, or about 0.75 of a ministry. Should it receive a ministry, or will a larger party be more entitled to it? This is precisely the question an apportionment method answers.

I focus on so-called *divisor* apportionment methods, which have been used to apportion seats in the U.S. House of Representatives to the fifty states, based on their populations. These methods are also used to allocate seats to political parties in most parliaments, based on the votes the parties receive in an election.

Here I propose using these methods to allocate *ministries* (sometimes called portfolios) to political parties, based on the numbers of seats the parties have in parliament. Beyond determining the number of ministries to which a party is entitled, however, I analyze the use of divisor apportionment methods to

---

[3] But *which* parties decide to form a government may well depend on how compatible they are in a coalition; the procedures discussed in chapter 8 are intended to find the most compatible coalitions.

TABLE 9.1
Webster and Jefferson Apportionments

Party	A	B	C	D	E	Total
Seats	75	65	30	21	10	201
Percent	37.3	32.3	14.9	10.4	5.0	100.0
Quota ($d = 13.4$)	5.60 (6)	4.85 (5)	2.24 (2)	1.57 (2)	0.75 (1)	15.0 (16)
Webster ($d = 13.8$)	5.43 (5)	4.71 (5)	2.17 (2)	1.52 (2)	0.72 (1)	(15)
Jefferson ($d = 10.8$)	6.94 (6)	6.02 (6)	2.78 (2)	1.94 (1)	0.93 (0)	(15)

determine the sequence in which parties choose the ministries they desire, tying this application to more theoretical research on fair division and sequential-allocation procedures. Unlike O'Leary, Grofman and Elklit (2005), I propose emendations in the apportionment and allocation procedures used in Northern Ireland and Denmark, such as allowing the parties to trade ministries after their initial selections.[4]

To begin the analysis, I illustrate the two most commonly used divisor methods—one proposed by Daniel Webster and the other by Thomas Jefferson—that have been used to apportion parliaments (Balinski and Young, 1982, 2001). Assume that five parties, A, B, C, D, and E, hold the numbers and percentages of seats shown in the first and second rows of Table 9.1.

To determine the number of ministries to which each party is entitled, a divisor d is chosen. Initially, let $d = 201/15 = 13.4$, the number of government coalition members per ministry. Dividing this number into the number of seats of each party gives its *quota*, or the exact number of ministries to which it is entitled (third row). For example, A is entitled to 5.60 ministries, whereas E is entitled to 0.75 of a ministry (as discussed earlier).

Because it is persons that become ministers, fractional numbers of ministries cannot be given to parties. Hence, one needs to find a way to round the quotas, either up or down, to integers.

Suppose the quotas are rounded in the usual manner: Numbers with decimals greater or equal to 0.50 are rounded up, and those with smaller decimals are rounded down. Observe that these rounded quotas for the five parties, shown in parentheses in the third row of Table 9.1, sum to 16, which is one ministry more than the number (15) to be allocated.

This excess-ministry problem can be solved by slightly increasing d from 13.4 to 13.8 (all values of d in the interval [13.64, 14.00] work), as shown in the

---

[4] But see the recommendations in McGarry and O'Leary (2006a, 2006b), based on their assessment of the Northern Ireland experience since 1999. In these articles, the authors seek to determine the degree to which a theory of conflict and cooperation in ethnically, religiously, and linguistically divided societies ("consociationalism") explains what has occurred in Northern Ireland in recent years. They conclude that both proponents and critics of this theory need to revise their views in light of the evidence from Northern Ireland.

fourth row. Thereby A's quota drops from 5.60 to 5.43, reducing its allocation from 6 to 5 seats, without changing the allocations of the other parties. Finding a d such that the allocations for all parties, when rounded, sum to the total number of ministries to be allocated is an apportionment method proposed by Daniel Webster in 1832 and independently by André Sainte-Laguë of France in 1910 (Balinski and Young, 1982, 2001). For convenience, I will refer to it as the Webster method.

Another well-known divisor method, proposed by Thomas Jefferson in 1792 and by Viktor d'Hondt of Belgium in 1878, uses a different rounding procedure. It finds a d such that only the integer portions of the quotas sum to the total number of ministries. Put another way, the quotas are all rounded down, whatever their fractional remainders.

In the example in Table 9.1, d needs to be decreased quite substantially from 13.4, which gives integer portions that sum to 12, to 10.8 (all values of d in the interval [10.72, 10.83] work). When this is done, one obtains new allocations that, when rounded down, sum to 15, as shown in parentheses in the last row of Table 9.1.

Notice that the Jefferson method favors the two largest parties (A and B get 6 rather than 5 ministries each), whereas the Webster method favors the two smallest parties (D and E get, respectively, 2 and 1 ministries rather than the 1 and 0 ministries they get under the Jefferson method). In general, the Jefferson method favors large parties, which is a bias that Balinski and Young (1982, 2001) consider appropriate in apportioning seats to parties in a parliament.

By giving parties an incentive to combine to obtain more seats, the Jefferson method encourages coalitions. Thereby it helps to thwart the fractionalization of parties common in proportional-representation (PR) systems (Rae and Taylor, 1970), especially PR systems in which the threshold for parliamentary representation is low (e.g., 0.67 percent in The Netherlands and 1.5 percent in Israel). By contrast, Balinski and Young (1982, 2001) argue that the Webster method is better for apportioning seats to states in the U.S. House of Representatives because it is the least biased of the five divisor methods (more on these shortly), favoring neither large nor small states.

However, all five divisor methods—each of which uses a different rounding procedure—have been criticized because they *violate quota* (Brams and Straffin, 1982). They can give a state more than its quota rounded up or less than its quota rounded down. As a case in point, B's quota is 4.85 ($d = 13.4$), but the Jefferson method gives it six ministries.[5]

---

[5] An apportionment method proposed in 1792 by Alexander Hamilton satisfies quota (Balinski and Young, 2001). But it is not useful for purposes of allocating cabinet ministries, because, unlike the divisor methods, it does not yield a sequence of choices by the parties (illustrated next for the Webster and Jefferson methods). For a recent review of apportionment methods, see Edelman (2006a), who proposes a nondivisor method that is mathematized in Edelman (2006b).

My purpose is not to argue for a particular apportionment method but, instead, to show how the two divisor methods used in the allocation of seats to parliaments—those of Webster and Jefferson—lead to a natural order in which the parties select ministries. For this purpose, let

$$s_i = \text{number of seats party } i \text{ receives;}$$

$$a_i = current \text{ allocation of ministries to party } i.$$

Each apportionment method asks which party most "deserves" the next ministry to be allocated, which gives insight into how ministries can be allocated in sequence rather than all at once (as given by the previously discussed rounding conventions).

As one criterion of deservingness, the Webster method gives the next ministry to the party that maximizes $W_i = s_i/(a_i + 1/2)$, or the number of seats per ministry if party i receives the next 1/2 ministry.[6] Clearly, if no ministries have yet been allocated, A gets to make the first choice, because $s_i/(1/2) = 2s_i$ is greater for A than for the four smaller parties.

But which party gets to make the next choice? Because $W_B = 65/(1/2) = 130$ and $W_C = 30/(1/2) = 60$ are greater than $W_A = 75/(3/2) = 50$, B and C make the next two choices. But then A gets to choose a second time before D makes a first choice because $W_A = 50$ is greater than $W_D = 21/(1/2) = 42$. Continuing in this manner, the Webster method gives the following sequence of fifteen party choices (the slash between E and C indicates a tie):

Webster: ABCABDABAE/CBABD   (5 A's, 5 B's, 2 C's, 2 D's, 1 E).

The Jefferson method gives the next seat to the party that maximizes $J_i = s_i/(a_i + 1)$, or the number of seats per ministry if party i receives the next ministry. As with the Webster method, if no ministries have yet been allocated, A gets to make the first choice because $s_i/1 = s_i$ is greater for A than for the four smaller parties.

---

[6] To be sure, receiving half a ministry is not possible; the 1/2 in the denominator of W_i reflects the rounding of fractions equal to or greater than 1/2 under the Webster method. Under the Jefferson method, as I will show, the 1/2 is replaced by 1. These constants, when added to a_i in the denominators of W_i (and later J_i), lead to *stable allocations*: no transfer of a ministry from one party to another reduces the inequality in representation among parties, based on different measures of inequality. The three divisor methods, in addition to those of Webster and Jefferson, that produce stable allocations are all based on different criteria of deservingness; all favor smaller parties in varying degrees (Balinski and Young, 2001; Marshall, Olkin, and Pukelsheim, 2002). Their deservingness measures (I have dropped the subscript i used in W_i and J_i) are $s/[a(a + 1)]^{1/2}$ for Hill or "equal proportions," $s/[(2a(a + 1)/(2a + 1)]$ for Dean or "harmonic mean," and s/a for Adams or "smallest divisors." Whereas the Webster method is used in four Scandinavian countries, and the Jefferson method in eighteen other countries, none of the other three divisor methods is used. The nondivisor Hamilton method, also called "largest remainders" (see note 5), is used in nine countries (Blais and Massicotte, 2002; Cox, 1997).

But what happens next? After A gets its first ministry, $J_A = 75/2 = 37.5$, which is less than $J_B = s_B = 65$ but not $J_C = s_C = 30$. Thus, B makes a first choice after A does, but then A gets to make a second choice before C makes a first choice. Continuing in this manner, the Jefferson method gives the following sequence of party choices (the slash between C and A indicates a tie):

Jefferson: ABABCABDABC/ABAB   (6 A's, 6 B's, 2 C's, 1 D, 0 E's).

Comparing the sequences of Webster and Jefferson, notice that the two largest parties, A and B, come up earlier as well as more frequently under Jefferson, whereas the opposite is true of the two smallest parties, D and E; C appears twice under both apportionment methods. The fact that E does not appear at all under the Jefferson method does not mean that it would necessarily go unrewarded. Its reward, however, might be something other than a ministry; or ministerial portfolios might be added so that E receives one.[7]

9.3. SOPHISTICATED CHOICES

The choice between Jefferson, which favors large parties, and Webster, which is more neutral, will depend on the degree to which one wishes (i) to encourage coalitions by giving more ministries to large parties, or (ii) to render allocations more proportional to the size of parties. Besides the choice of an apportionment method, however, political parties can choose between acting sincerely or sophisticatedly.

The latter choices exploit the parties' possible knowledge of one another's preferences, which we illustrate with a simple 2-party example. Suppose there are two parties, {A, B}, and four ministries to be allocated, {1, 2, 3, 4}. In addition, suppose A is somewhat larger than B, so the choice sequence is ABAB. Assume the parties rank the ministries from best to worst as shown in Example 9.1a. This might be the case, for example, if ministry 1 is the defense ministry, and B's interest is primarily in domestic politics.

Example 9.1a. Rankings

A	B
1	2
2	3
3	4
4	1

[7] If one believes that small parties should be favored, a divisor method that does so, like Adams' method (Balinski and Young, 1982, 2001), could be used. But whatever the method used, not all sequences—including some that might be considered fair—are feasible, which is a matter I return to later.

A party makes *sincere choices* if, when it selects a ministry, it chooses the best one available, based on its preference ranking. Define a pair of consecutive choices by each party as a *round*. Then the sincere choices of A and B on the 1st round (single underscore) and 2nd round (double underscore) are as shown in Example 9.1b.

Example 9.1b. Sincere Choices (ABAB)

	A	B
1st round →	1	2
	2	3
2nd round →	3	4
	4	1

Now assume the parties have *complete information*—each knows not only its own ranking but also the other party's ranking.[8] In this situation, choosing sincerely may not be rational. In fact, it is not in this example.

To see this, note that because A knows that B does not want ministry 1, A can choose ministry 2 on the 1st round and not fear the loss of ministry 1. If A's and B's goals are to benefit themselves, B has no defense against A's strategy and can do no better than choose ministry 3 on the 1st round. The resulting choices are as shown in Example 9.1c.

Example 9.1c. Sophisticated Choices (ABAB)

	A	B
2nd round →	1	2
1st round →	2	3
	3	4
	4	1

---

[8] Alternatively, parties might have *incomplete information*, which can be modeled by assuming their leaders have beliefs about the probability distributions of the preferences of other party leaders, and these beliefs are known. This is the approach that has been taken in modeling the exercise of peremptory challengers in jury selection when each side has incomplete information about prospective jurors (Roth, Kadane, and DeGroot, 1977; Brams and Davis, 1978). While incomplete information might make party leaders more circumspect about trying to manipulate choices in the manner I discuss next, it is unlikely that it would eliminate the problems of Pareto-nonoptimality and nonmonotonicity that are analyzed in section 9.4. Indeed, incomplete information may aggravate these problems; moreover, incomplete information about party preferences may also exacerbate the problems that a mediator or arbitrator would have in achieving envy-free or equitable allocations. These are discussed in section 9.7 and later chapters and may, even with complete information, be difficult if not impossible to realize.

These choices are *optimal*: by making different choices on a round, neither party can ensure itself of a better allocation and could, in fact, do worse.

For the general situation of two parties' making sequential choices with complete information, there is an elegant algorithm for optimal play (Kohler and Chandrasekaran, 1971). It works "from the bottom up" as follows. B's *final* choice will be the ministry ranked last by A. Cross that ministry off both parties' preference lists, thereby reducing each list by one ministry.

A's *final* choice is the ministry that is ranked last on B's reduced list. Cross that ministry off. Continue in this fashion, with each party's next-higher choice being the ministry that is last in the reduced list of the other party, until all ministries are chosen.

To illustrate this algorithm, consider Example 9.2 (3rd-round choices are shown in boldface). Working backward from B's last choice of ministry 6 on the 3rd round, one can show that A will select its third choice of ministry 3 on the 1st round. In toto, A obtains its top three ministries of {1, 2, 3}—or 123 for short—which is preferable to its sincere allocation of 125, whereas B does worse than its sincere allocation of 346 by obtaining 465.

Example 9.2. Sophisticated Choices (ABABAB)

	A	B
2nd round →	1	3
3rd round →	**2**	4
1st round →	3	1
	4	**6**
	5	5
	6	2

An *assignment A* is *efficient* (or *Pareto-optimal*) if there is no different assignment A' such that every party that receives a different allocation in A'

- can match a new ministry it gets in A' to a different old ministry it gets in A
- for each such match, weakly prefers the new ministry in A'
- for at least one match, at least one party strictly prefers the new ministry in A'.

Note that this definition implies that the two assignments that are compared must each have the same number of ministries. In the earlier examples, it is not hard to show that both the sincere and the sophisticated assignments are Pareto-optimal.[9]

---

[9] This is true in comparing individual items, but it need not be true in comparing bundles of items. For example, if parties A and B both rank four ministries in the order 1 2 3 4, and A

To illustrate a Pareto-nonoptimal assignment, suppose that A's allocation is 234 and B's is 165 in Example 9.2. Then

- A prefers its sophisticated allocation 123, because it strictly prefers 1 to 4.
- B prefers its sophisticated allocation 465, because it strictly prefers 4 to 1.

Hence, by trading ministries 1 and 4, both parties can improve their allocations.

If there are only two parties, both sincere and sophisticated choices always lead to Pareto-optimal assignments, as the preceding examples illustrate. If there are more than two parties, assignments are always Pareto-optimal if the parties are sincere (Brams and Straffin, 1979). If the parties are sophisticated, however, two problems arise.

9.4. THE TWIN PROBLEMS OF NONMONOTONICITY AND PARETO-NONOPTIMALITY

Most government coalitions include more than two parties. If there are only a handful, as is usually the case, it is likely that party leaders know well one another's rankings of ministries, especially if there are relatively few ministries to allocate.

To illustrate the problems that can arise when parties have complete information and are sophisticated, assume that there are exactly three parties, and their leaders must choose among six ministries. If the size ordering of the parties is A > B > C, but they are close in size, the choice sequence will be ABCABC under both the Webster and Jefferson methods.

Suppose the three parties rank the six ministries as shown in Example 9.3a. Although there is no simple algorithm, like that of Kohler and Chandrasekaran

---

receives 14 and B receives 23, this assignment is *not* what might be called *bundle* Pareto-optimal if A prefers the bundle 23 and B prefers the bundle 14. Unequal allocations may also be bundle Pareto-optimal. In Example 9.1, both players will prefer the unequal assignment (134, 2) to the sophisticated assignment (12, 34) if A prefers 34 to 2 and B prefers 2 to 34. If each party receives three or more items, bundle Pareto-nonoptimality can occur under the divisor apportionment methods. For example, assume A and B both rank six ministries in the order 1 2 3 4 5 6, and the sequence is ABABAB. Then the sincere/sophisticated outcome is (135, 246) to (A, B). But if A prefers 24 to 15, and B 15 to 24, this allocation is not bundle Pareto-optimal, because there is a trade (24 to A for 15 to B) that would make both parties better off, even though it does not involve switching their entire allocations. Henceforth, I assume Pareto-optimality to be as defined in the text, involving item-by-item comparisons but not the comparison of bundles. Note that if an assignment is Pareto-nonoptimal, it is bundle Pareto-nonoptimal (the bundles that the parties would benefit from trading are individual items; see the subsequent text for an illustration) but not necessarily the reverse.

(1971), for determining sophisticated choices when there are three or more parties, it is not difficult to show that the sophisticated choices in this example are in fact sincere—namely, (A, B, C) receive (12, 65, 43), as indicated in their 1st-round and 2nd-round choices in Example 9.3b.

Example 9.3a. Rankings

A	B	C
1	6	4
2	2	3
3	1	6
4	5	5
5	4	1
6	3	2

Example 9.3b. Sincere and Sophisticated Choices (ABCABC)

	A	B	C
1st round →	1	6	4
2nd round →	2	2	3
	3	1	6
	4	5	5
	5	4	1
	6	3	2

How one shows that these choices are sophisticated is by constructing a game tree in which A can choose among its top three choices. Given these, B can choose among its top three choices that remain after A's choice; and C can choose among its top three choices that remain after A's and B's choices.

On the 1st round, a party need never dip below its third choice, because the *best* allocation it can hope for if it does so is its first and fourth choices—14 in the case of A. If A selected ministry 1 on the 1st round, 14 would be the *worst* allocation it could obtain. Hence, choosing ministry 1 on the 1st round *weakly dominates* A's choosing ministries 4, 5, or 6 on this round. This choice is never worse, and sometimes better, than choosing ministries 4, 5, or 6.

Thus in Example 9.3a, this means that only $3 \times 3 \times 3 = 27$ branches of the game tree need be analyzed rather than $6 \times 5 \times 4 = 120$ branches if all available choices are considered. I will not display any game trees here but instead give intuitive explanations of why *backward induction* on the trees, which involves players' reasoning backward from their last possible choices to anticipate one another's earlier choices in order to make their own later optimal choices, produce the sophisticated outcomes to be presented (these calculations have been done on a computer).

To illustrate the problems of Pareto-nonoptimality and nonmonotonicity, suppose that C becomes the largest party and A the smallest in Example 9.4a, so the previous choice sequence is reversed and becomes CBACBA. Then the parties' sincere choices are as shown in Example 9.4a. Somewhat surprisingly, middle-size party B, whose two positions (second and fifth) in the sequence do not change, now gets its two best choices, whereas A, as one would expect, does worse (C does the same—it cannot improve on its two best ministries under ABCABC, even though it chooses earlier).

Example 9.4a. Sincere Choices (CBACBA)

	C	B	A
1st round →	<u>4</u>	<u>6</u>	<u>1</u>
2nd round →	<u>3</u>	<u>2</u>	2
	6	1	3
	5	5	4
	1	4	<u>5</u>
	2	3	6

This time, however, the worst-off party (A) under CBACBA can escape from its poor allocation by choosing insincerely on the 1st round. By selecting ministry 3 rather than ministry 1 initially (choosing ministry 2 does not help), A gets 13 when the other two parties are sincere. But the resulting allocation for C, 45, is poor; it can respond optimally by choosing ministry 6 on the 1st round (choosing ministry 3 does not help), in which case it gets 46 when the other parties respond optimally by choosing sincerely thereafter.[10] The upshot is the sophisticated choices shown in Example 9.4b.

---

[10] Why does choosing ministry 3 (C's second preference) not help? If B subsequently chooses 6, it is optimal for A to choose 4, in which case C ends up with 35, which is worse for it. On the other hand, if B subsequently chooses 2 (it will never choose 1 initially), it is optimal for A to choose either 1 or 5, in which case C ends up with 43, which is better for it. In other words, C may do either worse or better than 46, depending on what B does next. What B does next, however, does not change its 62 allocation, so it will be indifferent. I assume in a situation with multiple Nash equilibria that the equilibrium selected is that which maximizes a party's minimal allocation. Since 64 when C chooses 6 initially is better than 35 that C may receive if it chooses 3 initially, it will choose 6 initially. (It will not sincerely choose 4 initially for similar reasons.)

Example 9.4b. Sophisticated Choices (CBACBA)

	C	B	A
2nd round →	4	6	1
	3	2	2
1st round →	6	1	3
	5	5	4
	1	4	5
	2	3	6

The switch from choice sequence ABCABC (Example 9.3b) to CBACBA . (Example 9.4b), when the largest party (A) becomes the smallest and the smallest party (C) becomes the largest, leads to two anomalies when players are sophisticated:[11]

1. *Nonmonotonicity*: C does worse going earlier—when it moves up from positions 3 and 6 in sequence ABCABC to positions 1 and 4 in sequence CBACBA (a total of four position changes ahead).
2. *Pareto-nonoptimality*: All players—not just C—do worse when the choice sequence changes.

Brams and Straffin (1979), using a different example, showed that under-lying (2) is an *n*-person Prisoners' Dilemma, whereby all players are worse off when they make sophisticated choices in a different sequence.[12] To be sure, if a change in a choice sequence hurts all players, as is true going from sophisticated choices in Example 9.3b to sophisticated choices in Example 9.4b, it necessarily creates a nonmonontonicity problem for the party or parties whose positions move up in the sequence but do worse as a result of the change.

But there may also be a nonmonotonicity problem when the sequence change leads to *another* Pareto-optimal assignment, as illustrated in Example 9.5 in which the choice sequence is ABCABC. Like C in Example 9.4b, A begins by choosing its third-best ministry (3). This time, however, the third player to choose (C) is also not sincere, selecting ministry 5 on the 1st round in order to obtain its best ministry (2) on the 2nd round.

---

[11] Neither of these problems can occur if there are only two parties, however many choices they have. Nonmonotonicity is precluded by the Kohler and Chandrasekaran (1971) algorithm, because it gives priority to a party that makes an earlier selection; Pareto-optimality is shown by Brams and Straffin (1979).

[12] They also showed that the sophisticated outcome, compared with the sincere outcome, can make all players worse off without a change in the sequence, which is not true in Example 9.3b.

Example 9.5. Sophisticated Choices (ABCABC)

	A	B	C
2nd round →	1	6	2
	2	3	3
1st round →	3	5	5
	4	4	6
	5	2	1
	6	1	4

Now when the sequence changes to ABCCBA, as shown in Example 9.6, the new sophisticated outcome is also Pareto-optimal. Notice that because the 2nd-round sequence is CBA, C immediately gets a 2nd-round choice after its 1st-round choice.[13] The switch to ABCCBA causes A to select its second best-ministry (2) on the 1st round, but the parties are sincere thereafter.

Example 9.6. Sophisticated Choices (ABCCBA)

	A	B	C	
	1	6	2	
1st round →	2	3	3	
	3	5	5	← 2nd round
	4	4	6	
	5	2	1	
	6	1	4	

Observe that when C's second choice in sequence ABCABC moves up two positions in sequence ABCCBA, its allocation worsens from 25 to 35; also, when A's second choice moves down two positions after the change, its allocation improves from 13 to 12 (B's allocation, 64, is the same in both assignments). Because position changes *both* up and down in the choice sequence lead to nonmonotonicity, I call this *two-sided* nonmonotonicity.

This phenomenon cannot occur if the new assignment is Pareto-nonoptimal, as in Example 9.4b, because *no* party can benefit after the change. This example illustrates that one-sided nonmonotonicity is possible in a sequence given by an apportionment method.

[13] One should be mindful that no apportionment method can give choice sequence ABCCBA, which is an issue I discuss in section 9.5.

9.5. POSSIBLE SOLUTIONS: TRADING AND DIFFERENT SEQUENCING

One solution to the Pareto-nonoptimality problem would be to allow the parties to trade ministries after they make sophisticated choices. However, no simple two-party trades can save all parties from the Pareto-nonoptimal outcome in Example 9.4b. It takes a three-way trade in which A gives ministry 3 to C, C gives ministry 6 to B, and B gives ministry 2 to A to restore the sincere/sophisticated Pareto-optimal choices of ABCABC (Example 9.3b), wherein A and C get their two best ministries and B gets its best and fourth-best ministries.

The nonmonotonicity of Example 9.4b is actually more extreme than I indicated. Compare the choice sequence AABBCC with CACBAB, wherein C moves up a total of seven positions. It turns out that the first sequence gives the same sincere/sophisticated outcome as does ABCABC in Example 9.3b, and the second sequence the same sophisticated outcome as CBACBA in Example 9.4b. The fact that C does worse when it chooses first and third than when it chooses fifth and sixth—and gets its two best ministries in the latter case!—is startling, suggesting that there are circumstances when "the last shall come first."[14]

In fact, however, such an extreme case of nonmonotonicity cannot occur under the apportionment methods. Thus in the case of CACBAB, there are *no* party sizes that can lead to this sequence under either the Jefferson or the Webster methods. Although the subsequence CACBA—comprising the first five positions in sequence CACBAB—are feasible if C is the largest party, A the next-largest, and B the smallest, B cannot be in the sixth position; C must come next.[15] Similarly, choice sequence AABBCC also cannot occur under the Jefferson or Webster methods.

The infeasibility of certain choice sequences, because apportionment methods preclude them, raises a serious question about their use in parties' choosing cabinet positions sequentially, even when there is no monotonicity problem. Take the sequence ABCCBA, which I used to illustrate two-sided nonmonotonicity in Example 9.6. This sequence is not feasible under *any* apportionment method because the smallest party, C, implied by the initial subsequence,

---

[14] The switch from AABBCC to CACBAB is the most extreme example in which nonmonotonicity for C can occur. If the latter choice sequence were CCAABB, C can guarantee its two best ministries, precluding nonmonotonicity. If one wants to keep the ordering of A and B the same as in AABBCC, the most extreme switch illustrating monmonotonicity is the switch to CAACBB, whereby C moves up six positions rather than seven.

[15] Under the Jefferson method, the third and fourth positions of C and B imply that C more deserves to receive a second ministry than B deserves to receive a first ministry, so $s_C/2 > s_B$ or $s_C > 2s_B$. But for B to be in the sixth position, its deservingness in receiving its second ministry must be greater than C's deservingness in receiving its third ministry, so $s_B/2 > s_C/3$, or $s_C < (3/2)s_B$, which is a contradiction.

ABC, can never get a second choice before the two larger parties, A and B, get their own second choices.

But ABCCBA might be the fairest sequence if the three parties are roughly equal in size. It compensates for the boost given to parties A and B, at the start, by giving party C two consecutive choices later, which Brams and Taylor (1999, ch. 3) call *balanced alternation*. Yet all the apportionment methods rule out this sequence, however close in size A, B, and C are, as long as they are not equal.

One could, of course, simply mandate the use of balanced alternation in a situation like this. Such a mandate would acknowledge that because A and B benefit from choosing first and second—especially when all parties agree on which ministries are most important—C needs to be compensated with two consecutive choices thereafter.[16]

9.6. A 2-PARTY MECHANISM

So far I have shown that if Pareto-optimality is lost when sophisticated parties select ministries in some prescribed order (which need not be the product of an apportionment method), it can be restored. However, trades that do so may not be easy to arrange, especially if they require a coordinated exchange of ministries among more than two parties.

While this is a practical difficulty, there is a more fundamental theoretical problem with the apportionment methods themselves. If one party is slightly larger than another, every apportionment method prescribes that the larger party choose before the smaller party *on every round*. This advantage can grow and grow and would, therefore, seem unfair to the (slightly) smaller party. But because of nonmonotonicity, giving the smaller party earlier choices may actually hurt it—as well as other parties if the sophisticated outcome is Pareto-nonoptimal.

Is there a way around these problems? If the players are sincere, these problems disappear. But it is naïve to suppose that political party leaders do not know a good deal about their competitors' preferences for ministries. Almost surely they do; why would they not exploit this information in making their own optimal choices?

There is one situation, however, in which party leaders can be induced to be sincere. If they are not, the sincere outcome can still be implemented, as a

---

[16] If the parties are sincere so choices are monotonic, even balanced alternation will not necessarily produce the most balanced outcome. To illustrate, assume A's ranking is 1 2 3 4 and B's is 2 3 4 1. Choice sequence ABAB gives (13, 24) to (A, B), whereas choice sequence ABBA gives (14, 23), which is obviously one-sided in favor of B. In this instance, B's earlier second choice throws off the "balance." A notion of balance is formalized in Herreiner and Puppe (2002), wherein procedures for finding it are discussed.

subgame-perfect equilibrium, which is in equilibrium in every subgame, through swaps of ministries.

The mechanism I present next works for two parties, making it optimal for each to choose its most-preferred ministry on every round—lest it have to swap ministries later to reach precisely this (sincere) outcome. Here are the rules that render sincere choices a subgame-perfect equilibrium when two parties take turns selecting ministries.[17]

1. At the time of selection, a party may offer to swap a ministry that it selects for one previously chosen by the other party.
2. This offer is placed on the table until all ministries are selected.
3. After all ministries are selected, start with the most recent swap offer and go back to all previous offers. If the party offered a swap agrees to it, a swap is made.[18]

To illustrate this mechanism, consider again Example 9.1c and the sophisticated choices the parties make that result in assignment (12, 34) to (A, B) when the choice sequence is ABAB. If the players have complete information about one another's preferences, then B would know that A is insincere when it selects ministry 2 on the 1st round. When this happens, suppose that B selects, on its next turn, A's most-preferred ministry 1 (i.e., the one A skipped over) and offers to swap it for ministry 2, which B prefers.

At the conclusion of the 1st round, ministries 3 and 4 are still available. On the 2nd round, if A and B sincerely choose ministries 3 and 4, respectively, it will not be in B's interest to offer A a swap of ministry 4 for ministry 3. Hence, this offer will not be made; moreover, even if it were, it would not be accepted by A.

But assignment (23, 41) to (A, B) is Pareto-nonoptimal. Clearly, it would be in A's interest to accept B's offer to swap ministries 1 and 2, yielding the Pareto-optimal assignment of (13, 24), which is the sincere outcome.[19] This

---

[17] "Taking turns" need not be *strict alternation*—ABAB to allocate four ministries—that all apportionment methods prescribe if A is less than twice as large as B; AABA and AAAB are also feasible as A becomes larger and larger compared to B. Whereas the balanced-alternation sequence, ABBA, is never feasible under an apportionment method, I do not rule it out as a choice sequence, especially if two parties are close in size. Thus in Example 9.1, the sophisticated outcome under ABBA is (12, 34) to (A, B), in which A selects ministry 2 on the 1st round and B follows by selecting ministries 3 and 4. Applying the two-party mechanism if A insincerely chooses ministry 2 on the 1st round, B would next select ministries 1 and 3 and offer to swap ministry 1 for A's ministry 2. This swap would be rational and restore the sincere assignment of (13, 24) to (A, B) under the rules described in the text.

[18] The purpose of rule 3 is to ensure that there is not a multiplicity of offers on the table, all "up for grabs." By imposing an ordering for swapping, the rule precludes bargaining over whose offer takes precedence. Instead, it induces parties to respond to "skipping" as soon as they are hurt by it (see text).

[19] Patently, A would prefer a swap of ministries 1 and 3 to restore Pareto-optimality, yielding (12, 34) rather than (13, 24) as the Pareto-optimal outcome. But rule 3 rules out this swap, illustrating

outcome gives each party its first and third choices, compared with the sophisticated outcome in which A gets its two best ministries and B its two middling ministries.

Similarly, in Example 9.2, the sincere assignment of (125, 346) to (A, B) is more "balanced" than the sophisticated assignment, (123, 465), which gives A its top three ministries and B its second, fourth, and fifth-best ministries. The former can be induced in the following manner: When A starts by deviating from its sincere choice of ministry 1 and instead chooses ministry 3 on the 1st round, B chooses ministry 1 and offers to trade it for ministry 3. If A tries to counter by choosing ministry 4 and offering to trade it for ministry 1, B will reject this trade since it gains by waiting for the trade of ministries 1 and 3.

Still, B would like ministry 4 if A should choose it on the 2nd round. It can obtain it by choosing ministry 2 and offering to trade it for ministry 4. This trade will be accepted by A, because A knows that B will reject the trade of ministries 4 and 1.

If A and B sincerely choose ministries 5 and 6 on the final round, the allocation to the players before trades will be (345, 126). Now consider all the trade offers. First, A will accept the offer of ministry 2 for ministry 4. Next, B will reject the offer of ministry 4 for ministry 1. Finally, A will accept the offer of ministry 1 for ministry 3. This leads to the sincere outcome, (125, 346).

I now show that if the sophisticated outcome differs from the sincere outcome, it will be in the interest of at least one player (say, B) to offer to swap a ministry with the other player (A). In turn, it will be in the interest of A to accept this offer. The resulting outcome will be the sincere outcome, restored.

More precisely, B will offer to swap a ministry with A when (i) A is not sincere, and (ii) at least one of the ministries A skips over is a ministry that B ranks lower than the ministry that A chose. Thus, A and B have opposite preferences for a ministry that A chose and a ministry that A skipped over. Then B will choose a skipped-over ministry on which the players have opposite preferences and offer a swap to A, which it will be rational for A to accept. After all such offers are made and accepted by both parties, the sincere outcome will be restored. A formal proof of this result is given in the appendix.

Although one might think that this argument could be extended to more than two parties, Example 9.7, in which the sophisticated choices are indicated for choice sequence ABCABC, shows why this is not the case. These choices presume that A prefers the pair 23 to the pair 14, which it can achieve by selecting ministry 1 on the 1st round and ministry 4 on the 2nd round. Instead,

---

why ex-post trading may not restore the sincere outcome, especially if one player, like A in this example, can dictate the trade that will be made because it has greater bargaining power.

by selecting ministry 2 initially, which is an insincere choice, the sophisticated assignment becomes (23, 15, 46) to (A, B, C).

Example 9.7. Sophisticated Choices (ABCABC)

	A	B	C
	1	<u>1</u>	1
1st round →	2	3	2
2nd round →	<u>3</u>	2	<u>4</u>
	4	4	3
	5	<u>5</u>	5
	6	6	<u>6</u>

What can C do to restore the sincere outcome, (14, 35, 26), which it prefers
to the sophisticated outcome?[20] After A selects ministry 2 and B selects ministry 1 on the 1st round, C can no longer select ministry 1 that A skipped over
and offer to swap it for A's ministry 2. Although C could select ministry 3 that
both A and B would like as 2nd-round choices, either A or B could respond by
selecting ministry 4 and offering to swap it for ministry 3, which C prefers.
Hence, C can do no better than sincerely select ministry 4 on the 1st round,
whence all the players will make sincere choices on the 2nd round (in general,
players can do no better than be sincere on a last round).

To summarize, I have shown that there is a two-party mechanism that can
induce the parties to be sincere. But if one party is not, the other party can
respond by offering a swap that, when accepted, restores the sincere outcome. Thereby, sincerity is rendered optimal, whether the parties are sincere initially or swaps are used to restore the sincere outcome, which is a
subgame-perfect equilibrium in a game of complete information. This mechanism, however, does not generalize to an n-person mechanism that restores
sincerity.

9.7. ORDER OF CHOICE AND EQUITABILITY

The party that benefits from a sophisticated outcome may not be the first-
choosing party (A), as was true in all the previous two-party examples. In
Example 9.8, B obtains its top two ministries, and A its first and third choices,
when the choice sequence is ABAB and the parties are sophisticated (notice
that it is B, not A, that skips over its first choice, ministry 4, on the 1st round).

---

[20] Because B prefers the sophisticated outcome, it would have no incentive to help C in this
restoration.

Example 9.8. Sophisticated Choices (ABAB)

	A	B
1st round →	$\underline{1}$	$\underline{\underline{4}}$
	2	$\underline{2}$
2nd round →	$\underline{\underline{3}}$	1
	4	3

In this example, the two-party mechanism enables A to restore the sincere outcome, turning the tables on B by ensuring its two top choices and B's first and third choices. A does so by selecting ministry 4 on the 2nd round, after B insincerely selects ministry 2 on the 1st round, and by offering to swap it for ministry 2. It is rational for B to accept A's offer, which turns the Pareto-nonoptimal assignment of (14, 23) into the Pareto-optimal sincere assignment of (12, 43) to (A, B).

In this example, the sincere and sophisticated assignments are the *only* two Pareto-optimal assignments that give two ministries to each party.[21] Because the sincere assignment favors A and the sophisticated assignment favors B, there is no Pareto-optimal assignment that benefits both parties "equally."

Call an assignment *equitable* if two or more parties (i) receive an equal number of ministries, and (ii) rank them exactly the same (Brams and Taylor, 1996, 1999). In Example 9.1, the sincere assignment of (13, 24) to (A, B) is equitable.

If there is not an equitable assignment, typically one party will do better than the other, as measured by their comparative rankings.[22] Perhaps it is appropriate in Example 9.8 that the sincere outcome—or its restoration via the two-party mechanism—benefits A when the choice sequence is ABAB. However, if the choice sequence is the balanced-alternation sequence ABBA, the sincere outcome is (13, 42), favoring B. Moreover, there is nothing that A can do to turn the tables on B in this example, because the sincere and sophisticated outcomes coincide.

Manifestly, the order of choice may matter. But independent of order, there may be no equitable outcome that leaves both players equally satisfied, based on their rankings. Worse, one player may envy another player for receiving a set of items that it prefers, which is a subject I explore in depth in chapter 10.

---

[21] Brams and King (2005) prove that an assignment is Pareto-optimal if and only if it is the product of sincere choices in some sequence, which I discuss in chapter 10. Of the six different sequences that give two ministries to each party, three (ABAB, AABB, BAAB) lead to the sincere (12, 43) outcome that benefits A, and three (ABBA, BABA, BBAA) lead to the sophisticated (13, 42) outcome that benefits B.

[22] In Example 9.8, giving 14 to one party and 23 to the other might also be considered fair, but this assignment is Pareto-inferior to both (12, 43) and (13, 42) to (A, B).

9.8. SUMMARY AND CONCLUSIONS

The divisor methods of apportionment provide a way of allocating cabinet ministries to political parties that reflect their entitlements. These entitlements, which are based on the size of parties and perhaps other prerogatives,[23] determine their deservingness, and thereby their priority in making sequential choices.

Different apportionment methods define deservingness differently. Thus, the Jefferson method favors large parties, whereas the Webster method is more neutral.

But more than offering promise, the Jefferson method was used for the first time in November 1999 to allocate ten cabinet ministries to the "grand coalition" of four major parties on the Northern Ireland Executive Committee, the main governing body of Northern Ireland today. There is little doubt that the parties were quite well informed about one another's top preferences. Furthermore, there is evidence that at least some parties in Northern Ireland behaved strategically, forgoing sincere choices if they believed they could pick them up later. By contrast, strategizing in Denmark takes place mainly through the forming of party coalitions prior to an election in order to increase the seat shares of their members as well as their priority in the selection process (O'Leary, Grofman, and Elklit, 2005).

If trades of ministries had been permitted (they were not in either country), it is not evident whether any would have been made. But because the sophisticated choices of parties may lead to a Pareto-nonoptimal outcome, there seems no good reason not to permit trades. In professional sports drafts in the United States, trading is not only commonplace before and after the draft but sometimes involves future draft choices—even before the players that will be in the draft are known—as well as monetary payments.[24]

That early choices are not necessarily beneficial, nor late choices harmful, shows how sophistication can nullify, or even reverse, the favorableness of one's position in a choice sequence. It seems highly doubtful, however, that political parties would actually give up early choices for later ones *because* they foresee a monotonicity problem.

More serious, I think, is the problem caused by parties that differ only slightly in size. If there are only two large parties, as was the case in Northern

---

[23] In Northern Ireland, where single transferable vote (STV) is used in elections, the party that receives the most first-place votes under STV receives priority in the event of a tie as to which party is awarded the next ministry (O'Leary, Grofman, and Elklit, 2005). For information on voting processes in other countries, see Colomer (2001) and Cox (1997); a comprehensive analysis of electoral systems, including their evolution, use, and effects in different countries, is given in Colomer (2004).

[24] It is, therefore, not strictly indivisible goods that are being divided, which usually facilitates the division process but does not necessarily eliminate envy, as I show in chapter 12.

Ireland (together they received 60 percent of the ministries), being fair to both parties translated into giving *each* party one first minister and one junior minister ("dual premiers"). Indeed, these four positions were above and beyond the six (of ten) specialized ministries the two largest parties were awarded by the Jefferson method; altogether, they ended up with ten of the fourteen ministries, or 71 percent.

In most parliamentary systems, being the prime minister is far more important than being any other minister (Laver and Hunt, 1992; Warwick, 2001; Warwick and Druckman, 2001). If there are two large parties, therefore, it would seem sensible to use balanced alternation, or ABBA, in selecting the first four ministries (BAAB would follow as the next sequence of choices under balanced alternation). For example, if A received the prime ministership, and foreign affairs and finance were considered the most prestigious specialized ministries, B would get both of these before A would get a second choice under ABBA.

In the case of two parties, I gave a mechanism that ensured that sincere choices are optimal—either initially or after rational swaps are made—given the parties have complete information and can propose such swaps. This mechanism is needed even in the case of the balanced-alternation sequence ABBA, wherein B can do no better than be sincere in its two consecutive selections (see note 17). But I leave open how close in size two or more parties must be to justify use of a balanced-alternation sequence, like ABBA for two parties or ABCCBA for three parties.

Speaking pragmatically, I believe the use of divisor apportionment methods can abet cabinet selection by putting responsibility in the hands of party leaders. By fixing the number and order of party choices, these methods would eliminate bargaining *among* parties—and the costs associated with such bargaining—but bargaining *within* parties about which ministries to choose, and who gets what ministry, would continue.

If party leaders are more satisfied by a method that is considered fair, the government coalition is likely to be more stable. Satisfaction can further be enhanced by using the two-party bargaining mechanism that induces sincere choices and balanced alternation—when each is appropriate—and by allowing trades of ministries among parties after their selection. Thereby, some of the problems of cabinet selection can be ameliorated if not solved.

APPENDIX

To formalize the 2-party mechanism discussed in section 9.6 for implementing sincere outcomes, which are Pareto-optimal (Brams and Straffin, 1979), assume the selection of items proceeds according to the following rules:

1. Two parties, A and B, alternate choosing single items from some set.
2. After the first party to select an item has made its choice, either party can make an offer to swap the item it chooses for an item already chosen by the other party.
3. This offer is placed on hold until all items have been chosen.
4. Starting with the most recent offer and going back to the first offer made by a party, each party accepts of rejects the offer it receives.
5. If a party accepts an offer, a swap is made; otherwise, there is no swap.

Proposition 9.1. *If either A or B makes an insincere choice by "skipping over" one or more sincere choices (see section 9.6), the other party can induce an outcome for itself at least as good as the sincere outcome and sometimes a better one.*

Proof. Let A be the party at any point in play that makes an insincere choice from among the items still available. Let x be the item that A chose insincerely.

Now consider B's highest-ranked item, y, among those items still available. If the preferences of A and B are strict (i.e., there are no ties), there are two possibilities:

1. B ranks y higher than x, or $y >_B x$. Because B prefers an item among those still available (y) to the item that A chose insincerely (x), B has no reason to offer a swap for x.

This is not to say that a swap would not restore the sincere outcome. For example, assume A ranks four items 1 2 3 4 and begins by choosing item 3. Assume B ranks the items 1 3 2 4. If B chooses item 2 after A chooses item 3 and offers a swap of 2 for 3—after which A and B sincerely choose items 1 and 4, respectively—it will be rational for A to accept B's offer, which will restore the sincere outcome of (12, 34) to (A, B). But B could have done better by not offering this swap and instead choosing item 1 initially, giving (23, 14) when A and B sincerely choose items 2 and 4, respectively, on the next round.

2. $x >_B y$. Because B prefers an item that A chose insincerely (x) to its highest-ranked item among those still available (y), B has reason to offer a swap for x.

This is not to say that B should offer to swap y for an item that A skipped over. For example, assume that A ranks four items 1 2 3 4 and begins by choosing item 4. Assume B ranks the items 4 2 1 3. If B chooses item 2 after A chooses item 4 and B offers a swap of item 2 for item 4, the sincere outcome, (12, 43), will be restored if the players' two choices on the next round are sincere. On the other hand, if B offers to swap its lower-ranked item 3 for item 4, the outcome will be (13, 42), which B prefers.

There will always be swaps that restore the sincere outcome if, because of insincere choices, A possesses item i and B possesses item j but the players rank these items in the "opposite" order (e.g., A ranks them in the order $j \ldots i$, and B in the order $i \ldots j$). By swapping these items, the players can obtain preferred items, and, in so doing, restore the sincere outcome. But such a swap will not be rational if the offering player can obtain a better outcome by not making an offer in case 1, or offering a different swap in case 2, as I have illustrated.

In case 2, it is true that if A skips over only one item when it chooses x, then the best offer that B can make is to swap A's sincere choice for x, which will restore the sincere outcome. But if A skips over more than one item, it may be possible for B to do better, as I showed, by not offering to swap its most-preferred item among those still available. In sum, B can guarantee for itself an outcome at least as good as the sincere outcome, and sometimes a better one, if A skips over one or more preferred items. ∎

10

Allocating Indivisible Goods:
Help the Worst-Off or Avoid Envy?

10.1. INTRODUCTION

Probably the most-discussed trade-off in economics is between efficiency and equity (LeGrand, 1991). While efficiency is arguably best achieved by unfettered competition in the marketplace that maximizes total wealth, the marketplace may be extremely unfair in distributing this wealth, especially to people who lack certain skills or social support (Roemer, 2000).[1]

The efficiency versus equity trade-off has analogues in other fields. In politics, centralized government may be less democratic, but decentralized government may be more wasteful and corrupt (though some analysts would claim just the opposite). In law, economic justice may best be served by giving people rights to a good education, decent housing, and other economic-enhancing opportunities, but these rights may vitiate their incentives to strive to their fullest extent possible, encouraging shirking instead.

In this chapter, I assume that a fair allocation is *efficient* (or *Pareto-optimal*): there is no other distribution that can help some players without hurting others, or at least not making them worse off. Assuming efficiency, I then focus on the trade-offs in the *distribution* of items, which, as in chapter 9, I assume are indivisible and which may range from the marital property in a divorce to religious sites in an international dispute.[2]

The indivisible items in chapter 9 were cabinet ministries, and I showed that sequential apportionment methods are vulnerable to manipulation. Moreover,

---

Note: This chapter is adapted from Brams and King (2005) with permission of Sage Publications Ltd.

[1] The inadequacy of the market in making choices that protect the environment is discussed in Adger, Paavola, Huq, and Mace (2006).

[2] If there is only one item, as in the case of the baby disputed by two mothers in the famous Bible story (I Kings 3:16–28), the allocation problem is exacerbated. King Solomon resolved the mothers' dispute by proposing to divide the baby in two (as if it were divisible), but his real purpose was more devious—to try to distinguish the true mother from the impostor (Brams, 1980, 2003, pp. 118–123). For different solutions to King Solomon's problem, see Glazer and Ma (1989), Perry and Reny (1999), Olszewski (2003), and Bag and Sabourian (2005).

if there are three or more players and they are sophisticated, the outcome chosen may be inefficient.

But efficient allocations need not be equitable. As I showed in section 9.7, the blame for an inequitable allocation does not necessarily lie with the apportionment method used. Instead, there may not *exist* an efficient allocation that is equitable, regardless of what method is used.

Likewise in this chapter, I show that there may not exist an efficient allocation that both helps the worst-off and avoids envy. However, I do not focus on algorithms for dividing up a set of indivisible items, such as the "descending demand procedure" (proposed by Herreiner and Puppe, 2002). Rather, I emphasize properties of fair division, showing which are compatible and which are not. Once a choice of compatible properties has been made, then finding efficient algorithms that yield them is certainly an important task to undertake.

As in chapter 9, I assume that each player can strictly rank all items to be divided from best to worst (ordinal preferences) but not attach numerical values, or cardinal utilities, to the items (cardinal preferences).[3] While this assumption eliminates the practical difficulty of comparing large numbers of subsets of items (or bundles)—2^{20}, or over a million, if there are twenty items—it also may produce indeterminacy.

For example, if there are four items, $\{1, 2, 3, 4\}$, and a player ranks them from best to worst in the order 1 2 3 4, it is impossible to say from the ordinal rankings alone whether this player prefers the bundle $\{1, 4\}$ to the bundle $\{2, 3\}$ or vice versa.[4] Realistically, this is information players may not possess—even of *their own* preferences if the bundles are quite large—so it is reasonable to assume only simple rankings and derive fairness consequences from these.

---

[3] This is the framework that is used in Brams and Fishburn (2000), Edelman and Fishburn (2001), and Brams, Edelman, and Fishburn (2001, 2003), which contain numerous references to the literature on fair division of indivisible goods. But these papers do not emphasize, or give general results on, the fundamental trade-off encapsulated in the subtitle of this chapter. While the present chapter is self-contained, I mention, as background reading on the fair-division and distributive-justice literature, ten recent books by economists, mathematicians, and philosophers relevant to the topics discussed herein: Broome (1991), Young (1994), Moulin (2003), Kolm (1996), Roemer (1996), Brams and Taylor (1996, 1999), Robertson and Webb (1998), Binmore (2005), and Barbanel (2005), some of which discuss the fair division of divisible goods like cake (more on this in later chapters); see also Brams (2006a) for a brief overview and Fleischacker (2004) for historical backgrouund. For an extensive bibliography, the Web site EqualityExchange is worth perusing: http://mora.rente.nhh.no/projects/EqualityExchange/.

[4] By contrast, if this player associated cardinal utilities of [5, 3, 2, 1] with items (1, 2, 3, 4), one could say that it prefers $\{1, 4\}$ (utility: $5 + 1 = 6$) to $\{2, 3\}$ (utility: $3 + 2 = 5$), assuming its preferences are separable and its utilities are additive. If they are not, then a framework in which players can rank both individual items and bundles of items is required; trade-offs in this framework are analyzed in Beviá (1998). Unlike the present framework, however, Beviá assumes that there is a divisible good (money), which facilitates "smoothing out" allocations. I analyze the fair division of divisible and indivisible goods in chapter 14.

To facilitate comparisons among efficient distributions of items, I postulate two criteria of fairness (to be refined shortly):

1. Help the worst-off player by maximizing the minimum rank of items that any player receives (*maximin*).
2. Avoid envy by preventing, insofar as possible, any player from receiving a set of items it considers inferior to a set received by another player (*envy-freeness*).

Because the maximin criterion takes into account only the lowest-ranked items of players, I offer a second maximin criterion, called *Borda maximin*, that better reflects the overall satisfaction of the players (this concept is defined in section 10.2). Also, because efficient, envy-free allocations may not exist, I concentrate on allocations that do not *ensure* envy, based on ordinal preferences, rather than those that are definitely envy-free (this distinction will be illustrated in section 10.4).

The chapter proceeds as follows. In section 10.2 I define maximin and Borda maximin allocations and illustrate differences between these concepts. In section 10.3 I characterize efficient allocations and show that at least one maximin and one Borda maximin allocation must be efficient. At the same time, there may be maximin and Borda maximin allocations that are inefficient.

In section 10.4 I illustrate with an example that maximin and Borda maximin allocations may be disjoint. Then I prove the main result—that maximin and Borda maximin allocations may be envy-ensuring. Indeed, if the preferences of players for items are unrestricted, the conflict between the two maximin criteria, on the one hand, and the avoidance of envy on the other, is inescapable—with one exception. This exception occurs when there are just two players; if each receives at least two items, maximin and Borda maximin allocations never ensure envy, but they do allow for it, which I call an *envy-possible* allocation and illustrate in section 10.4.

In section 10.5 I provide a sufficient condition for an efficient allocation not to ensure envy, but it is not necessary. Nevertheless, I show that there always exists at least one efficient envy-unensuring allocation if each player receives at least two items. On the other hand, there may be no envy-free allocation; moreover, even if one exists, it may be inefficient.

In section 10.6 I illustrate that maximin and Borda maximin allocations may require allocating unequal numbers of items to the players. Thereby "inequality" may be a virtue rather than a vice. Furthermore, equal allocations do not necessarily maximize the sum of the Borda scores of all players, which may be viewed as an indicator of overall well-being.

To understand better the quantitative dimensions of the conflict between the two maximin criteria and envy-ensuringness, I briefly report in section 10.6 on a computer analysis of the relative frequencies of this conflict for different populations. The principal finding is that maximin allocations ensure envy quite often, but Borda maximin allocations do so only rarely.

In section 10.7 I conclude with a discussion of both the obstacles and the opportunities for efficient fair division of a set of indivisible items. In theory, the obstacles are large. Not only is the envy-ensuringness of maximin and Borda maximin allocations unavoidable, except in the case of two players receiving at least two items, but even in this case there is no guarantee of envy-freeness—envy depends on the players' preferences for subsets of items. On the brighter side, while maximin and Borda maximin allocations cannot expunge envy, there is always an efficient allocation that is not envy-ensuring. I give an algorithm for finding it and also briefly discuss the applicability of the theoretical results to real-life cases.

10.2. MAXIMIN AND BORDA MAXIMIN ALLOCATIONS

I begin by defining and illustrating the concepts of maximin and Borda maximin allocations. A *maximin allocation* of items to players is an allocation that renders the lowest-ranked item that any player receives as high as possible. To illustrate, consider two players, A and B, with the following preferences for four items {1, 2, 3, 4}, ranked from best (on the left) to worst (on the right):

Example 10.1
> A: 1 2 3 4
> B: 2 3 4 1

If A receives items {1, 3} and B receives items {2, 4}, the lowest-ranked item that either player receives is its 3rd-best; this is also true if A receives items {1, 2} and B receives items {3, 4}. Because every other allocation in which A and B receive two items each involves A's receiving item 4 or B's receiving item 1 (their 4th-best items), the two aforementioned allocations are the only maximin ones.

In addition, there are two unequal allocations in which no player receives a 4th-best item: A–{1}, B–{2, 3, 4}; and A–{1, 2, 3}, B–{4}. To include these allocations in the maximin set seems highly questionable, however, because the player receiving its top three items can hardly be considered worse off (because it receives a 3rd-best item) than the player receiving only its best item.[5]

---

[5] To avoid this problem, one could restrict maximin comparisons to those that provide as "even" a distribution of the set of items as possible. For example, if there were three players and eight items, this would be a distribution that gives one player two items and the other two players three items each. Among all such distributions, a maximin distribution would be one that maximizes the lowest-ranked item received by the player who receives only two items. Because even this comparison is dubious, I restrict future maximin comparisons to allocations in which players receive exactly the same number of items.

A *Borda maximin allocation* of items to players is an allocation that maximizes the minimum modified Borda score that any player receives. The *modified Borda score* of a player is the sum of the individual scores it receives for each item it is allocated, wherein the score for a player's lowest-ranked item is 1 point, the score for its next-lowest-ranked item is 2 points, and so on up to its top-ranked item.[6]

To illustrate the calculation of Borda maximin, consider the six equal allocations in Example 10.1. In order to simplify notation, let the allocation of {1, 3} to A and {2, 4} to be B be represented by (13, 24); and let the resulting Borda scores of $4 + 2 = 6$ to A, and $4 + 2 = 6$ to B, be represented by [6, 6]. This representation is given as allocation 1 in the list below; the five other allocations to (A, B), and their Borda scores, are also shown:

1. (13, 24) [6, 6]	3. (14, 23) [5, 7]	5. (24, 13) [4, 4]
2. (12, 34) [7, 5]	4. (23, 14) [5, 3]	6. (34, 12) [3, 5]

Allocation 1 is the Borda maximin allocation, because it gives each player 6 points, which is greater than the minimum number of points that each of the five other equal allocations gives the players. Moreover, every unequal allocation, which gives one player either one or zero items, necessarily gives that player a lower Borda score than 6 (indeed, 4 or less). Hence, allocation 1 is the unique Borda maximin allocation among *all* allocations (equal or unequal) in Example 10.1.

Allocation 1 is also a maximin allocation—no player gets lower than a 3rd-best item—but it shares this status with allocation 2. Sometimes, however, the set of Borda maximin and maximin allocations may not overlap, as I will illustrate in section 10.4.

It is worth noting that unequal allocation (1, 234), with Borda scores of [4, 9], maximizes the *sum* of Borda scores, giving a total of 13 points to both players. Thus, Borda maximin allocations do not necessarily lead to what I call *Borda maxsum* allocations. This is a point I will return to in sections 10.5 and 10.6, where I will show that Borda maximin allocations also may give players unequal numbers of items.

To recapitulate, maximin selects allocations 1 and 2; Borda maximin singles out allocation 1. Sometimes these two maximin criteria may give entirely different allocations, and they need not be Borda maxsum allocations.

---

[6] The one-unit difference of cardinal scores is a simplification: preferences are rarely linear and equally spaced. While this cardinalization of ranks reflects the scoring method of the Borda count voting system, the reflection is not exact: Borda count voting scores give the lowest-ranked alternative 0 points, the next lowest-ranked alternative 1 point, and so on up to an award of $n - 1$ points to the highest-ranked alternative if there are n alternatives. I start with 1 point for the lowest-ranked item, rather than 0, to ensure that every item adds positive value to a player's set of items. While Borda scores are but one of an infinite number of possible cardinalizations of ordinal preferences, all seem vulnerable to the problems I discuss (see note 10).

10.3. CHARACTERIZATION OF EFFICIENT ALLOCATIONS

I next characterize all efficient allocations. As I will show, there is at least one maximin and one Borda maximin allocation that is efficient. However, both maximin and Borda maximin allocations may be inefficient, as I will illustrate later.

To make precise a notion of efficiency in the present ordinal framework, I need the following definitions. Assume two sets of items, S and S', have the same cardinality (i.e., number of members) but not the same items. For an individual player, a set S of items *dominates* set S' if, for every item in S but not in S', there is a different item in S' but not in S which that player ranks lower.

Consider two different allocations, A and A', to *all* players. A *Pareto-dominates* A' if, for every player, its allocation in A dominates its allocation in A'. In this situation, A' is *Pareto-dominated* by A. An allocation is *efficient* if it is not Pareto-dominated.[7]

Thus in Example 10.1, allocation 1 Pareto-dominates allocation 4, because both A and B prefer their allocations in the former to those in the latter, rendering allocation 4 Pareto-dominated. Allocations 2 and 3, as well as allocation 1, are efficient, because they are not Pareto-dominated by any other allocations.

By contrast, allocations 5 and 6, as well as allocation 4, are Pareto-dominated, because both A and B prefer their allocations in 1, which gives each player its 1st and 3rd-best choices, to their allocations in 4, 5, and 6. It is easy to check that allocation 2 Pareto-dominates allocations 4 and 5, and allocation 3 Pareto-dominates allocations 5 and 6.

I next characterize efficient allocations, offering a simple test for efficiency. When players choose items one at a time in some sequence, a *sincere choice* by a player is its choice of its top-ranked item from those not yet chosen.

Proposition 10.1. *An allocation of items is efficient if and only if it is the product of sincere choices by the players in some sequence.*

Proof. Sufficiency: Consider a set of items I, containing p items, and an arbitrary sequence of players of length p. Let A be the allocation resulting from sincere choices by all the players, selecting in the order of the sequence. Suppose A' is any other allocation. I claim that A is not Pareto-dominated by A'.

To show this, let I^* be the nonempty subset of items in I that are allocated to different players in A and A'. Let i be the item in I^* that was selected first in A, and let X be the player that selected i. Because X selected i over all other items

---

[7] It is worth emphasizing that efficiency, as defined here, does not take into account information about preferences over bundles of items. Thus, for instance, if two players, A and B, rank four items in the order 1234, the two allocations, (14, 23) and (23, 14), to (A, B) are efficient. But if both A and B prefer the first allocation to the second, then the second allocation is inefficient, based on this bundle information, which I do not assume.

in I^*, X must prefer i to all other items in I^*. Thus, there is no item that X receives in A', but not in A, that it prefers to i.

Player X's allocation in A', therefore, does not dominate its allocation in A. Because A' was arbitrary, A is not Pareto-dominated by any other allocation.

Necessity: Now consider a situation in which there are m players and a set I of p items that must be allocated. Let A be an arbitrary efficient allocation of the items to the players. I next present an algorithm that is well defined and produces a sequence of players, called O (for order), for which sincere choices produce A.

Step 1. Select any player from among those players that received a top-ranked item in A.

Step 2. Place this player first in the sequence O.

Step 3. Remove the selected player's top-ranked item from the rankings of all players.

Step 4. Repeat steps 1–3, each time placing the selected player in the next position of the sequence, until all items are removed and a sequence O of length p is completed.

I will illustrate this algorithm with an example following the completion of this proof.

I claim the algorithm is well defined: it will continue uninterrupted until a sequence of players of length p is generated. Suppose, for the sake of contradiction, that this is not the case. Specifically, suppose that the process stops before O is fully generated to a length of p players. This can occur only if, at some point, a situation occurs in which no player receives its top-ranked item (among those items remaining).

Consider an arbitrary player X_1, and denote by x_1 its top-ranked item at the current moment. By the supposition in the preceding paragraph, one knows that X_1 does not receive item x_1 in A. Let X_2 be the player that does receive this item in A. One knows from the supposition that x_1 is not player X_2's currently top-ranked item. Denote by x_2 the top-ranked item in X_2's current ranking. One knows that X_2 did not receive item x_2 in A. Denote by X_3 the player that did receive item x_2. Continue this process, identifying players X_1, X_2, X_3, . . . and items x_1, x_2, x_3, ...

Because the number of players is finite, one eventually arrives at a duplication in the listing X_i of players. Suppose that the first duplication occurs in the naming of player X_n, and suppose X_n is X_j for some $j < n$. Now consider the set of players X_j, X_{j+1}, X_{j+2}, . . . , X_{n-1}, and a trade between these players in which X_j receives the item x_j from X_{j+1}, X_{j+1} receives the item x_{j+1} from X_{j+2}, . . . , X_{n-1} receives item x_{n-1} from X_j.

This trade will be beneficial to all players involved, because each player will receive a higher-ranked item in exchange for a lower-ranked item. Thus,

A is Pareto-dominated by the allocation resulting from this trade. This contradicts the assumption that *A* is efficient. Hence, the algorithm always produces a sequence of players of length *p*, without stopping, and so is well defined.

Finally, I claim that efficient allocation *A* does in fact result from sincere choices with respect to the sequence *O* produced by the algorithm. Suppose, for the sake of contradiction, that some allocation other than *A* results from sincere choices involving sequence *O*. Let X be the first player appearing in the sequence that sincerely selects an item, say *x*, not allocated to it in *A*. At that moment, *x* would be the top-ranked item of X; otherwise, X would not select it sincerely. But, under the rules of the algorithm, X would have appeared at that position in the sequence *O* only if its top-ranked item at that moment, namely *x*, had been among the set of items that X receives under *A*. This contradicts the supposition that *x* is not among the set of items that player X receives in *A*. Thus, *A* must result from sincere choices using sequence *O*.

Because *A* was chosen arbitrarily, every efficient allocation results from sincere choices with respect to some player selection order *O*. ■

I illustrate the algorithm given in the proof of Proposition 10.1 with an example.

Example 10.2
    A: 1 2 3 4 5 6
    B: 5 6 2 1 4 3
    C: 3 6 5 4 1 2

Consider the allocation (12, 56, 34) to (A, B, C). Clearly, it is not Pareto-dominated by another allocation of two items each to the players, because either A or B, each of which gets their top two choices, would be hurt by such an allocation. Furthermore, no unequal allocation Pareto-dominates allocation (12, 56, 34), because any player receiving one or zero items would be worse off, even if the one item was its top choice. Thus, allocation (12, 56, 34) is efficient.

Now apply the algorithm to Example 10.2. Observe that all three players receive their top-ranked items in the postulated allocation. Following step 1, choose any one of these players, say B, which received its top-ranked item 5 in the postulated allocation. Following step 2, B becomes the first player in sequence *O*. According to step 3, remove item 5 from the rankings of all the players, which results in the following reduced rankings:

    A: 1 2 3 4 6
    B: 6 2 1 4 3
    C: 3 6 4 1 2

Continuing in this manner by arbitrarily selecting players that received their top-ranked choices in the postulated allocation—but now with already chosen items removed—a sequence of players of length 6 is generated. There may be different orders O that lead to the same allocation, including sequences that have players choosing two items in a row. For example, sequence AABBCC leads to the same allocation as ABCABC.[8]

I will not list all sequences for this example and indicate which produce what efficient allocations—this task is more manageable in the next example. Example 10.3 raises the question of whether the allocation that is generated by the most sequences is, in some sense, better than other efficient allocations.

Example 10.3
    A: 1 2 3
    B: 1 3 2
    C: 2 1 3

There are three efficient allocations—(1, 3, 2), (2, 1, 3), and (3, 1, 2)—in which each player receives one item. The first is the product of sincere choices in three sequences, ABC, ACB, and CAB; the second in one sequence, BAC; and the third in two sequences, BCA and CBA. It is apparent that different sequences may generate the same efficient allocation.

There are also sequences in which one player takes more than one turn, such as AAB or AAA. By Proposition 10.1, these also generate efficient allocations when the players make sincere choices, but they are obviously unfair to the player or players that receive no items.

It may seem that the more sequences that generate an allocation, the fairer it is. As a case in point, consider allocation (1, 3, 2), which is generated by three of the six sequences in which the players each receive one item. This allocation is the unique maximin and Borda maximin allocation.

But this is not the case in Example 10.1, in which the unique maximin and Borda maximin allocation to (A, B) is (13, 24), which is the product of two sequences (ABAB and BAAB). In addition, one sequence (AABB) leads to efficient maximin allocation (12, 34). However, there are three sequences

---

[8] One might think that sequences in which there is a strict alternation of players, such as ABCABC, will produce fairer allocations than those in which a player gets two turns in a row, such as AABBCC, in which C, for example, gets only a 5th and 6th choice. But as Brams and Taylor (1999) show, a sequence called "balanced alternation," which in the present example is ABCCBA and gives late-choosing player C two turns in a row, may be superior. This question is complicated if the players have complete information about all players' rankings and, thereby, can make strategic calculations. In this situation, they may have good reason not to be sincere, as Brams and Straffin (1979) show for Example 10.2 in the context of teams that choose players in professional sports drafts. Indeed, the sophisticated strategies of players when the sequence is ABCABC produces a worse allocation for *everybody* than when the sequence is AABBCC, because the former sequence induces a 3-person Prisoners' Dilemma in Example 10.2. This example illustrates how strategic choices may not lead to efficient allocations; also, they may be nonmonotonic, whereby choosing earlier in a sequence actually leads to a worse outcome for a player than choosing later. I showed examples of these problems in chapter 9.

(ABBA, BABA, BBAA) that generate the third efficient allocation, (14, 23), which is neither maximin nor Borda maximin.

Fortunately, for every profile there is always an efficient maximin and an efficient Borda maximin allocation:

Proposition 10.2. *There is at least one maximin and at least one Borda maximin allocation in the set of efficient allocations.*

Proof. I first prove Proposition 10.2 for Borda maximin allocations. Suppose A_0 is an arbitrary Borda maximin allocation for a given preference profile. If A_0 is efficient, the proof is completed. If A_0 is inefficient, then A_0 is Pareto-dominated by another allocation, say A_1. Hence, for some players the allocations they receive in A_1 dominate their allocations in A_0, though the minimum Borda score will remain the same (otherwise, A_0 would not be a Borda maximin allocation).

Thus, the Borda scores for these players increase in going from A_0 to A_1. But for every other player, the allocation in A_1 is the same as that in A_0. Hence, the Borda scores of these players remain unchanged. Therefore, A_1, like A_0 itself, must be a Borda maximin allocation. Furthermore, the *sum* of the players' Borda scores for A_1 is greater than that for A_0.

If A_1 is an efficient allocation, the proof is completed. If not, repeat the preceding argument to obtain another Borda maximin allocation, A_2, that has a Borda score sum greater than that of A_1. If A_2 is efficient, the proof is completed. If A_2 is inefficient, repeat the preceding argument again.

Continuing in this manner, create a sequence, $A_0, A_1, A_2, A_3, \ldots$ of Borda maximin allocations with increasing Borda score sums. Because such sums are bounded from above, no such infinite sequence can occur. Thus, one of the A_i must be efficient. Hence, there exists an efficient Borda maximin allocation for every preference profile.

The argument for the existence of efficient maximin allocations proceeds along similar lines. Using the Borda score sum, note that an allocation A_1 that Pareto-dominates another allocation A_0 will raise the ranks of items of some players, and hence their Borda scores, though the minimum rank will remain the same (otherwise, A_0 would not be a maximin allocation). ∎

I suggested earlier the *possibility* that a Borda maximin allocation may be inefficient in the proof of Proposition 10.2, but in none of the previous examples did I exhibit such an inefficient allocation. Accordingly, I now prove their existence.

Proposition 10.3. *There exist maximin and Borda maximin allocations that are inefficient.*

Proof. In Example 10.3, the allocation (2, 3, 1) to (A, B, C) is inefficient by Proposition 10.1. No player receives its top-ranked item, so (2, 3, 1) is not the product of sincere choices, whatever the sequence of player choices. Also note that a trade between players A and C that benefits each results in allocation (1, 3, 2), which is efficient. Both the inefficient allocation and the efficient allocation are maximin and Borda maximin, because there are no other allocations that give a worst-off player better than a 2nd-best item, or a higher Borda score than 2. ∎

Henceforth, I will focus on *efficient* maximin and Borda maximin allocations, which by Proposition 10.2 always exist.

10.4. MAXIMIN AND BORDA MAXIMIN ALLOCATIONS MAY BE ENVY-ENSURING

In this section I will prove that maximin and Borda maximin allocations provide no insurance against envy. But first I show that the two maximin criteria may either converge or diverge in their choice of fair allocations:

Proposition 10.4. *Maximin and Borda maximin allocations may be disjoint or overlapping.*

Proof. To prove the first part of the proposition (disjointness), consider

Example 10.4
    A: 1 2 3 4 5 6 7 8 9
    B: 3 4 2 6 8 7 1 5 9
    C: 5 8 2 7 9 1 3 4 6

There are three efficient maximin allocations, giving each player, at worst, a 5th-best item. These allocations to (A, B, C) and their Borda scores are as follows:

1. (124, 368, 579)   [23, 20, 20]
2. (123, 468, 579)   [24, 19, 20]
3. (125, 346, 879)   [22, 23, 19]

Three sequences of sincere choices that generate allocations 1, 2, and 3, respectively, are AABCABBCC, AAABBBCC, and AABBABCCC. Showing that these allocations are the only maximin allocations is tedious, so I do not include a demonstration here.[9]

Observe that the minimum Borda scores for these allocations are 20 for allocation 1, 19 for allocation 2, and 19 for allocation 3, so it would seem that allocation 1 is the Borda maximin allocation. However, it turns out that there is a fourth allocation, which is not maximin (because A and B receive 6th-best

---

[9] A computer program is available on request to check all unproved computational claims.

items), in which the worst-off player receives a higher minimum Borda score (21 points):

    4. (126, 347, 589)   [21, 21, 22]

A computer calculation verifies that this is the unique Borda maximin allocation.[10]

That the maximin and Borda maximin allocations can overlap is shown by Example 10.1, in which allocation 1 is both maximin and Borda maximin, making it presumably the "fairest" allocation in this example.[11] ∎

But what is the fairest allocation among the three maximin, and the disjoint Borda maximin, divisions in Example 10.4? Like allocation 1 in Example 10.1, allocation 1 in Example 10.4 maximizes the minimum Borda score *among the maximin allocations*; but unlike Example 10.1, nonmaximin allocation 4 is the Borda maximin allocation.[12]

Whereas Borda maximin helps the worst-off player by maximizing its overall satisfaction, maximin's focus is narrower; it maximizes the rank of only the lowest-ranked item that any player receives, independent of whatever other items this player possesses. It is not surprising, therefore, that these criteria may clash over what allocation is fairest.

---

[10] This example well illustrates how the cardinalization of the ordinal ranks can matter. (By "cardinalization" I mean the attribution of cardinal scores to ranks that preserves player orderings by giving the most points to a player's highest-ranked item, the next-most points to its next highest-ranked item, etc.) Thus, instead of using Borda scores, assign each player's top five items its Borda score *plus* two additional points, but keep the old Borda scores for each player's bottom four items (under the presumption that these items, as a group, are worth two points less than the top five). Then the Borda maximin allocation given in the text now has a minimum score of 25, whereas efficient maximin allocation 1 has a minimum score of 26, giving it the edge over all the others. I hypothesize that any cardinalization different from Borda, like the one just illustrated, will still lead to most of the difficulties identified in this chapter—in particular, it will not preclude envy-ensuring allocations.

[11] In Example 10.1, there are two maximin allocations, one of which is Borda maximin. Below I give an example of the opposite—two Borda maximin allocations, one of which is maximin.

    A: 1 2 3 4 5 6   B: 2 1 4 6 3 5   C: 4 5 6 3 1 2

The Borda maximin allocations, in which each player receives a minimum Borda score of 9, are as follows:

    1. (13, 24, 56) [10, 10, 9]   2. (13, 26, 45) [10, 9, 11]

While only allocation 1 is maximin—giving each player an item ranked no lower than 3rd-best—it is allocation 2 that maximizes the sum of the players' Borda scores (i.e., is Borda maxsum), suggesting that allocation 2 is competitive with allocation 1, even though allocation 2 is not maximin.

[12] What explains this difference? Note that allocation 4 gives each player its two best items and a 6th-best item, whereas allocation 1 gives only player A its two best items (and a 5th-best item). While players B and C get their single best items in allocation 1, they also get only their 4th-best and 5th-best items, which lowers their Borda scores to 20 points rather than the minimum of 21 points that allocation 4 gives the players in Example 10.4.

More surprising is that both a maximin and a Borda maximin allocation may ensure envy. In our ordinal framework, call an allocation *envy-ensuring* if, for some player A, A *assuredly envies* another player B, because B's allocation dominates A's allocation with respect to A's preferences.

To illustrate this concept, assume that A and B both rank four items in the order 1 2 3 4. If A receives items 24 and B receives items 13, this allocation is envy-ensuring: A envies B, because A prefers B's item 1 to its own item 2, and B's item 3 to its own item 4.

By contrast, if A receives items 14 and B receives items 23, this allocation is *envy-possible*; the presence of envy depends on the preferences of A and B for *sets* of items. In particular, one player will envy the other if and only if A prefers 23 to 14, or B prefers 14 to 23. Thus, the presence of envy in an envy-possible allocation will depend on the cardinal utilities the players attach to sets of items (see note 2 for an example).

If each player receives only one item, as might occur in a gift exchange, an allocation may be envy-ensuring even though the players all have different rankings of the items, as illustrated by Example 10.3. In this example, there are three allocations to (A, B, C) that are efficient—(1, 3, 2), (2, 1, 3), and (3, 1, 2)—wherein at least one player receives its best item. None is envy-free, however, because the player that receives item 1 in each allocation (A or B) is envied by at least one of the other two players. Allocation (1, 3, 2), in which no player receives its worst item, is the unique efficient maximin and Borda maximin allocation, but it is envy-ensuring because B envies A.

Before showing that maximin and Borda maximin allocations cannot, in general, prevent assured envy (at an ordinal level), consider the one exceptional case—when there are exactly two players, and each receives two or more items. In Example 10.1, recall that allocation (12, 34) is Borda maximin (each player gets a Borda score of 6); the fact that this allocation does not ensure envy—in fact, it is envy-free—is no accident.

Proposition 10.5. *Assume there are $n = 2$ players and $k \geq 2$ items that each player receives. Then an efficient Borda maximin allocation is never envy-ensuring.*

Proof. Suppose, for the sake of contradiction, that there exists an efficient Borda maximin allocation A that is envy-ensuring. In addition, suppose players A and B receive sets X and Y, containing k items each. For concreteness, suppose that B envies A.

Let $Y = \{y_1, y_2, \ldots, y_k\}$, where B prefers y_i to y_j if $i < j$. Because B envies A, set X dominates set Y for B. Thus, there exists an ordering of the elements in X, say $X = \{x_1, x_2, \ldots, x_k\}$, such that B prefers x_i to y_i for all i. For B, therefore, the Borda score of X exceeds that of Y by at least the number of items in Y:

$$\text{Borda}_B X \geq \text{Borda}_B Y + k. \tag{10.1}$$

Because the original allocation is efficient, it must also be the case that A prefers x_i to y_i for all i; otherwise, A would be Pareto-dominated by another allocation. Hence,

$$\text{Borda}_A X \geq \text{Borda}_A Y + k. \tag{10.2}$$

Because there are only two players, we know

$$\text{Borda}_B X + \text{Borda}_B Y = \text{Borda}_A X + \text{Borda}_A Y. \tag{10.3}$$

From (10.1) and (10.2) it follows that

$$\text{Borda}_B Y - \text{Borda}_B X + k \leq \text{Borda}_A X - \text{Borda}_A Y - k. \tag{10.4}$$

Adding (10.3) and (10.4),

$$2\text{Borda}_B Y + k \leq 2\text{Borda}_A X - k$$

$$\text{Borda}_A X \geq \text{Borda}_B Y + k. \tag{10.5}$$

Let $d_i = \text{Borda}_A\{x_i\} - \text{Borda}_A\{y_i\}$ for $i = 1, 2, \ldots, n$. Let d_i^0 be the minimal value of d_i for $i = 1, 2, \ldots, n$. Because there are $2k$ items altogether, $d_i^0 \leq k$, with equality holding only if and only if A prefers all items in X to all items in Y. Now consider two cases:

Case 1. Suppose $d_i^0 < k$. Let A' be the allocation that results when A and B swap items x_i^0 and y_i^0; and let X' and Y' be the allocations to A and B, respectively, in allocation A'. Clearly, $\text{Borda}_B Y' > \text{Borda}_B Y$, because B is receiving a higher-ranked item in the trade. Furthermore, because A ranks x_i^0 fewer than k units above y_i^0 (because $d_i^0 < k$), $\text{Borda}_A X' > \text{Borda}_A X - k \geq \text{Borda}_B Y$ from (10.5). Thus, both players receive a higher Borda score from A' than that received by B from A. This contradicts the assumption that A is a Borda maximin allocation.

Case 2. Suppose $d_i^0 = k$. In this case, A prefers all items in X to all items in Y. Hence, if $k \geq 2$, inequality (10.1) is strict. This implies that inequality (10.5) is strict as well:

$$\text{Borda}_A X > \text{Borda}_B Y + k. \tag{10.6}$$

As in case 1, let A' be the allocation that results when A and B swap items x_i^0 and y_i^0, and let X' and Y' be the allocations to A and B, respectively, in allocation A'. Clearly, $\text{Borda}_B Y' > \text{Borda}_B Y$, because B is receiving a higher-ranked item in the trade. Furthermore, because $d_i^0 = k$, $\text{Borda}_A X' = \text{Borda}_A X - k > \text{Borda}_B Y$ from (10.6). Thus, both players receive a higher Borda score from A' than that received by B from A. This contradicts the assumption that A is a Borda maximin allocation. ∎

Next I turn to the main negative findings. To avoid trivialities, assume that there are enough items so that each player can obtain at least one item. If this were not the case, then any player receiving no item would envy those players that receive some items. In addition, assume that the number of items is such that they can be equally divided among all the players.

I begin by showing that there always exists a preference profile in which an efficient *maximin* allocation is envy-ensuring (Proposition 10.6). I then show that the same conflict arises if the allocation is an efficient *Borda maximin* allocation (Proposition 10.7), except when $n = 2$, as I just demonstrated (Proposition 10.5). Finally, I show that Borda maximin allocations that are envy-ensuring may or may not be unique (Proposition 10.8).

Proposition 10.6. *Assume the number of items to be allocated is an integer multiple k of the number of players n. There is always an efficient maximin allocation that is envy-ensuring for some preference profile.*

Proof. Start with the n-person, n-item case (i.e., $k = 1$). Assume players P_1, P_2, $P_3, \ldots, P_n$ rank items as follows:

Example 10.5

P_1: 1 2 3 4 5 ... $n - 1$ n
P_2: 1 2 3 4 5 ... $n - 1$ n
P_3: 3 1 2 4 5 ... $n - 1$ n
P_4: 4 1 2 3 5 ... $n - 1$ n
.
.
.
P_n: n 1 2 3 4 ... $n - 2$ $n - 1$

It is apparent that an efficient maximin allocation is one in which all players get their first choices—except *both* P1 and P2, which have the same ranking—one of which (say, P_2) must get its second choice. But then P_2 will envy P_1 in the allocation (1, 2, 3, ..., n).

This example can readily be extended to the n-player, kn-item case ($k \geq 2$), which I illustrate with a 4-player, 8-item example, in which items have been grouped into two-sets (e.g., 12, which indicates a ranking of item 1 ahead of item 2):

Example 10.6

A: 12 34 56 78
B: 12 34 56 78
C: 56 12 34 78
D: 78 12 34 56

In the efficient maximin allocation (12, 34, 56, 78), which is the product of sincere choices in sequence AABBCCDD (among others), B envies A.[13] ∎

---

[13] There is another efficient maximin allocation, (34, 12, 56, 78), which is the product of sincere choices in sequence BBAACCDD (among others), in which A envies B. But because A

Proposition 10.7. *Assume the number of items to be allocated is an integer multiple k of the number of players n. If n ≠ 2, there is always an efficient Borda maximin allocation that is envy-ensuring for some preference profile. If n = 2, such a division can exist only if k = 1.*

Proof. Consider the n-person, n-item case (i.e., $k = 1$) analyzed in the proof of Proposition 10.6. The efficient maximin allocations in Example 10.5, whereby each player receives its best choice—except for one of either P_1 or P_2—is also a Borda maximin allocation that ensures envy.[14]

I now offer a construction when $k \geq 2$ that yields Borda maximin allocations that are envy-ensuring for $n \geq 3$ (recall from Proposition 10.5 that there exist no such allocations when $n = 2$). I begin the construction with three players, A, B, and C, and later show how additional players can be added:

1. A and B have the same preference ranking of items; let numbers 1, 2, 3, . . . , n identify these items, ranked from best to worst by both players.
2. A receives the first $k - 1$ odd-numbered items, 1 3 5 . . . $2k - 3$; its kth item is item $3k - 1$, its next-to-last ranked item.
3. B receives the first k even-numbered items, 2 4 6 . . . $2k$.
4. C's 1st, 3rd, 5th, . . . , $2k - 1$st-ranked items are those that A receives (in numerically increasing order). C's 2nd, 4th, 6th, . . . , $2k$th-ranked items are those that neither A nor B receives (in numerically increasing order); C receives these items. C ranks the items that B receives below its $2k$th-ranked item (in numerically increasing order).

The simplest example that illustrates this construction, which I call CON, is for $n = 3$, $k = 2$:

Example 10.7
  A: 1 2 3 4 5 6
  B: 1 2 3 4 5 6
  C: 1 3 5 6 2 4

CON gives allocation (15, 24, 36), with Borda scores of [8, 8, 8]. Note that because the Borda scores of B and C are given by their 2nd and 4th-ranked

---

and B have the same preferences, this is isomorphic to the one given in the text. Henceforth, I ignore isomorphic allocations (i.e., when two or more players have the same preferences).

[14] But envy-ensuringness does not extend to allocation (12, 34, 56, 78) in Example 10.6, wherein each player receives two items. The Borda maximin allocation in this example is (14, 23, 56, 78), which gives the players—in particular, A and B—minimum Borda scores of 13 points. These scores are greater than the minimum Borda scores of 11 points each that the envy-ensuring maximin allocation, (12, 34, 56, 78), gives A and B. Example 10.6, therefore, does *not* show that there exists a Borda maximin allocation that is envy-ensuring when the players get more than one item each.

items, each player receives $5 + 3 = 8$ points. To ensure that A also receives 8 points, I ask how many points it must receive, in addition to the 6 points it receives from obtaining item 1, to give it 8 points. The answer of 2 points means that it must receive its 5th-ranked item (item 5), which is what step 2 of CON specifies.

To show that A always can be given a kth item that equates its Borda score with the (equal) Borda scores of B and C, suppose that A's kth item is *not* that which equalizes its Borda score but instead the kth odd-numbered item (item 3 in Example 10.7). Then its Borda score would be k points greater than those of B and C, which receive items at even-numbered ranks. But because there are $k + 1$ items below A's kth odd-numbered item (items 4, 5, and 6 in Example 10.7), I can drop k items below this item (i.e., go from item 3 to item 5) to award A—instead of its kth odd-numbered item—one that equalizes the Borda scores of all three players. The award of this item to A does not conflict with the items that C receives, because C's even-ranked items are determined after A and B receive their items, according to step 4 of CON.

To show that CON yields an efficient allocation, note that the allocation is the product of sincere sequence ABCBAC in Example 10.7. In general, sequence ABCABCABC . . . BAC, which gives A the first $k - 1$ odd-numbered items, B the first k even-numbered items, and C the remaining items, except for item $3k - 1$, which A takes before C does on the last round.

Before showing that CON yields a Borda maximin allocation, observe that the allocation it gives in Example 10.7 ensures that C envies A. Because, in general, the k items that C receives at ranks 2, 4, 6, . . . are preceded by the k items that A receives, C will envy A. Later I will show that this result can be extended to the case of more than three players.

I begin with the $n = 3$ case, in which each player receives k items, where $k \geq 2$. The allocation to players A, B, and C, according to CON, is

$$A = (1\ 3\ 5 \ldots 2k - 3\ \ 3k - 1, \quad 2\ 4\ 6 \ldots 2k - 2\ \ 2k,$$
$$2k - 1\ \ 2k + 1\ \ 2k + 2 \ldots 3k - 2\ \ 3k).$$

The Borda scores of the players, in which A receives its 1st, 3rd, 5th . . . -best items, except for its kth item that it ranks next-to-last, and B and C receive their 2nd, 4th, 6th . . . , kth-best items, are

$$s(A) = 3k + 3k - 2 + 3k - 4 + \cdots + 3k - (2k - 4) + 2$$

$$s(B) = s(C) = 3k - 1 + 3k - 3 + 3k - 5 + \cdots + 3k - (2k - 3) + 3k - (2k - 1).$$

Combining terms and simplifying, it is not difficult to show that each of these sums is $2k^2$, giving the three players identical Borda scores.

It is evident that a Borda maxsum allocation is achieved if and only if each item is allocated to a player that ranks it highest. In fact, this is true of all the items received by the players in CON, except for the kth item received by A;

this item, $3k - 1$, which A ranks next-to-last, contributes 2 points to A's Borda score. If C, which ranks this item the highest of the three players, had received it, it would have contributed $3k - (2k - 2) = k + 2$ points to C's Borda score (1 more point than the last term of $s(C)$ above). The addition of the k points to the sum of the Borda scores that the players receive from the CON allocation gives a Borda maxsum value of $6k^2 + k$ points.

Now for an allocation to achieve this Borda maxsum value, it is necessary for C to receive all the items shown in A above *plus* item $3k - 1$, which it ranks higher than A or B. If it does not receive one of these items, the reduction in the maxsum value will be at least k points. Postulating A' to be any other allocation, consider two cases:

Case 1. Suppose that C does not receive in A' all the items it receives in A. Then the maximal possible Borda sum will be $6k^2 + k - k = 6k^2$. In this case, an allocation cannot possibly give each player a Borda score strictly greater than $2k^2$ points.

Case 2. Suppose that C does receive in A' all the items it receives in A (as provided by CON). These items contribute $2k^2$ points to the Borda score of C. The remaining items, if allocated only to A or B, contribute $4k^2$ points to the Borda sum of the allocation. Such an allocation cannot give each player a Borda score strictly greater than $2k^2$ points.

Because no allocation in A' can provide a Borda score strictly greater than $2k^2$ to each player, I conclude that A, which gives each player exactly this score, is Borda maximin.[15]

It remains only to show that A can be extended to situations in which there are more than three players. Consider any number $n - 3$ of additional players D, E, F, ..., all of which rank $(n - 3)k$ additional items in such a manner that D ranks k of these items best, E a different set of k items best, F the next k items best, and so on. A, B, and C rank these additional items below items 1, 2, ..., $3k$, which they rank according to CON. More concretely, consider the following example, in which $n = 6$ and $k = 2$:

Example 10.8

A:	1	2 3 4 5 6	7 8	9 10	11 12
B:	1	2 3 4 5 6	7 8	9 10	11 12
C:	1	3 5 6 2 4	7 8	9 10	11 12
D:	7 8	9 10	11 12	1 2 3 4 5 6	
E:	9 10	11 12	7 8	1 2 3 4 5 6	
F:	11 12	7 8	9 10	1 2 3 4 5 6	

---

[15] There may be other Borda maximin allocations, including unequal ones, as I will show in Example 10.9. The one generated by CON is distinguished by its always ensuring envy, which other efficient allocations that equalize the Borda scores of players may not do.

I claim that the following allocation, in which A, B, and C receive their CON allocation for the first six items, and D, E, and F receive their two best items from among the remaining six, is Borda maximin:

$$(1\ 5, 2\ 4, 3\ 6, 7\ 8, 9\ 10, 11\ 12) \qquad [20, 20, 23, 23, 23]$$

In order to try to raise the minimum Borda score of A, B, and C above 20, and hurt D, E, and F as little as possible, one could give each of A, B, and C an item possessed by D, E, and F. But giving them the next-best item of each of the latter players would lower D, E, and F's Borda scores to 12, far below the 23 points they now receive and the present minimum Borda score of 20. This example can be generalized to show that a CON allocation to A, B, and C of the first $3k$ items, and an allocation to each of the remaining $n - 3$ players of its k best items, is not only Borda maximin but also envy-ensuring (C envies A). ∎

The next proposition shows that sometimes an escape from efficient, envy-ensuring Borda maximin allocations is possible.

Proposition 10.8. *Efficient Borda maximin allocations that are envy-ensuring may or may not be the only efficient Borda maximin allocations.*

Proof. Proposition 10.7 shows by construction that there is always a preference profile that renders an efficient Borda maximin allocation envy-ensuring if $n \neq 2$. But there may be other efficient Borda maximin allocations that are not envy-ensuring, as illustrated by the following example:

Example 10.9

```
A: 1 2 3 4 5  6  7  8 9 10 11 12
B: 1 2 3 4 5  6  7  8 9 10 11 12
C: 1 7 3 9 5 10 11 12 2  4  6  8
```

Because this preference profile is the product of CON, the allocation

1. (1 3 5 11,   2 4 6 8,   7 9 10 12)   [32, 32, 32, 32]

is Borda maximin and envy-ensuring. However, there are three other allocations, which give the same (equal) Borda scores, that are not envy-ensuring:

2. (1 2 6 11,   3 4 5 8,   7 9 10 12)   [32, 32, 32, 32]
3. (1 5 6 8,   2 3 4 11,   7 9 10 12)   [32, 32, 32, 32]
4. (1 2 4,   3 5 6 8 11,   7 9 10 12)   [32, 32, 32, 32]

In all four allocations, notice that C receives the same four items, but A and B receive different allocations. Only in allocation 4 do the players receive different numbers of items (A—3, B—5, C—4).

It is easy to show that allocations 2, 3, and 4 are not envy-ensuring. The fact that there are Borda maximin allocations different from that generated by

CON mitigates somewhat the conflict between Borda maximin and envy-ensuringness. However, this is not the case in both Example 10.7 and the following next-larger example generated by CON ($n = 3$, $k = 3$):

Example 10.10
    A: 1 2 3 4 5 6 7 8 9
    B: 1 2 3 4 5 6 7 8 9
    C: 1 5 3 7 8 9 2 4 6

The CON allocation, (138, 246, 579), which has Borda scores of [18, 18, 18], is the *only* (nonisomorphic) Borda maximin allocation. In both this example and Example 10.7, therefore, an escape from the conflict between Borda maximin and envy-ensuringness is impossible. Indeed, there are other examples, not generated by CON, in which every Borda maximin allocation is envy-ensuring:

Example 10.11
    A: 1 2 3 4 5 6
    B: 1 2 3 4 5 6
    C: 1 5 4 6 2 3

There are two allocations—(14, 23, 56) and (13, 24, 56), each giving a minimum Borda score of 8—which are Borda maximin as well as being maximin (no player gets worse than its 4th-best item). In addition, there is a third maximin allocation, (12, 34, 56), which is not Borda maximin. *All* these allocations ensure envy.[16] Example 10.11 can be generalized to all $n \times 2$ cases, in a manner analogous to the extension pattern used in Example 10.8, to show that all maximin and Borda maximin allocations are envy-ensuring. ■

I have shown in this section that there are always preference profiles that render maximin and Borda maximin allocations envy-ensuring (except when there are two players in the case of Borda maximin allocations). Indeed, there is a profile in which every maximin allocation is envy-ensuring, whatever the number of players n and the number of items k that each player receives. While there is always a Borda maximin allocation that ensures envy, except when $n = 2$, there may be other Borda maximin allocations that do not ensure envy.

That neither of the maximin criteria rules out assured envy highlights the dilemma of helping worst-off players—doing so may actually guarantee that they envy other players. As I will show next, escaping envy may require abandoning maximin or Borda maximin allocations.

---

[16] This 3-player, 6-item example is actually stronger than Example 10.7, which has maximin allocations, like (14, 23, 56), that are not envy-ensuring. In Example 10.11 there are none—all maximin and Borda maximin allocations are envy-ensuring.

10.5. FINDING ENVY-UNENSURING ALLOCATIONS

With the knowledge that both maximin and Borda maximin allocations can ensure envy, I turn next to the problem of finding efficient allocations that do not lead to this unhappy outcome. Fortunately, in all situations in which maximin and Borda maximin allocations are envy-ensuring, there is always at least one efficient allocation that is *envy-unensuring*—that is, either envy-possible or envy-free—and so does not ensure envy.

Proposition 10.9. *For any preference profile, there is always an efficient allocation that is envy-unensuring if there are at least twice as many items as players.*

Proof. Consider an arbitrary ordering of the players, $P_1P_2P_3 \ldots P_n$, such that each player appears exactly once in the ordering. Assume that each player makes a sincere choice of an item when its (i.e., the player's) turn comes up. These are the round 1 choices. On round 2, reverse the order of choice, and assume again that each player makes a sincere choice on this round when its turn comes up. If there are more items to be allocated, assume the players make sincere choices in any order.

This selection order appears below as the round 1 subsequence (terminated by the first slash), followed by the round 2 subsequence (terminated by the second slash) and, if necessary, an arbitrary ordering of players to complete the selection order (indicated by asterisks):

$$P_1P_2 \ldots P_n / P_nP_{n-1} \ldots P_1 / \text{***}.$$

The resulting allocation is efficient, which follows immediately from Proposition 10.1 because sincere choices are made throughout the sequence. In addition, I claim that this allocation is envy-unensuring. For $i < j$, P_i will choose, as its first item, one that it ranks higher than any item chosen subsequently by P_j. Thus, the allocation to P_j will not dominate that to P_i with respect to P_i's preferences, so P_i will not assuredly envy P_j. By the same token, P_j will not assuredly envy P_i, because P_j makes a second choice before P_i, precluding P_i's allocation from dominating P_j's allocation with respect to P_j's preferences. Since the order of players' making sincere choices in round 1 was chosen arbitrarily, the existence of envy-unensuring allocations is general. ■

The construction in the proof of Proposition 10.9 gives one method for finding an efficient envy-possible or envy-free allocation. It also makes perspicuous why each player must receive at least two items; if not, the reversing of the first subsequence to obtain the second subsequence, which gives all players undominated allocations, would not be possible.

While the construction provides a sufficient condition for an allocation to be efficient and envy-unensuring, it is not necessary—other constructions are possible. For example, assume there are three players and six items, and all the players rank the items exactly the same, which is a worst-case preference profile (i.e., one most likely to induce envy). Then the sequence starting ABCCB . . . , *without* an A following the last B, will result in an envy-unensuring allocation, whether the sequence ends with a third B (ABCCBB) or a third C (ABCCBC). Indeed, there are three other structurally different sequences, in which A makes only a first appearance in a sequence (ABBCCC, ABCCCB, ABCBCC), that yield allocations that are envy-unensuring, even in this worst-case scenario in which the three players have identical preferences.

In all five sequences, A's sincere first choice is sufficient to preclude A's envying, with certainty, any other player (because A's top-ranked item might be more valuable to it than all the other items combined). But how can one be sure that the other two players will not be envious? In fact, it is not difficult to show the following:

Condition for Envy-Unensuringness. *Given sincere choices, a sequence yields an envy-unensuring allocation, whatever the preference profile of the players, if, for each pair of players S and T, there is* not *a matching such that the first S to appear in the sequence is ahead of the first T to appear, the second S to appear is ahead of the second T to appear, and so on.*

No such one-to-one matching is possible if the number of appearances of the players is different, so the condition presumes the same number of appearances. In the envy-unensuring sequence given by the construction in the proof of Proposition 10.9 (e.g., ABCCBA for three players and six items), the players do appear the same number of times. Moreover, because in this sequence there is *not* the stated matching for each pair, the Condition for Envy-Unensuringness is satisfied. Thus in sequence ABCCBA, the first A precedes the first B, but the second A follows the second B.

The Condition for Envy-Unensuringness, while necessary and sufficient if the players' preferences are identical, is not necessary if their preferences are different. That is, there may be other sequences that result in envy-unensuring allocations when players make sincere choices.

In Example 10.1, for instance, sequence ABBA gives (14, 23), and sequence BAAB gives (13, 24), to (A, B);[17] the first allocation is envy-possible, and the second is envy-free. However, there is a third efficient division, (12, 34), induced by sequence AABB, that is envy-possible. Clearly, if the preferences of

---

[17] While ABBA and BAAB are structurally the same sequences, they produce different allocations because the preferences of the players are different, unlike the worst-case scenario I posited earlier in the text.

the two players were identical, this latter sequence would generate an envy-ensuring allocation.

To summarize, while the construction given in the proof of Proposition 10.9 suffices to give an efficient, envy-unensuring allocation, it may not be the only one to do so. This is true even if each player appears the same number of times in a sequence (e.g., as in AABB), which makes a matching possible but does not satisfy the Condition for Envy-Unensuringness.

A unique envy-free allocation that is maximin and Borda maximin, such as (13, 24) in Example 10.1, would seem to be the fairest of efficient allocations. Unfortunately, two problems can beset such envy-free allocations.

Proposition 10.10. *For a given preference profile, an envy-free allocation may not exist. If one exists, it may be inefficient and the only envy-free allocation.*

Proof. Consider

Example 10.12
    A: 1 2 3 4
    B: 1 3 4 2

wherein there are two efficient maximin and Borda maximin allocations, both of which are envy-possible:[18]

    1. (12, 34)   [7, 5]
    2. (23, 14)   [5, 6].

There is also one efficient, envy-ensuring allocation:

    3. (24, 13)   [4, 7]

In allocation 3, A envies B because A prefers B's item 1 to one of its items (item 2), and B's item 3 to the other of its items (item 4).

Using Proposition 10.1, it is not hard to show that these three allocations exhaust the efficient allocations that give each player two items. Because none of the inefficient or unequal allocations in this example is envy-free either, this example yields no envy-free allocation.

Consider next

Example 10.13
    A: 1 2 3 4 5 6
    B: 4 3 2 1 5 6
    C: 5 1 2 6 3 4

---

[18] This and the next two examples are drawn from Brams, Edelman, and Fishburn (2001), wherein they are discussed as "paradoxes of fair division." Here I present them as further obstacles in precluding envy via efficient or equal divisions.

The unique envy-free allocation to (A, B, C) is (13, 42, 56), whereby A and B get their 1st-best and 3rd-best items, and C gets its 1st-best and 4th-best items. Clearly, A prefers its two items to those of B (which are A's 2nd-best and 4th-best items) and those of C (A's two worst items). Likewise, B and C prefer their items to those of the other two players. Consequently, all three players prefer their items to those of the other two players, so the allocation is envy-free.

Compare this allocation with (12, 43, 56), whereby A and B receive their two best items, and C receives, as before, its 1st-best and 4th-best items. This allocation Pareto-dominates (13, 42, 56), because two of the three players (A and B) prefer their items in (12, 43, 56), whereas both allocations give player C the same two items (56). It is easy to see that allocation (12, 43, 56) is efficient; it is the product of sincere choices in sequence AABBCC as well as several other sequences.

So far I have shown that efficient allocation (12, 43, 56), which is envy-possible (because C may envy A), Pareto-dominates inefficient allocation (13, 42, 56), which is envy-free. To show that allocation (13, 42, 56) is uniquely envy-free, note first that an envy-free allocation must give each player its best item (because a player not receiving its best item may envy the player that does receive it). Thus, an allocation that does not give a player its best item is either envy-possible or envy-ensuring. Second, even if each player receives its best item, this allocation cannot be the only item it receives, because then that player might envy any player that receives two or more items, *whatever* these items are.

By this reasoning, then, the only possible envy-free allocations in Example 10.13 are those in which each player receives two items, including its top choice. In the six possible allocations in which each player receives its top choice and one other item, only the allocation (13, 42, 56) is envy-free. Therefore, the unique envy-free allocation is inefficient. ∎

I provided in this section a sufficient condition for an allocation to be envy-unensuring. In fact, the proof of Proposition 10.9 showed how to generate an efficient, envy-unensuring allocation. However, because an envy-unensuring allocation may not be envy-free but only envy-possible, I then investigated the existence of envy-free allocations. My main finding was negative: an envy-free allocation may not exist; even if one does, it may be inefficient (Proposition 10.10).

This shows that players may have to make a difficult choice, as in Example 10.13, between an efficient allocation that may cause envy and an inefficient one that does not. But if they choose efficiency and an allocation that enables them to avoid envy, none of these may be maximin or Borda maximin. Thus, eschewing envy may carry the unwanted baggage of being inefficient or not helping the worst-off.

10.6. UNEQUAL ALLOCATIONS AND STATISTICS

In this section, I show that the best way to achieve maximin and Borda maximin allocations may require that the players receive unequal numbers of items. I also present statistics on the relative frequency with which maximin and Borda maximin allocations, when they are equal, ensure envy. These statistics give one insight into how serious a conflict there is between helping the worst-off and avoiding envy.

In section 10.2 I argued that the maximin property is not meaningful in comparing allocations that give players different numbers of items (though I suggested an extension of this property that makes unequal allocations more comparable; see note 5). By contrast, because Borda maximin takes into account *all* items in a player's set, valuing them by the same standard, it renders comparable sets that contain different numbers of items.

One might suppose that Borda maximin as well as Borda maxsum will always favor equal allocations, but this is not the case.

Proposition 10.11. *Allocations that give unequal numbers of items to players may be the only ones that are Borda maximin; they may also be Borda maxsum.*

Proof. Consider

Example 10.14
    A: 1 2 3 4 5 6 7 8 9
    B: 3 1 2 4 5 6 7 8 9
    C: 4 1 2 3 6 5 7 8 9

There are exactly two *unequal* allocations, (12, 357, 4689) and (12, 3589, 467), that maximize the minimum Borda scores of players, which are [17, 17, 17] for both allocations.[19]

On the other hand, among *equal* allocations there are two allocations, (129, 357, 468) and (129, 358, 467), that maximize the minimum Borda scores of players, which are [18, 17, 16] for the first allocation and [18, 16, 17] for the second allocation. Because the worst-off player in the equal allocations receives fewer points (16) than the worst-off (and best-off) player in the two unequal allocations (17 points), the unequal allocations are Borda maximin among *all* allocations. Also, Borda scores for all the aforementioned allocations (equal and unequal) sum to 51, which, it can be shown, is the maximum sum among all possible allocations in Example 10.14 (i.e., Borda maxsum). ∎

---

[19] Unlike Example 10.9, in which the unequal Borda maximin allocation was accompanied by three equal ones, the unequal Borda maximin allocations in Example 10.14 are the only ones.

Whereas an unequal allocation may be both egalitarian and benefit-inducing (e.g., the Borda maximin and the Borda maxsum allocation in Example 10.14), Borda maxsum allocations may also be very inegalitarian (e.g., Example 10.1). These examples illustrate that a wide range of allocations is possible satisfying different normative criteria.

To obtain a better idea of the fundamental trade-off discussed in this chapter, a computer simulation was done to calculate the relative frequencies that efficient maximin and Borda maximin allocations are envy-ensuring, based on a sample of several hundred randomly generated preference profiles.[20] The percentages of *equal* efficient (n, k)-allocations that are (i) envy-ensuring, (ii) envy-ensuring and maximin, and (iii) envy-ensuring and Borda maximin are shown below:[21]

(n, k)-Allocation:	(2, 2)	(2,3)	(2, 4)	(2,5)	(3, 2)	(3, 3)	(4, 2)
Envy-ensuring	29	24	8	5	53	25	68
Maximin	22	8	4	3	37	18	52
Borda maximin	0	0	0	0	6	0	12

Reading down the columns, maximin, but especially Borda maximin, act like sieves to reduce the proportion of envy-ensuring allocations. In fact, from Proposition 10.5 it is known that there are no two-item Borda maximin allocations that are envy-ensuring (first four columns).

Reading across the rows in the two-player and three-player cases, observe that an increase in the number of items also reduces the proportion of allocations that are envy-ensuring. Although the (3, 3) percentage for Borda maximin is 0, Example 10.10 shows that this is not a null set—as is no other Borda maximin (n, k)-allocation in which there are three or more players equally dividing six or more items.

It is evident that envy-ensuringness is more worrisome the more players there are, and the fewer the number of items they must divide equally. Indeed, in the case in which there are n players that must divide exactly n items (not shown in the table), envy becomes a virtual certainty as the number of players increases. While there is welcome relief from envy when the allocation is Borda maximin for two players that receive at least two items each, beyond

[20] The size of the samples was governed by computer memory and computational capabilities, varying from 500 for all but the (n, k) cases of (3, 3) and (4, 2), for which the samples were 200.

[21] I focus on equal allocations, because maximin allocations are not meaningful for unequal allocations (see note 5). While Borda maximin allocations render unequal allocations comparable with equal ones, unequal Borda maximin allocations are rare, Examples 10.8 and 10.14 notwithstanding. More to the point of making comparisons, however, there seem to be no examples of unequal Borda maximin allocations that are envy-ensuring. For this to happen, a player would have to envy another receiving at least as many items; I have found no example of this and suspect one does not exist.

two players there is always a preference profile in which both maximin and Borda maximin allocations are envy-ensuring.

10.7. SUMMARY AND CONCLUSIONS

While maximin and Borda maximin allocations cannot prevent envy (except for Borda maximin when there are two players), the good news is that these allocations reduce the probability that it occurs. The other good news is that there is always an efficient allocation that is not envy-ensuring, given that each player receives at least two items, but it will not necessarily be a maximin, a Borda maximin, or a maxsum allocation.

That an efficient, envy-unensuring allocation may not satisfy either of the two maximin criteria may require that a wrenching choice be made between helping the worst-off and avoiding envy. This choice is even more difficult when each player is entitled to only one item, because in this case envy is most difficult to escape (each player must desire a different item). To the best of my knowledge, the fact that Rawlsian (Rawls, 1971) criteria like maximin and Borda maximin may ensure envy rather than prevent it has not heretofore been investigated in a framework in which simple preference rankings are assumed.

While this clash is reminiscent of Arrow's (1951, 1963) impossibility theorem, wherein the conflict among several reasonable conditions for aggregating individual choices into a social choice is unavoidable, fair division is not generally a problem of collective choice. Rather, it is usually a private issue, as when parents decide how a family estate will be divided among their children.

But larger entities, including countries, may also face such issues, in which case millions of people may be affected by how fair-division questions are resolved. A case in point is the breakup of Czechoslovakia into the Czech Republic and Slovakia in 1993, in which the two sides had to split a number of indivisible items, including certain military assets. Likewise, the division of both Germany and Berlin after World War II, while ostensibly involving the Allies' drawing borders over the divisible good of land, also included, by extension, indivisible goods like universities and opera houses that were situated on this land.

In such cases, allocations may be inefficient because of practical constraints that players place on the division. More surprising is that independent of any practical constraints, requiring that an allocation be envy-free may force it to be inefficient.

Although the symbolic value of giving players equal numbers of items, such as landing slots at an airport, may be important, equal allocations may violate the two maximin criteria and, as well, ensure envy. To be sure, unequal divisions can also force a violation of these criteria, so neither equality nor in-equality can be held up as an egalitarian ideal.

The computer simulation showed that Borda maximin allocations ensure envy less often than maximin allocations. The chances of assured envy also become less as the number of items increases, but they never fall to zero as long as there are not two players. While envy may be ineradicable if one desires to help the worst-off, it is not clear that abandoning the maximin criteria to avoid it is a better alternative.

11

Allocating a Single Homogeneous
Divisible Good: Divide-the-Dollar

11.1. INTRODUCTION

Much work in the mathematical social sciences is devoted to showing the conditions under which individually rational actions can lead to collectively inferior outcomes. This problem is epitomized by the game of Prisoners' Dilemma, in which each player has a dominant, or unconditionally best, strategy of not cooperating, but the resulting outcome, and the unique Nash equilibrium, is worse for both players than if they had both cooperated. The game is nonzero-sum in that both players may win (by cooperating) or lose (by not cooperating) simultaneously; what one wins is not necessarily equal to what the other loses, making the sum of their payoffs zero.

This clash between individual and collective interests is also illustrated by another nonzero-sum game, *divide-the-dollar* (DD), wherein two players, A and B, independently propose a division of a dollar into cents, with each demanding a certain amount. If the players are equally entitled to it, the question is how to induce the players to divide it equally, with minimal sanctions for not being egalitarian (i.e., bidding 50 cents each).

I stress "minimal," because it is desirable that egalitarian behavior, or something close to it, emerges as a consequence of the players' rational calculations, not be imposed by an outside party or be the product of dire threats. Of course, a dollar could be any *divisible good*, the sum of whose parts is the same however it is divided (i.e., it loses no value when it is split between two people).

A dollar is also a *homogeneous good*, whose component parts (e.g., cents) are valued the same by all players. This is not true of a *heterogeneous good*, like a cake of different flavors, whose various parts players may value differently (see chapter 13).

The usual rules of DD specify that each player receives whatever it bids if the sum of the bids does not exceed 100 cents; otherwise, the two players

Note: This chapter is adapted from Brams and Taylor (1994) with permission of Springer Science and Business Media; see also Brams and Taylor (1996, ch. 8).

receive nothing. An intriguing feature of this game is that every ordered pair $(x, 100 - x)$ for A and B, respectively, where $0 \leq x \leq 100$—as well as (100, 100)—is a Nash equilibrium, whereby neither player has an incentive to depart unilaterally from such a strategy lest it do worse by doing so.[1]

To illustrate, assume that the players propose (50, 50). If A raises its bid (say, to 51) or lowers it (say, to 49), it would do worse by receiving 0 in the first case and 49 in the second, assuming that B sticks with its bid of 50. Similarly, if the players propose (100, 100), each receives 0. But because neither can do better by raising or lowering its bid, (100, 100), like (50, 50), is stable in the sense of Nash.[2]

Unlike Prisoners' Dilemma, in which the players' dominant strategies lead to the inefficient Nash equilibrium, the problem in DD is that there are 102 pure-strategy Nash equilibria as well as many mixed-strategy equilibria (Myerson, 1991, p. 112).[3] In the absence of dominant strategies, it is difficult for the players to coordinate their bids in order to select one of the 101 $(x, 100 - x)$ efficient Nash equilibria—and avoid the (100, 100) inefficient Nash equilibrium—or nonequilibrium bids like (50, 60), that give the players payoffs of 0.

To be sure, only one of the plethora of efficient equilibria in DD gives the *egalitarian outcome* of 50 cents to each player, which is the unique symmetric Nash equilibrium in pure strategies that is also efficient. While this egalitarian outcome would seem the evident focal point of the players, there is little besides its "prominence," as Schelling (1960, pp. 56–58) puts it, to commend it as *the* rational choice.

Simple as DD is, this variable-sum game—in which the sum of the payoffs to the players varies between 0 and 100, depending on their bids—highlights the problem in which two players can both benefit if they make "reasonable" bids. However, if either player is too greedy, both may end up getting nothing.

DD has been much discussed in the game theory literature, but solving the coordination problem, much less justifying the egalitarian outcome, has

---

[1] DD was introduced by Nash (1953) and is sometimes called the "demand game"; see van Damme (1991, pp. 145–150) for further analysis than is given here of its game-theoretic properties. In the related "ultimatum game," the proposals of the two players are not independent. A makes a demand for a certain amount of money, with the remainder going to B if it accepts the ultimatum; if not, both players get nothing. See Güth and Tietz (1990) and Forsythe, Horowitz, Savin, and Sefton. (1994) for theoretical and experimental results on the ultimatum and related games.

[2] Bids that each exceed 100, such as (101, 101), also constitute a Nash equilibrium. But they are clearly less attractive than the other Nash equilibria, because each *guarantees* a player a payoff of 0 whatever the other player does. Consequently, I assume that no players would ever make bids above 100 in DD.

[3] *Mixed strategies* involve choosing, according to some probability distribution, different bids at different times, such as 50, say, 80 percent of the time, and 60 20 percent of the time. All mixed-strategy equilibria are inefficient (van Damme, 1991, p. 146).

proved difficult.[4] An attempt to rationalize salient choices by postulating conditions under which players, when confronted with multiple equilibria in a coordination game, reconceptualize their choices in a nonrational "involuntary phase," and solve a different (and more tractable) game in a "reasoning phase," is given in Bacharach (1993).

The approach taken here is different. I explicitly postulate new rules and show how they resolve the coordination problem by singling out the egalitarian outcome as uniquely rational.[5] Because the payoff scheme embodied in the new rules does not penalize the players for greediness to the degree that DD does, the egalitarian outcome can be obtained without the need to make incredible threats.

These rules, which describe variants of DD when there are two or more players, involve the following:

- changing the payoff structure of DD to reward the lowest bidders first (DD1);
- adding a second stage that provides the players with new information yet restricts their choices at the same time (DD2);
- both changing the payoff structure and adding a second stage (DD3).

I illustrate the possible applicability of the different procedures to a real-world allocation problem (setting of salaries by a team), in which there may be entitlements. I conclude by assessing the strengths and weaknesses of the different procedures and their solutions.

11.2. DD1: A REASONABLE PAYOFF SCHEME

DD treats harshly bidders whose total request exceeds 100. Is it possible to render its payoff structure more reasonable, in some sense, and also to induce *egalitarian behavior*—bids of 50 by each player—as well as the egalitarian outcome of 50 cents to each player?

I use the term "reasonable" to rule out payoff schemes that induce egalitarian behavior by means of punishment for noncompliance to some preset standard. For example, a payoff scheme that gives each player 50 cents if it bids 50, and nothing otherwise, trivially induces egalitarian behavior. It accomplishes this feat, however, by dictating that some standard (i.e., bids of 50 by

---

[4] Philosophers, such as Lewis (1969) and Ullman-Margalit (1977), have argued that the coordination problem gets solved by the establishment of conventions or norms, but the rational foundations of this literature have been challenged (Gilbert, 1989, 1990).

[5] The consequences of another set of rules, which allows play to progress to a new round if the players make incompatible offers, is analyzed in Chatterjee and Samuelson (1990).

each player) must be met, lest the players get nothing,[6] which is the kind of Hobbesian solution I wish to rule out.

When a Leviathan steps in, the resulting coercive solution deprives people of the interesting and morally difficult choices they would otherwise face in real-life situations. The barbaric nature of such solutions also limits their practical use and effectiveness, especially insofar as they proscribe certain behaviors. Thus, the death penalty as a type of punishment has questionable deterrent value, perhaps in part because it has been only fitfully administered; it is no longer used in many countries. In the United States, the prohibition of alcoholic beverages from 1920 to 1933 by the Eighteenth Amendment is an example of a ban that proved easy to violate and was revoked.

Call a payoff scheme in DD for n players *reasonable* if it satisfies the following five conditions:

1. Equal bids are treated equally.
2. No player receives more than what it bids.
3. If 100 units are sufficient to give every player what it bids, then these bids are the amounts disbursed to the players.
4. If 100 units are insufficient to give every player what it bids, then the 100 units are, nevertheless, completely disbursed to the players.
5. If all bids are greater than the egalitarian level of $100/n$, where n is the number of players, then the highest bidder does no better than the lowest bidder.

The question now becomes: Can one alter the payoff structure of DD in a reasonable way so that the egalitarian outcome is a solution, in some sense, in the corresponding game? (Note that the payoff scheme of DD violates condition 4.) Of course, "solution" could involve any one of a number of things, including dominant strategies, the iterated elimination of dominated strategies, and Nash equilibria.

In fact, the iterated elimination of *weakly dominated strategies*—strategies that are never better, and sometimes worse, than another strategy—is the concept I shall employ and illustrate shortly.[7] Such strategies always yield a Nash

---

[6] This payoff scheme is even harsher than DD, because it penalizes underbidding (i.e., bidding less than 50), whereas DD gives players at least their underbids, not 0, unless they bid 0. Clearly, there is a continuum of schemes that punish players for deviating, in varying degrees, from preset standards, but I do not consider them further for reasons given in the text. An experimental and theoretical literature that assumes players make bids according to the rules of DD, but have only incomplete information about the resources to be divided, is also worth noting; see, for example, Rapoport and Suleiman (1992) and references therein.

[7] This is the same solution concept used in Abreu and Matsushima (1994), who show in a different context how the severity of punishment can be mitigated by implementation via this concept. See also Jackson, Palfrey, and Srivastava (1994) and Sjöström (1994).

equilibrium. Like all reasonable schemes, this equilibrium does not result in egalitarian behavior, but it does lead to an egalitarian outcome:

Proposition 11.1. *Egalitarian* behavior *is weakly dominated under every reasonable payoff scheme for n = 2. However, there is a reasonable payoff scheme, which works for all n, that yields the egalitarian* outcome *as the result of unique bids remaining after the iterated elimination of weakly dominated strategies.*

The proof of Proposition 11.1 is given in the appendix to this chapter. The proof of the second part of the theorem relies on showing that there is a reasonable payoff scheme that yields the egalitarian outcome as the result of unique bids remaining after the iterated elimination of weakly dominated strategies. Of particular interest here is the fact that the procedure works for more than two people.[8] The payoffs are made according to the following

Payoff Scheme (PS). One starts with the lowest bidder—or, more generally, with the group tied for lowest bid—and pays them what they bid if there is enough money to do so. If there is not enough money, the money available is divided evenly among this group. One next moves to the group tied for the second-lowest bid and proceeds in exactly the same way, but now working with only the money that is left after the group of lowest bidders has been paid. One continues in this fashion until the money is exhausted, after which no one else is rewarded, or all the players receive what they bid.[9]

Example 11.1

Suppose that four players (A, B, C, D) bid (10, 60, 60, 80). Under PS, A would be paid first and receive its full 10 bid, leaving 90 remaining. B and C are tied for second-lowest bid (60) and are thus paid next. Since only 90 is left (and not the 120 that would be needed to pay them both what they bid), each

---

[8] Moulin and Shenker (1992) independently proposed a related scheme ("serial cost sharing"). Showing that it uniquely satisfies certain properties (including coalitionproofness), they applied it to cost sharing and surplus sharing. Their allocation mechanism specifies how much players share in the costs of developing a technology, based on how much they demand of its use, whereas I assume no distinction between inputs (costs of development) and outputs (benefits of technology). In the present setup, the bids and payoffs are in the same currency (i.e., money), and doing well means getting a larger monetary payoff. By contrast, in the Moulin-Shenker scheme, doing well means paying lower costs in the development of a technology that will later be shared according to the players' demands. For further results on serial cost sharing, see Moulin (1994) and work cited therein. Anbarci (2001) shows that a modification of condition 4 and elimination of condition 5 induces egalitarian behavior in an iterated strict-dominance equilibrium, but unlike DD1, his scheme may reward greed.

[9] Demange's (1984) bidding procedure employs a similar payoff scheme, but her scheme becomes operative by default—only when a player objects to the highest bidder's proposed allocations to all players. Her procedure generalizes Crawford's two-person procedure to *n* players; see Young (1994, pp. 143–145) for further discussion.

receives 45. The money has now been exhausted, so A, which bid 80, gets nothing. If, instead, the bids had been (40, 50, 80, 80), the payoffs would have been (40, 50, 5, 5).

I call DD with PS—instead of the usual payoff scheme—DD1, which defines a new procedure and hence a new game (i.e., with different rules of play). It is worth pointing out that there is still a multiplicity of equilibria under DD1. For example, egalitarian bids of 25 by the four players, or by only some players with the others bidding 26, are also Nash equilibria (see the appendix). But it is easy to see that no bids other than 25 or 26 can constitute a Nash equilibrium, because either a higher bidder would always have an incentive to lower its bid, or a lower bidder would have an incentive to raise its bid.

Thus, all the other equilibria that PS generates are only slight perturbations of the Nash equilibrium of bidding $b = 26$ by the four players, which reinforces the latter's robustness as the solution of DD1. Nonetheless, the fact that this solution requires 75 iterations to find (50 iterations would be required if there were only two players choosing $b = 51$, with the number of iterations approaching 100 as the number of players increases), is troublesome. I will return to this issue after analyzing DD3, but next I describe and analyze DD2.

11.3. DD2: ADDING A SECOND STAGE

Proposition 11.1 shows that there does not exist a reasonable payoff scheme for DD that yields egalitarian behavior, although DD1 comes very close in the sense of inducing equal bids (of 26 each in Example 11.1) that give the egalitarian outcome. This raises a new question: Can one change the rules of DD so that the revised rules, together with a reasonable payoff scheme, yield egalitarian behavior?

My starting point is a new two-player game, DD2, which provides for a 2nd stage but retains the old (unreasonable) payoff scheme of DD. In adding a 2nd stage, I introduce a new solution concept, "dominance inducibility."

An outcome is *dominance inducible* by a player if that player has an opening move, or a choice, that, when known to the other player, yields the outcome as the result of the successive elimination of weakly dominated strategies in the subgame describing what takes place after the opening move.[10]

---

[10] Moulin (1979) called an outcome *dominance solvable* if it is the result of the successive elimination of weakly dominated strategies, which is true of the egalitarian outcome of DD1. By contrast, dominance inducibility presupposes a prior choice by a player that makes the subsequent subgame—in particular, that played in stage II of DD2—dominance solvable. As I show next, players in the stage II subgame always have dominant strategies, so the elimination of dominated strategies is immediate, not iterative.

The rules of play of DD2, which is applicable only to two players, are as follows:

1. In stage I, players A and B make *initial* bids, x and y, which are made public.
2. In stage II, A and B make their *final* bids, b_1 and b_2, but the players are restricted in their choices to just x or y.
3. Based on the final bids in stage II, which are made public, the usual rules of DD determine the payoffs. If $b_1 + b_2 \leq 100$, A and B receive b_1 and b_2, respectively; otherwise, each receives 0.

Rule 2 may be interpreted as allowing a player to

• *affirm* its stage I bid; or
• *usurp* the other player's stage I bid.[11]

Of course, in stage I each player is free to choose any feasible stage I bid between 0 and 100. Consequently, the addition of stage II to DD enormously increases the number of possible strategies available to each player from 101 to 101×2^{100} because, for each of the 101 possible bids at stage I, each player may choose either its own bid, or the other player's bid, at stage II. (Because in exactly one of each of the 101 possible binary-choice situations at stage II the other player's bid will be the same as the first player's, each has 100 instead of 101 "my bid or your bid" decisions to make at stage II and hence 2^{100} *distinct* choices at this stage.)

This explosion of strategies would appear to complicate play of DD greatly. Surprisingly, however, just the opposite is the case: the addition of stage II singles out the egalitarian outcome as the only one that is dominance inducible by either player, as shown by

Proposition 11.2. *In DD2 for two players, the egalitarian outcome is dominance inducible if either player bids 50. Moreover, it is the only outcome that is dominance inducible.*

Proof. Assume that A bids $x = 50$ and B bids y in stage I. I will speak of 50 and y as being *on the table* at this point because they are made public. The subgame describing what takes place once these bids are on the table, and therefore common knowledge, involves each player's independently choosing either 50 or y, with the payoffs as in DD. If $y = 50$, then "choose 50" is the only choice available (since there is only one number, 50, on the table at this point), and this choice dominates all others by default. The remaining two possibilities, $y > 50$ and $y < 50$, give the following payoff matrices:

---

[11] "Usurp" is not meant to imply that a bid is taken away from the player usurped; if a player's bid is usurped, it can still affirm its own bid or usurp the other player's bid.

Case 1. $y > 50$

                  B

	$x = 55$	$y > 50$	
$x = 50$	(50, 50)	(0, 0)	← Dominant strategy (weak)
$y > 50$	(0, 0)	(0, 0)	

A (left label)

              ↑

Dominant strategy (weak)

Case 2. $y < 50$

                  B

	$x = 50$	$y < 50$	
$x = 50$	(50, 50)	(50, y)	← Dominant strategy (strong)
$y < 50$	(y, 50)	(y, y)	

A (left label)

              ↑

Dominant strategy (strong)

The dominant strategies in stage II associated with $x = 50$ in these two cases demonstrate that the egalitarian outcome is dominance inducible by either player (A in the foregoing cases, which chooses $x = 50$ in stage I).

To see that no other outcome is dominance inducible by either player, note first that if a play of x_0 by A always leads to outcome (x_0, y_0) via the successive elimination of weakly dominated strategies, then $x_0 = y_0$; otherwise, some stage I bids by B would preclude the outcome (x_0, y_0) from even occurring, except when $x_0 = 0$, which is clearly dominated for A. Moreover, an outcome of (x_0, y_0) for $x_0 > 50$ is impossible. So it suffices to show that for $x_0 < 50$, the outcome (x_0, y_0) is not dominance inducible by, say, A.

To see this, consider the scenario wherein A bids $x_0 < 50$ at stage I and B bids $100 - x_0$ at stage I. Then, in the subgame describing stage II, B does strictly better by affirming $100 - x_0$ (when A affirms x_0)[12] than by usurping x_0. Thus, "choose x_0" is not the result of the successive elimination of weakly dominated strategies in the subgame unless $x_0 = 50$. ∎

In fact, it turns out that if $x \neq 50$ and $y \neq 50$, there are four qualitatively different types of games that can occur, only two of which have a Nash equilibrium in pure strategies:

[12] If A usurps $100 - x_0$, both players get 0.

Game 1. x > 50 and y > 50.

The payoffs of the players are (0, 0) at every outcome in the game matrix associated with stage II, ruling out these choices as part of a Nash equilibrium unless $x = 100$ and $y = 100$. If $x \neq 100$ and $y \neq 100$, then both players could have done better if A had bid $100 - y$, or B had bid $100 - x$, at stage I.

Game 2. x > 50 and y < 50 and x + y ≤ 100.

This game is illustrated by $x = 60$ and $y = 40$. The payoff matrix for stage II is as follows:

B

		$x = 60$	$y = 40$
	$x = 60$	(0, 0)	(60, 40)
A			
	y = 40	(40, 60)	(40, 40)

Strategies associated with outcomes (40, 60) and (60, 40) are the pure-strategy Nash equilibria in this stage II game, which gives it some resemblance to the game of Chicken.[13] But if, say, $y = 35$, then the strategies associated with (35, 60) would not be a Nash equilibrium, because A could have done better by choosing a stage I bid of 40 and affirming it in stage II.

Game 3. x > 50 and y < 50 and x + y > 100.

The payoffs to the players are (0, 0) at every outcome in the stage II game matrix, except the (y, y) payoff associated with the players' joint choice of $y < 50$. But the strategies associated with this outcome do not constitute a Nash equilibrium, because A could have done better by unilaterally defecting to a strategy of bidding 50 at stage I and then affirming. This new strategy would give A a payoff of 50, as opposed to $y < 50$, whether B affirms $y < 50$ or usurps 50.

Game 4. x < 50 and y < 50.

The payoffs to the players are exactly what they bid at every outcome, but no strategies associated with these outcomes are in equilibrium because each player could have done better by raising its stage I bid.

In effect, the addition of stage II to DD, making DD2 a two-shot rather than a one-shot game, solves the coordination problem the players have under DD. Although (50, 50) is the prominent Nash equilibrium in DD, as I indicated

---

[13] It would be Chicken if the preferences of the players were strict over the four outcomes, and the (40, 40) outcome in the matrix were next-best for both players. This would be the case by making it, say, (45, 45) in the game matrix. But the rules of DD2 do not allow the payoffs at this outcome to be different from the players' y bids.

earlier, it is not otherwise compelling. What Proposition 11.2 establishes is that *either* player can induce the rational choice of (50, 50) by making, at stage I, an initial bid of 50.

Note that this solution does not require that the players communicate with each other, much less make commitments or binding agreements, so play of DD2 is, in the parlance of game theory, *noncooperative*. The communication, as it were, occurs when the stage I bids are made public and thereby become common knowledge. Given that just one player bids 50 at this stage, the other player can do no better than also bid 50 at stage II.

There is a natural generalization of DD2 to the case where there are $n > 2$ players: all n players bid independently at stage I, and then each can, at stage II, choose to affirm its bid or usurp any of the other bids. If the sum of the stage II bids is at most 100, then each player receives what it bid. Otherwise, all players receive payoffs of 0. But unlike the situation when $n = 2$, when $n > 2$ the solution given by Proposition 11.2 does not generalize, as shown by

Proposition 11.3. *In DD2 for $n > 2$ players, no outcome is dominance inducible by any player.*

Proof. As in the proof of Proposition 11.2, any outcome that is dominance inducible by a player must be of the form $(x_0, x_0, \ldots, x_0)$, where x_0 is the stage I bid of that player. Let y_0 be the bid corresponding to the egalitarian outcome. If x_0 were less than y_0, then a bid of y_0 by any other player and a choice by every player to usurp y_0 would yield an outcome strictly better than $(x_0, x_0, \ldots, x_0)$. If x_0 were greater than y_0, then the outcome $(x_0, x_0, \ldots, x_0)$ would be impossible. Finally, if $(x_0, x_0, \ldots, x_0)$ is the egalitarian outcome, then one need only consider the case in which B bids $x_0 - 1$ at stage I, C bids $x_0 + 1$ at stage I, and everyone else bids x_0 at stage I. That is, given these stage I choices, it is now easy to construct scenarios at stage II in which (i) $x_0 + 1$ is a strictly better choice for A than is x_0; (ii) x_0 is a strictly better choice for B than is $x_0 - 1$; and (iii) $x_0 - 1$ is a strictly better choice for some third player C than is $x_0 + 1$. Hence, "choose x_0 at stage II" is not the result of the successive elimination of weakly dominated strategies. ∎

Although DD2 does not employ a reasonable payoff scheme and breaks down when there are more than two players, the egalitarian outcome it induces does not require the successive elimination of many dominated strategies. Instead, this outcome can be induced immediately by a stage I bid of 50 by either player. The inclusion of stage II, which induces egalitarian behavior as well as an egalitarian outcome, gives DD2 an additional advantage: it makes the procedure *rectifiable* by enabling one player to "correct" its initial bid. Thus, if one player overbids or underbids and the other player bids 50 in stage I, the first player can, in effect, correct its mistake by usurping the bid of 50 by the other player in stage II.

11.4. DD3: COMBINING DD1 AND DD2

The third variant of DD I propose combines the idea of two stages from DD2 with the idea of successively paying off the lowest bidders from DD1. I call this variant DD3, whose rules of play are described next.

In stage I, each of n players independently makes an initial bid. Once these bids are on the table and therefore known to all players, each player has a choice of affirming its own bid or usurping any of the other bids. (Thus, I allow for the possibility that all players choose to affirm or usurp the same bid in stage II.) Once the players choose their stage II bids, payoffs are made according to PS, exactly as in DD1. That is, one starts with those tied for lowest bid and proceeds to pay off as many bidders as one can until either the money is exhausted or all the players receive what they bid. The main result is the following:

Proposition 11.4. *In DD3 for $n \geq 2$ players, the egalitarian outcome is dominance inducible by any player. Only $n - 1$ successive eliminations of dominated strategies are necessary to arrive at bids leading to this outcome.*

Proof. For simplicity, I again consider only the case $n = 4$ with egalitarian outcome (25, 25, 25, 25). Suppose that P1 bids 26 at stage I. I will show that three eliminations of weakly dominated strategies yield "choose 26 from among the four or fewer bids on the table" as the sole remaining strategy choice for all four players at stage II. I first establish two claims:

Claim 1. A choice of 26 at stage II weakly dominates a choice of any $y < 26$ at stage II.

Proof. A choice of 26 guarantees a payoff of at least 25 to the player making this bid. Of course, a choice of y cannot yield a payoff greater than y, so 26 is strictly better than any bid $y < 25$, and at least as good as 25. However, if $y = 25$ and all the other bids are zero and affirmed, then 26 is a strictly better stage II choice than 25.

Claim 2. Suppose that n (or fewer) stage I bids are on the table, one of which is 26. In addition, suppose that at stage II

- no player will choose any $y < 26$; and
- no player will choose any bid among the k largest of those on the table.

Let y be the $(k + 1)^{st}$ largest of the stage I bids. Then a choice of 26 weakly dominates a choice of y at stage II.

Proof. A choice of y, as described in the claim, can yield a payoff of at most 25, and this occurs when everyone else also chooses y. (That is, if anyone else chooses some $y' < y$, then a choice of y yields a payoff of at most $(100 - y')/3 \leq (100 - 26)/3 < 25$.) As before, a choice of 26 guarantees a payoff of at least 25. Thus, a choice of 26 is at least as good in every scenario as a choice of y.

Moreover, if everyone else chooses 26, then a choice of 26 yields a strictly better payoff (26) than does a choice of y (≤ 22).

The proposition now follows immediately from claims 1 and 2. To begin with, the strategies $0, \ldots, 25$ are eliminated in the first reduction (by claim 1); the highest of the four stage I bids is eliminated in the second reduction (this is the $k = 0$ version of claim 2); the next-highest of the stage I bids is eliminated in the third reduction ($k = 1$ in claim 2); and, finally, the third-highest of the four stage I bids is eliminated in the fourth reduction ($k = 2$ in claim 2). The proposition said $n - 1$ reductions (instead of n reductions, as the four just described suggest), but it is easy to see that either the first or the last of the reductions is vacuous. (That is, if some bid is less than 26, then one cannot have three greater than 26.) ∎

DD3 would appear to provide the best of both possible worlds: a reasonable payoff scheme for n players, as in DD1; and the additional information given by stage II of DD2, which speeds up dominance inducibility to only $n - 1$ successive eliminations. But is stage II, an admitted complication, really necessary if DD1 also does the job, albeit in more steps? Before comparing the different solutions, I introduce an additional parameter (entitlements) and indicate how the procedures might actually be applied.

11.5. THE SOLUTIONS WITH ENTITLEMENTS

In this section I illustrate an application of some of the ideas discussed, but in a more general setting than that developed so far. Specifically, assume that

- some players are entitled to more of the amount being divided than others; and
- these entitlements must be reflected in the solution (rather than its being an egalitarian apportionment).

Surprisingly, if there are only two players, each of which has a different entitlement, DD2 fails to provide these, even when there is complete information about these entitlements. To illustrate this result, consider the following example.

Example 11.2
Assume that the (male) president (P) of a company plans to award pay raises to two (female) vice presidents (VP1 and VP2), but he does not think they deserve the same raise. He announces that VP1, whom I call S, deserves only a small (s) raise, and VP2, whom I call B, deserves a big (b) raise.[14]

---

[14] When players have different rights or claims, a variety of allocation rules have been explored using cooperative game theory; see, for example, O'Neill (1982), Aumann and Maschler (1985), Young (1987), Chun (1988), Curiel, Maschler, and Tijs (1988), Chun and Thomson (1992), Bossert (1993), Dagan and Volij (1993), and Fleurbaey (1994), in which solutions are derived from axioms. When different players (e.g., men and women seaching for marriage

So as not to appear dictatorial, however, P says that he will not implement these raises but will instead ask each VP how much she deserves. She may say medium (m) as well as s or b, but there are budgetary limits. Whereas the company can afford one raise of s and one of b, or two of m, requests for more (i.e., one m and one b, or two b's) are unacceptable. In fact, consistent with the rules of DD2 as well as DD, P says that if the VPs make either of the latter pair of (unacceptable) requests, they will get no raises, making the payoff scheme unreasonable.

Despite P's announcement of which VP is more deserving, the unfavored S has no incentive to request only s. To see this, consider the following 3×3 payoff matrix, in which each player may bid s, m, or b in stage I and choose either her bid or the other player's bid in stage II (the payoffs at the conclusion of stage II are shown, whereas the strategies are the stage I bids):

		|	s	m	b
		|		B	
	s	|	(s, s)	(m, m)	?
S	m	|	(m, m)	(m, m)	(m, m)
	b	|	?	(m, m)	$(0, 0)$

These payoffs, except for the question marks, are an immediate consequence of what the players would bid (and receive) in stage II to maximize their payoffs. For example, when S bids s and B bids m in stage I, S can do no better than usurp m in stage II when B affirms m, resulting in (m, m).

As for the question marks, if the players affirm their stage I bids in stage II, their payoffs will be (b, s) in the lower left cell and (s, b) in the upper right cell. But given P's announcement of the deservingness of S and B, the latter payoff assumption, which matches P's announcement, seems much more reasonable than the former. Hence, I assume payoffs of (s, b) in the upper right cell but leave open what the players' payoffs are in the lower left cell.

Even with this cell a question mark, S's strategy of m weakly dominates s. With s eliminated, B's strategy of m weakly dominates her strategy of b in the reduced 2×3 matrix, leaving a further reduced 2×2 matrix. Given the question mark, no further reductions are possible. Nevertheless, I conclude that neither player will bid what P announces to be her entitlement (s for S and b for

---

partners, college students seeking roommates) have different preferences for each other, the problem of finding a stable matching is explored theoretically in Gusfield and Irving (1989) and Roth and Oliveira Sotomayor (1990); the latter work also contains an empirical analysis of the results of applying a matching algorithm to the placement of new physicians in hospitals for their internships and residencies (both the physicians and the hospitals can indicate their preferences for each other, but the hospitals' preferences take precedence, in a certain sense, giving them what might be called a greater entitlement in finding a stable matching).

B), because precisely these strategies are eliminated in the successive reductions. Hence, P's announcement about the deservingness of his VPs has no bite in inducing them to make the "right" (i.e., his proposed) choices.

To sum up, if there are entitlements and they are common knowledge, DD2 does not yield them in the example. DD1, on the other hand, is more successful:

Proposition 11.5. *Suppose $e_1, \ldots, e_n$ are positive integers (entitlements) summing to k, and suppose that n players are allowed to submit bids (claims) $b_1, \ldots, b_n$. Define the greed of player i to be the number $g_i = b_i - e_i$. Assume under DD1 that players are paid off in the order determined by greed (least greedy to most greedy), with ties resulting in an allocation that is proportional to the entitlements of those involved in the tie. [Thus, if two players have entitlements of 8 and 12, and they bid 10 and 14 each, then they would divide a remaining pot of, say, 50 as (8/20)(50) = 20 for the first, and (12/20)(50) = 30 for the second.] Then iterated domination of weakly dominated strategies results in a bid of $e_i + 1$ by player i for each i.*

Sketch of proof. Arguments similar to those in claims 1–8 of the proof of Proposition 11.1 (see the appendix to this chapter) show that any bid $x \le e_i$ by player i is weakly dominated by a bid of $e_i + 1$. Now assume that some bid $x > e_i + 1$ remains after the iterated elimination of weakly dominated strategies. Among all such x's (for all players), choose one making $g_i = x - e_i$ as large as possible. Thus, a bid of x will leave player i as most greedy, or perhaps tied for most greedy. It then follows easily that a bid of x yields player i a payoff of at most e_i. Thus, a bid of $e_i + 1$ in place of x by player i would never be worse, and it would be strictly better if every other player bid one unit more than its entitlement. ∎

As with the egalitarian outcome without entitlements under DD1, no player receives its bid of $e_i + 1$ but instead the exact entitlement of e_i. Thus, optimal bids are perturbations of the egalitarian outcome, but they are only slight, rendering outrageous requests under DD1 nonoptimal.

Example 11.3

As a possible application of DD1 (or DD3), consider a team of players that works closely together (e.g., in a company or on the athletic field). Assume that the team must set raises for its members, based on their previous performance, that are to be taken from a preset pool of money.

Each player (assume they are all female) may request any amount for herself up to the size of the pool. At the same time, each player makes a recommendation of a pay raise for every *other* player, with the sum of her own request and her recommendations for all the other players equal to the pool.[15]

---

[15] De Clippel, Moulin, Plassmann, and Tideman (2006) rule out self-requests but do, as here, allow recommendations of others.

The recommendations for each player—by every other player—are averaged to determine each player's entitlement.

DD1 is then applied, with allocations made in the order of the closest matches between the entitlements and the requests of each player. Given that the players make honest assessments of their teammates, then they have an incentive to make honest assessments of themselves.[16]

To see why, assume that everybody requested the entire pool for herself. This is obviously not a Nash equilibrium, because every player would then have an incentive to lower her request slightly and thereby receive almost the entire pool, leaving almost nothing for anybody else. This logic eventually carries the players toward the ratings they think the other players will give them. Provided players are honest in their assessments of others (there seems to be no good reason why they should not be in the absence of collusion), players can do no better than try to reflect the others' assessments, slightly perturbed, *in their own requests.*

Of course, incomplete information may prevent a perfect match. Nevertheless, I believe DD1 or DD3 would be viewed as fair by the players, because these procedures benefit players whose self-ratings agree with those of others. If there is exaggeration and posturing, it would more likely come in bargaining over the size of the pool available for salary raises rather than in the individual requests by the players.

11.6. SUMMARY AND CONCLUSIONS

Divide-the-dollar (DD), which punishes players by giving them nothing if their bids exceed 100, has a multiplicity of Nash equilibria. Although the egalitarian outcome is prominent, it is not otherwise distinguishable in noncooperative play. Furthermore, DD does not satisfy five conditions of reasonableness that preclude punitive behavior, which is often difficult to enforce even if severe punishment is "on the books." Practically speaking, why threaten such punishment if the threat is likely to be empty and there are, moreover, "softer" ways of inducing reasonable behavior?

An alteration in the payoff structure of DD, whereby the players who bid the least are paid off first (DD1), can induce the egalitarian outcome via iterated elimination of weakly dominated strategies. Like all payoff schemes that satisfy the reasonableness conditions, however, the players' optimal bids are not egalitarian but slightly greater.

---

[16] A related scheme for land taxation also induces players to make honest self-assessments (Niou and Tan, 1994): if they undervalue their property, it will be bought by somebody else; if they overvalue it, their taxes will be too high.

If the rules of DD are revised to add a second stage but leave intact the old payoff structure, the resulting procedure is DD2. Based on dominance inducibility, it gives both the egalitarian outcome and egalitarian behavior if there are only two players, but it fails if there are more players.

Combining the second stage of DD2 with the payoff scheme of DD1 gives DD3. Under this procedure, which, like DD1, is reasonable, the successive elimination of weakly dominated strategies is greatly speeded up, which makes its solution more transparent than that of DD2. Like DD2, it can be implemented if only one player makes an egalitarian bid in stage I because of dominance inducibility; and it is rectifiable, because a player who (mistakenly) does not make an egalitarian bid in stage I can usurp the egalitarian bid of another player in stage II.

In addition to not generalizing to $n > 2$, DD2 does not hold up well when entitlements are introduced, even for two players, whereas DD1—and by extension, DD3—does. Provided the players on a team are sincere in evaluating each other's merits, DD1 and DD3 encourage honest assessments of self-worth, only slightly perturbed upward.

The main advantage that these procedures have over letting others, such as a boss or teammates, be the sole determinants of one's salary increase is that they encourage personal responsibility. One cannot simply blame others for a faulty evaluation if one's own estimate partially determines the result.

Both DD1 and DD3 incorporate in their payoff functions a person's request in such a way as to reward him or her for a searching and accurate self-assessment. This makes it rational for a person to gather information about others' perceptions of his or her performance before making a request. By adjusting one's request to others' perceptions, DD1 and DD3 induce one to see oneself as others see one. Psychologically speaking, this is probably good not only for fostering more realistic attitudes but also for promoting better team performance.

APPENDIX

Proposition 11.1. *Egalitarian* behavior *is weakly dominated under every reasonable payoff scheme for n = 2. However, there is a reasonable payoff scheme, which works for all n, that yields the egalitarian outcome as the result of unique bids remaining after the iterated elimination of weakly dominated strategies.*

Proof. To prove the first part of Proposition 11.1, it suffices to show that, in a two-person game, if the payoff scheme satisfies the five conditions of reasonableness in section 11.2, then a bid of 51 weakly dominates a bid of 50. To show this, I divide an opponent's bids into four cases and demonstrate that, in all cases, a bid of 51 yields at least as good an outcome as does a bid of 50 and, in at least one case, a strictly better outcome.

Case 1. Opponent bids $b \leq 49$.

In this case, a bid of 51 yields an outcome of 51 by condition 3. By the same condition, a bid of 50 yields an outcome of 50, so a bid of 51 is strictly better.

Case 2. Opponent bids $b = 50$.

In this case, a bid of 51 yields an outcome of at least 50 by conditions 2 and 4. On the other hand, a bid of 50 cannot yield a better outcome by condition 2.

Case 3. Opponent bids $b = 51$.

In this case, a bid of 51 yields exactly 50 by conditions 1 and 4. As before, a bid of 50 can yield no more than 50 by condition 2.

Case 4. Opponent bids $b \geq 52$.

In this case, a bid of 51 yields an outcome of at least 50 by conditions 4 and 5. A bid of 50 cannot yield a better outcome by condition 2. This completes the proof of the first part of Proposition 11.1.

To prove the second part of Proposition 11.1, I show that there is a reasonable payoff scheme that yields the egalitarian outcome as the result of unique bids remaining after the iterated elimination of weakly dominated strategies. The payoffs are made according to the following payoff scheme in section 11.2.

Payoff Scheme (PS). One starts with the lowest bidder—or, more generally, with the group tied for lowest bid—and pays them what they bid if there is enough money to do so. If there is not enough money, the money available is divided evenly among this group. One next moves to the group tied for second-lowest bid and proceeds in exactly the same way, but now working with only the money that is left after the group of lowest bidders has been paid. One continues in this fashion until the money is exhausted, after which no one else is rewarded, or all the players receive what they bid.

It is easy to check that PS satisfies the five conditions of reasonableness given in section 11.2. The claim now is that the iterated elimination of weakly dominated strategies results in unique bids that yield the egalitarian outcome.

For clarity of exposition, I work with the case $n = 4$ and show that the outcome (25, 25, 25, 25) results from simultaneous bids of 26 by all four players. Moreover, these bids are the only ones left after a sequence of seventy-five successive eliminations of weakly dominated strategies. The arguments all extend to the case where there are n players and kn units ($n, k \geq 2$), where k ("cents" in DD) is the egalitarian bid of each player (25 in this example).

The result I now prove is an immediate consequence of the following ten claims.

Claim 1. A player never receives more than what it bid.
Proof. This is obvious from the description of PS.

Claim 2. Suppose that in making the payoffs, there are exactly t players not yet paid. Suppose that A is one of these, and none of the other $t-1$ has a lower bid than A. Then A receives at least the minimum of two quantities—what it bid, or $1/t$ of the money that is left over.
Proof. If A's bid is the lowest of those remaining, it receives either this bid or, if this bid is greater than the amount of money that is left over, the money that is left over. If A's bid ties for lowest with one or more of the remaining bids, the tied players receive either their bids or, if the sum is greater than the amount that is left over, they split this amount. Thus, the claim is true whether or not there is a tie.

Claim 3. A bid of 26 guarantees a payoff of at least 25.
Proof. Let s denote the number of players bidding less than 26. Then $4-s$ bid 26 or more. By claim 1, at most $25s$ cents are needed to pay off the s players bidding less than 26. Thus, at least $100-25s$ cents are left over. By claim 2, A receives at least the minimum of 26 and $[1/(4-s)][100-25s] = 25$.

Claim 4. A bid of $x \leq 25$ is never better than a bid of 26.
Proof. This follows from claims 1 and 3.

Claim 5. If $x \leq 25$, then there are scenarios in which a bid of 26 is strictly better than a bid of x.
Proof. If everyone else bids 0, then a bid of 26 yields a payoff of 26, whereas a bid of x yields a payoff of x.

Claim 6. For $0 \leq x \leq 25$, a bid of 26 weakly dominates a bid of x.
Proof. This follows from claims 4 and 5—that is, a bid of 26 is at least as good as, and sometimes better than, a bid of 25 or less.

Claim 7. For $26 \leq x < y \leq 100$, neither x nor y weakly dominates the other.
Proof. If A bids x, and the other three players also bid x, then these bids yield payoffs of 25 to everyone. If A bids y, this yields A a payoff of at most 22 (since each of the other three will receive at least 26). Thus, there is a scenario wherein x is a strictly better bid than y. For a scenario in which y is strictly better than x, consider the situation in which A bids y, B bids x, and everyone else bids 0. Then x yields x and y yields y when $x + y \leq 100$.

Claim 8. In the first reduction caused by each player's elimination of dominated strategies, the strategies $0, 1, \ldots, 25$ are precisely the ones that are eliminated.
Proof. This follows from claims 6 and 7.

Claim 9. Assume that $26, \ldots, j$ are the only strategies that have not yet been eliminated, where $27 \le j \le 100$. Then j is weakly dominated by x for every x such that $26 \le x < j$.

Proof. The only way j could be better than x is if the sum of the other three bids were less than $100 - x$. (Thus, more than x is available, so a bid of "x" yields x and a bid of "j" yields more than x.) But $x \ge 26$, so $100 - x \le 74$, whereas the sum of the other three bids is at least 78. Thus, x is at least as good as j in every scenario. Moreover, if everyone else bids x, then x yields a strictly better outcome (25) than does j (≤ 22).

Claim 10. With the same assumptions as in claim 9, if $26 \le x \le y \le j - 1$, then neither x nor y weakly dominates the other.

Proof. As in the proof of claim 7, it is easy to see that x produces a better outcome if everyone else bids x, whereas y produces a better outcome if everyone else bids j. ∎

12

Allocating Multiple Homogeneous Divisible Goods: Adjusted Winner

12.1. INTRODUCTION

Most disputes—divorce, labor-management, merger-acquisition, and international—involve only two parties, but they frequently involve several homogeneous goods that must be divided, or several issues that must be resolved. In this chapter I describe a procedure called *adjusted winner* (AW) that has been applied to disputes ranging from interpersonal to international (Brams and Taylor, 1996, 1999).

To introduce AW, I discuss in section 12.2 two properties of fair division, proportionality and envy-freeness (the latter I discussed in a different context in chapter 10). The well-known cake-cutting procedure of "I cut, you choose" satisfies these properties, but I argue that it suffers from being inequitable. Possible ways to improve on this procedure will be discussed in chapter 13.

In section 12.3 I illustrate AW with a hypothetical example and discuss its properties. To illustrate its possible real-life application, I apply it to the long-standing Egyptian-Israeli dispute, which was settled by the Camp David accords of 1978. These accords were formalized by a peace treaty in 1979, which terminated one of the most enduring conflicts since World War II.

I present in some detail the issues at Camp David in section 12.4. In section 12.5 I discuss how each side might have valued getting its way on each issue. AW, it turns out, would have given each side nearly two-thirds of what it wanted and probably reflected quite well the actual agreement that was reached.

I suggest that AW might well have expedited this agreement, perhaps by two or three years after the conclusion of the Yom Kippur War in 1973. In other disputes, AW might save the disputants not just from impasse but from the possibility that a stalemate may turn violent.

Camp David serves as a springboard for discussing several practical aspects of using AW—including different methods for assigning points to issues by the disputants—in section 12.6. In section 12.7 I offer some observations

Note: This chapter is adapted from Brams and Togman (1996) with permission; see also Brams and Taylor (1999, ch. 6).

about the fairness of the Camp David agreement, both actual and that achieved by AW. I conclude that the principles of fairness embodied in AW seem especially important in settling international disputes, wherein due process and the rule of law common in democratic societies often do not hold sway.

12.2. PROPORTIONALITY, ENVY-FREENESS, AND EFFICIENCY

There are several important criteria by which to judge fairness. One is that all parties to a dispute are entitled to a fair share of a heterogeneous divisible good, like a cake, parts of which each party may value differently. For example, one party may like the cherry in the middle, whereas another party may like the nuts on the side.

The simplest notion of a fair share is a proportional share. That is, each of n parties in a dispute is entitled to at least $1/n$ of the cake, as he or she views it. Fair division procedures that guarantee a proportional share are said to satisfy the property of *proportionality*.

Another criterion of fairness, and one that is more difficult to satisfy, is *envy-freeness*, which I analyzed in the context of indivisible goods in chapter 11. An envy-free division is one in which each party believes it received the most-valued portion, or a piece tied for most-valued. One way of conceptualizing such a division is to imagine an allocation in which no party believes it could do better by trading its portion for someone else's portion.

To illustrate how fair-division procedures can ensure proportionality and envy-freeness, consider "I cut, you choose," or cut-and-choose. Suppose players A and B wish to divide a heterogeneous cake between themselves. If A cuts the cake into two pieces, and B is allowed to choose whichever piece it prefers, each party can ensure both proportionality and envy-freeness by adhering to the following strategies:

1. By cutting the cake into two pieces that it considers to be of equal value, A can guarantee itself what it believes to be half the cake, regardless of which piece B chooses. Similarly, by choosing first, B can guarantee itself what it believes to be at least half the cake. Thus, this procedure, in conjunction with these strategies, guarantees proportionality—each player thinks it obtained at least one-half of the cake.
2. When these strategies are followed, this procedure guarantees that neither A nor B will believe that the other player received a larger portion of the cake than what it received. Thus, this procedure produces an allocation that is envy-free.

Envy-freeness and proportionality are equivalent when there are only two players—that is, the existence of one property implies the existence of the other. However, there is no such equivalence when there are three or more players. For example, if each of three players thinks it received at least one-third of the cake, it may still be the case that one thinks another received a larger piece—say, one-half—so proportionality does not imply envy-freeness if there are three or more players.

Envy-freeness, on the other hand, does imply proportionality, for if none of the players envies another, each must believe it received at least one-third of the cake. Thus, envy-freeness is the stronger notion of fairness.

Cut-and-choose is also efficient if both parties place a positive value on all parts of the cake. That is, there is no other allocation that is better for one and at least as good for the other. This is because if one party gets a smaller piece that it values less, the other party must get a larger piece that it values more.

12.3. ADJUSTED WINNER (AW)

Although the simplicity of cut-and-choose is appealing, many disputes are not over a single heterogeneous good. AW is designed for disputes in which there are two parties and multiple homogeneous goods (or issues), each of which is divisible.

Under AW, each of two players is given 100 points to distribute across two or more goods. After the players make their point assignments independently, the goods are then allocated to them according to the procedure described next.

I illustrate this procedure with a simple example. Suppose that A and B must divide three goods, G_1, G_2, and G_3, between themselves. Based on the importance they attribute to obtaining each good, assume they distribute their points in the following manner (the larger number of points allocated to each good by A or B is underscored):

	G_1	G_2	G_3	Total
A's announced values	6	67	27	100
B's announced values	5	34	61	100

Initially, A and B receive the goods to which they have assigned more points. Thus, A is awarded G_1 and G_2, giving it $6 + 67 = 73$ of its points; and B is awarded G_3, giving it 61 of its points.

If A's total points were equal to B's at this juncture, the procedure would end. However, this is not the case: A receives more of its points than B receives of its points, so A is the initial winner.

The next step is to transfer a good or goods from A to B so as to give both players the same point totals. This is called an *equitability adjustment*. The first good to be transferred is that with the lowest ratio of A's points to B's. In this example, G_1 has a lower ratio ($6/5 = 1.20$) than G_2 ($67/34 \approx 1.97$).

Even transferring all of G_1 to B leaves A with a slight advantage (67 of its points to $5 + 61 = 66$ of B's). Hence, one turns next to the good with the second-lowest ratio, G_2, transferring only that fraction of G_2 necessary to give A and B the same number of points.

Let x denote the fraction of G_2 that A will retain, with the rest transferred from it to B. Choose x so that the resulting point totals are equal for A and B. This occurs when A's points (left side of the equation below) are equal to B's points (right side) after the equitability adjustment:

$$67x = 5 + 61 + 34(1 - x).$$

This equation yields $x = 100/101 \approx 0.99$. Consequently, A ends up with 99 percent of G_2, for a total of 66.3 of its points, whereas B ends up with all of G_1, all of G_3, and 1 percent of G_2, for the same total of 66.3 of its points.

This outcome can be shown to satisfy several important properties (for details, see Brams and Taylor, 1996, 1999):

1. *Envy-freeness.* If A is truthful, it is assured of obtaining a minimum of 50 points, even if B knows A's point allocation and chooses an optimal manipulative strategy to maximize its (B's) points.
2. *Efficiency.* There is no other allocation that would give one player more of its points without giving the other less.
3. *Equitability.* A's valuation of its portion is exactly the same as B's valuation of its portion.

By comparison, cut-and-choose, while envy-free and efficient, is not equitable. This is because the chooser, by choosing what it thinks is the larger piece, does better in its eyes than the cutter, who obtains exactly 50 percent, does in its eyes.

In order for AW to satisfy the properties I have just described, two important conditions must be met: linearity and additivity. *Linearity* means that the added value, or marginal utility, of obtaining more of a good is constant (instead of diminishing, as is usually assumed). Thus, for example, $2x$ percent of G_1 is twice as good as x percent for each player.

Additivity means that the value of two or more goods is equal to the sum of their points. Put another way, obtaining one good does not affect the value of obtaining another, or winning on one issue is separable from winning on another. Thus, goods or issues can be treated independently of each other, with packages of goods no more—nor less—than the sum of their individual parts. I will return to the additivity of issues when I discuss the Camp David agreement.

I have illustrated how AW works to solve a hypothetical dispute and indicated the properties it satisfies. But how useful would it be in resolving actual conflicts? Short of having parties to real-world disputes utilize the procedure, perhaps the best way to evaluate its potential usefulness is to look at what, hypothetically, would have occurred had the parties to an actual conflict applied AW.

12.4. ISSUES AT CAMP DAVID

On September 17, 1978, after eighteen months of negotiation and a thirteen-day summit meeting, President Anwar Sadat of Egypt and Prime Minister Menachem Begin of Israel signed the Camp David accords. Their final bargaining was not easy. By the third day of their summit meeting, the animosity between the two leaders had grown so great that they refused to meet with each other face-to-face, so the remaining ten days of negotiations had to be conducted through intermediaries.

Six months later, the accords provided the framework for the peace treaty that the two nations signed on March 26, 1979. This epochal agreement shattered the view of many observers that the thirty-year-old Arab-Israeli conflict was probably irreconcilable.

There are a number of factors that make the Camp David negotiations an excellent case for examining the potential usefulness of AW. First, there were several issues over which the Egyptians and Israelis clashed. These issues can be considered as if they were goods to be divided fairly under AW, except to obtain a good translates into getting one's way, or winning, on an issue.

Second, most of the issues were to some degree divisible, rendering the equitability-adjustment mechanism of AW applicable to the issue that must be divided. Third, there is now considerable documentation on the positions of the two sides on each issue, based on detailed accounts of the negotiations at Camp David by several of the participants as well as outside observers. The empirical evidence enables one to make reasonable point assignments to each issue, based on the expressed concerns of each side.

The Camp David accords, of course, need to be seen in the context of the wrenching conflict that existed between the Arab countries and Israel from the time of the latter's creation in 1948. The Arab states, including Egypt, did not recognize Israel's right to exist and continually sought to annihilate it. However, Israel was victorious in the 1948–49 war, the 1956 Sinai conflict, and the Six-Day War of 1967. As a result of the 1967 war, Israel conquered and laid claim to substantial portions of territory of its Arab neighbors, including the Sinai Peninsula, the West Bank, the Gaza Strip, and the Golan Heights.

In 1973 Egypt and Syria attempted to recapture the Sinai Peninsula and the Golan Heights, respectively, in the Yom Kippur War but were repelled by Israel. Henry Kissinger's shuttle diplomacy in 1973–74 helped bring about two disengagement agreements between the warring sides, but a permanent resolution of their conflict remained elusive.

When Jimmy Carter became the U.S. president in January 1977, he deemed the attenuation of the Middle East conflict one of his top priorities. This conflict had contributed to major increases in the world price of oil; the fallout of these increases had been inflation and slowed economic growth.

From Carter's perspective, stable oil prices required an end to the turmoil in the Middle East (Quandt, 1986, p. 32). Furthermore, Carter believed that the prevailing disengagement was unstable; some sort of permanent settlement was necessary to prevent still another Arab-Israeli war and the potential involvement of the United States. Thus, after assuming the presidency, Carter almost immediately began to use his office to press for peace in the Middle East.

The original U.S. plan was to involve all the major parties, including the Palestine Liberation Organization (PLO), in the negotiations. But as talks proceeded, it became clear that the most likely resolution that could be reached would be one between Egypt and Israel. Indeed, Sadat at one point sent the U.S. president a letter urging that "nothing be done to prevent Israel and Egypt from negotiating directly" (Carter, 1982, p. 294).

By the summer of 1978, it seemed to Carter that a summit meeting was necessary to bridge the remaining gap between Egypt and Israel. He invited Sadat and Begin to meet with him at Camp David.

When the Egyptian and Israeli leaders convened at Camp David, there were several major issues on which the two sides sharply disagreed. These issues can be grouped into six categories. Much of the dispute centered on different territorial claims regarding the Sinai Peninsula, the West Bank, the Gaza Strip, and Jerusalem. I describe each side's most serious concerns regarding each issue next.

1. The Sinai Peninsula

This large tract of land was conquered by Israel during the Six-Day War in 1967 and remained under its control after the Yom Kippur War. In many ways it was the most important issue dividing the two sides in the negotiations. For Israel, the Sinai provided a military buffer that offered considerable warning in case of a possible Egyptian attack. Israel had set up military bases in the peninsula, including three modern airbases of which it was very protective.

Israel had also captured oil fields in the Sinai that were of significant economic value. Furthermore, Israel had established civilian settlements in the Sinai that it was loath to give up. At one point at Camp David, Begin told a member of the American negotiating team, "My right eye will fall out, my right hand will fall off before I ever agree to the dismantling of a single Jewish settlement" (Brzezinski, 1983, p. 263).

For Egypt, the Sinai was of such great importance that no agreement could be achieved that did not include Egyptian control over this territory. Almost all observers of the negotiations concur that, among all his goals, Sadat "gave primacy to a full withdrawal of Israel's forces from the Sinai" (Stein, 1993, p. 81). He let the United States know at the earliest stages of the negotiations that while he would allow some modifications of the pre-1967 borders, the Sinai must be returned *in toto* (Quandt, 1986, p. 50).

Roughly midway through the eighteen months of negotiation leading up to Camp David, Sadat began focusing almost exclusively on the Sinai in his discussions with both the Israelis and the Americans (Quandt, 1986, p. 177). From a material perspective, both its military significance and its oil fields made the return of the Sinai imperative for the Egyptians. But perhaps more importantly, the Sinai was highly valued by Egypt for symbolic reasons. For Egypt, "the return of the whole of Sinai was a matter of honor and prestige, especially since Sinai had been the scene of Egypt's 1967 humiliation" (Kacowicz, 1994, p. 135) in the Six-Day War.

2. Diplomatic Recognition of Israel

Since its creation in 1948, Israel had not been recognized as a legitimate and sovereign nation by its Arab neighbors. In fact, almost all Arab countries remained officially at war with Israel and, at least for propaganda purposes, sought its liquidation. For Israel, diplomatic recognition by Egypt, its most powerful neighbor, was an overriding goal.

But Israel wanted more than just formal recognition. Israeli leaders desired normal peaceful relations with Egypt, including the exchange of ambassadors and open borders (Brzezinski, 1983, p. 281). Such a breakthrough would help liberate Israel from its pariah status in the region.

Egypt balked at normalizing relations with Israel, in part because other Arab nations would vehemently oppose such measures. Sadat also believed that normal diplomatic relations would take a generation to develop because they would require such profound psychological adjustments (Telhami, 1990, p. 130).

In the actual negotiations, Sadat went so far as to assert that questions of diplomatic relations, such as the exchange of ambassadors and open borders, involved Egyptian sovereignty and therefore could not be discussed (Quandt, 1986, p. 130). Recognition of Israel became so contentious an issue that it presented one of the major obstacles to the signing of both the Camp David accords in 1978 and the formal peace treaty in 1979.

3. The West Bank and the Gaza Strip

For most Israelis, the West Bank and the Gaza Strip were geographically and historically integral to their nation—at least more so than was the Sinai. Indeed, the Israeli negotiating team held retention of these areas to be one of its central goals (Brzezinski, 1983, p. 236).

Begin, in particular, considered these territories to be part of Eretz Israel, or the Land of Israel, and not occupied foreign land. As one observer put it, "Begin was as adamant in refusing to relinquish Judea and Samaria [the West Bank] as Sadat was in refusing to give up any of Sinai" (Quandt, 1986, p. 66). By contrast, if Begin were to give up the Sinai, he was intent on getting some recognition of Israel's right to the West Bank and the Gaza Strip in return (Kacowicz, 1994, p. 139).

For Egypt, these two territories had little economic or geostrategic worth; Sadat did not focus much on them as the negotiations proceeded. However,

Egypt did face pressure from other Arab countries not to abandon the Palestinian populations in these territories. Sadat told his aides that he would not leave Camp David without some commitment from the Israelis to withdraw from the West Bank and Gaza (Telhami, 1990, p. 129). In fact, once he arrived at Camp David, Sadat informed Carter, "I will not sign a Sinai agreement before an agreement is also reached on the West Bank" (Carter, 1982, p. 345).

4. Formal Linkage of Accords and Palestinian Autonomy

One of the major issues of the negotiations was the extent to which an Egyptian-Israeli agreement should be tied to formal, substantive progress on the issue of Palestinian autonomy. Begin held that there should be no linkage. While Egypt and Israel might agree to some framework for the resolution of the Palestinian question, Begin claimed that this must be a separate matter, not part of a treaty between the two states (Quandt, 1986, p. 178).

Sadat seemed to be of two minds on this issue. On the one hand, he pushed for Israeli recognition of the Palestinians' right to self-determination as part of the treaty, holding that a bilateral agreement could not be signed before an agreement on general principles concerning a Palestinian state was reached. On the other hand, he pointed out that a truly substantive agreement on this issue could not be negotiated by the Egyptians alone. However, he opposed possible deferral of this issue to an Arab delegation, which he knew could sabotage an agreement.

5. Israeli Recognition of Palestinian Rights

From the Israeli perspective, recognizing the rights of the Palestinian people was difficult because of competing sovereignty claims between the Israelis and Palestinians. When President Carter declared in Aswân, Egypt, that any solution to the conflict "must recognize the legitimate rights of the Palestinian people," the Israelis reacted negatively (Quandt, 1986, p. 161). But because this recognition was not attached to any substantive changes (see issue 4), it was not viewed as excessively harmful to Israeli interests. In fact, Israeli foreign minister Moshe Dayan at one point sent a letter to the American negotiating team indicating that Israel would be willing to grant equal rights to Arabs in the West Bank (Quandt, 1986, p. 106).

From the Egyptian perspective, some form of Israeli recognition of the rights of Palestinians was deemed necessary. Even if the formulation was vague and largely symbolic, Sadat felt strongly that he needed at least a fig leaf with which to cover himself in the eyes of the other Arab countries (Brzezinski, 1983; Quandt, 1986, p. 188). Rhetorically, such a declaration would allow Egypt to claim that it had forced Israel finally to recognize the rights of the Palestinian population, an accomplishment that no other Arab state had been able to achieve. Furthermore, this formulation was appealing to Sadat because it would not require the participation of other Arab states.

6. Jerusalem

Control of Jerusalem had been a delicate issue since 1948. The United Nations demanded in 1949 that the city be internationalized because of competing religious and political claims. Until the Israelis captured and unified the city in 1967, it had been split between an eastern and a western section.

For Israel, Jerusalem was the capital of their nation and could not be relinquished. At Camp David, Dayan told the Americans that it would take more than a U.N. resolution to take the city away from Israel: "They would also need to rewrite the Bible, and nullify three thousand years of our faith, our hopes, our yearnings and our prayers" (Dayan, 1981, p .177).

As was the case with other territorial claims, Egypt faced pressure from other Arab nations to force Israeli concessions on this issue. An Egyptian representative impressed on the Israelis that a constructive plan for Jerusalem would "lessen Arab anxiety and draw the sting from Arab hostility" (Dayan, 1981, p. 49). However, Egypt did not push strenuously on this issue and, in fact, seemed willing to leave it for the future.

12.5. THE AW SOLUTION

How might AW have been used to resolve these issues as fairly as possible? In Table 12.1, I give plausible allocations of 100 points that Egypt and Israel might have made to the six issues.

These hypothetical allocations, to be sure, are somewhat speculative; it is impossible to know exactly how Israeli and Egyptian delegates would have distributed their points had they used AW. However, it should be noted that while different point allocations would produce different issue resolutions, this would not alter any of the properties that AW guarantees—envy-freeness, efficiency, and equitability.

The allocation of points in Table 12.1 is based on the preceding analysis of each side's interests in the six issues. Briefly, it reflects Egypt's overwhelming interest in the Sinai, Sadat's insistence on at least a vague statement of Israeli recognition of Palestinian rights to protect him from the wrath of other Arab nations, the Israelis' more limited interests in the Sinai, and Begin's strong views on Eretz Israel—that is, retaining the West Bank and Gaza Strip and control over Jerusalem. Notice that each side has a four-tier ranking of the issues: most important (55 points for Egypt, 35 for Israel), second-most important (20 points), third-most important (10 points), and least important (5 points).

TABLE 12.1

Hypothetical Israeli and Egyptian Point Assignments

Issue	Israel	Egypt
1. Sinai	35	<u>55</u>
2. Diplomatic Recognition	<u>10</u>	5
3. West Bank/Gaza Strip	<u>20</u>	10
4. Linkage	<u>10</u>	5
5. Palestinian Rights	5	<u>20</u>
6. Jerusalem	<u>20</u>	5
Total	100	100

Note: The larger number of points allocated to each item is underscored.

This hypothetical allocation represents a truthful, rather than a strategic, point distribution for each side. Although in theory it is possible to benefit from deliberately misrepresenting one's valuation of the issues, in practice this would be difficult. Indeed, the parties may succeed only in hurting themselves, as I will discuss later.

Initially under AW, Egypt and Israel each win on the issues for which they have allocated more points than the other side (underscored in the table). Thus, Egypt would be awarded issues 1 and 5, for a total of 75 of its points; Israel would be awarded issues 2, 3, 4, and 6, for a total of 60 of its points.

Since Egypt wins more of its points than Israel does of its points, some issue or issues must be transferred, in whole or in part, from Egypt to Israel in order to achieve equitability. Because the Sinai (issue 1) has a lower ratio ($55/35 \approx 1.57$) of Egyptian to Israeli points than the issue of Palestinian Rights (issue 5) does ($20/5 = 4.0$), the former must be divided, with some of Egypt's 55 points on issue 1 transferred to Israel, which allocates 35 points to this issue, to create equitability.

Let x denote the fraction of this issue that Israel will obtain. Setting Israel's points equal to Egypt's gives

$$60 + 35x = 20 + 55(1-x),$$

which yields $x = 1/6 \approx 0.17$. As a result, Israel is given about 17 percent of issue 1, plus all of issues 2, 3, 4, and 6, for a total of exactly 65.8 of its points. Egypt wins the remaining 83 percent of issue 1, along with all of issue 5, for the same total of 65.8 of its points. This final distribution is envy-free, equitable, efficient—and, as I will argue in the next section, practically strategy-proof.

It should be noted that AW, using the hypothetical point allocations of Table 12.1, produces an outcome that mirrors quite closely the actual agreement reached by Egypt and Israel. From Israel's perspective, it essentially won on issue 2, because Egypt granted it diplomatic recognition, including the exchange of ambassadors. Israel also got its way on issue 3, when Egypt "openly acknowledged

Israel's right to claim in the future its sovereign rights over the West Bank and Gaza" (Kacowicz, 1994, p. 139). Additionally, Israel won on issue 4, because there was no formal linkage between the Camp David accords—or the peace treaty later—and the question of a Palestinian state or the idea of Palestinian self-determination. And, finally, Jerusalem was not part of the eventual agreement, which can be seen as Israel's prevailing on issue 6.

Egypt prevailed on issue 5: Israel did agree to the Aswân formulation of recognizing the "legitimate rights" of Palestinians. That leaves issue 1, on which Egypt wins 83 percent according to the hypothetical division.

But was this a divisible issue on which an equitability adjustment could be made? In fact, the Sinai issue was multifaceted and thus lent itself to division. Besides the possible territorial divisions, there were also questions about Israeli military bases and airfields, as well as Israeli civilian settlements and the positioning of Egyptian military forces.

Egypt won on most of these issues. All the Sinai was turned over, and Israel evacuated its airfields, military bases, and civilian settlements, some forcibly. However, Egypt did agree to demilitarize the Sinai, and to the stationing of U.S. forces to monitor the agreement, which represented a concession to Israel's security concerns. Viewing this concession as representing roughly 17 percent of the total issue seems to be a plausible interpretation of the outcome.

One problem that arises for this hypothetical case relates to the separability of issues. An issue is *separable* if the value to a party of winning on that issue is independent of its winning on other issues. If issues are separable, then their points can be added, as assumed under AW. Winning on a set of two or more issues gives a value for the set equal to the sum of the points of the individual issues that the set comprises. In applying AW, a key question is whether the issues can be treated independently of each other.

In the case of Camp David, it can be argued that the recognition of Palestinian rights was not independent of territorial issues. For Sadat, in particular, recognition may have been more important *because* of his failure to win Israeli concessions on the West Bank, the Gaza Strip, or Jerusalem.

Although finding reasonably separable issues—whose points can be summed—is never an easy task, skillful negotiators can mitigate this problem. This happened in negotiations over the Panama Canal treaty, which was signed and ratified by the United States and Panama in 1977, when the two sides reached a consensus on ten different issues that split them (Raiffa, 1982). An analysis of the point allocations made to these issues showed that lumping them together would have reduced the point totals of each side, indicating that under AW two sides can do best by carving out as many separable issues as possible (Brams and Taylor, 1996).

At Camp David, it is likely that the two sides would have come up with a different set of issues than those considered here. This might have facilitated the application of AW if there had been, say, a dozen rather than a half-dozen issues. Nevertheless, the list I presented works tolerably well, at least to illustrate the potential of AW, with both sides obtaining nearly two-thirds of what they desired.

12.6. PRACTICAL CONSIDERATIONS

I next examine the Camp David case more closely to anticipate several difficulties, when applying AW, that can arise in trying to (1) minimize AW's vulnerability to manipulation and spite, (2) make point assignments, (3) render issues separable, (4) optimize timing, and (5) define issues.

1. Minimize AW's Vulnerability to Manipulation and Spite

A potential problem with applying AW is its manipulability, whereby one side may try to manipulate its point distribution in an attempt to increase its "winnings." Assume, for example, that Israel, anticipating that Egypt would put an overwhelmingly large number of points on the Sinai—enabling Egypt almost certainly to "win" on this issue—reduced its points on the Sinai from 35 to 20. (I will show the effects of the opposite strategy—increasing its allocation—shortly.) Also anticipating that Egypt would not put too many points on Palestinian rights, suppose that Israel increased its own points on this issue from 5 to 20 (corresponding to the amount it took away from the Sinai issue), hoping, possibly, to win on Palestinian rights.

Under this scenario, Israel initially is awarded issues 2, 3, 4, and 6, for a total of 60 of its points, the same as before. However, Egypt wins only issue 1, for a total of 55 of its points, because now there is a 20–20 tie on Palestinian rights.

Because Egypt trails in points 55 to 60 at the start, it is awarded the tied issue of Palestinian rights. However, because these 20 points would now put it ahead of Israel by 75 to 60 points, there must be an equitability adjustment on this issue. By giving Egypt 12.5 points (62.5 percent) and Israel 7.5 points (37.5 percent) on this issue, each side would seem to end up with a total of 67.5 points, slightly more than the 65.8 points each side formerly received.

But this improvement for Israel is illusory, because it is based on Israel's announced rather than true preferences. In fact, this maneuver backfires in two ways. First, insofar as Israel's point allocation in Table 12.1 reflects its true preferences, it actually ends up with fewer points. Instead of obtaining 37.5 percent of 20 points on Palestinian rights (its manipulative allocation), it actually obtains 37.5 percent of 5 points (its true allocation), or 1.875 points in addition to its initial 60 points, giving it a total of 61.875 points. This number is less, not more, than the 65.8 points it obtains by being honest in its announced allocation, whereas Egypt ends up with more (67.5 points).

The second way in which Israel's manipulative strategy backfires in this scenario is perhaps more costly. When both parties announce their true preferences,

Israel is awarded part of the Sinai issue according to the equitability-adjustment mechanism. However, in the manipulative scenario, because Israel reduced the number of points it put on the Sinai, Egypt wins this issue outright and need not make any concessions to Israel. In such a case, it could be assumed that Egypt would not have to demilitarize the Sinai, or allow the stationing of U.S. forces to monitor the agreement.

Although AW is manipulable in theory, in practice it is probably not manipulable unless a player has precise information about how the other side will distribute its points. Only then can the manipulator optimally allocate its points to exploit its knowledge. Short of having this information, however, a manipulative strategy like the one just described is dangerous. The manipulator may succeed only in hurting itself and helping the other side, the opposite of what it intended to do.

In fact, Israel would do better to increase the number of points it puts on the Sinai—say, from 35 points to 45 points—while putting 10 points rather than 20 points on Jerusalem. Now, after the equitability adjustment, Israel would win one-fourth rather than one-sixth of the Sinai. The problem with this maneuver is that if Egypt at the same time came down, say, from 55 points to 40 points on the Sinai—thinking the latter figure sufficient to ensure that it would win on this issue—Israel would win it instead. Egypt would lose it, which is exactly the opposite of what both parties want. Once again, this is a case of being too clever for one's own good.

A party to a dispute might, out of spite, try to manipulate its point distribution in an attempt to deny the other side a good or an issue it desires. Imagine, for example, that Egypt wanted to deny Israel diplomatic recognition, even though Egypt itself did not value this issue highly, by increasing the points it allocates to the diplomatic-recognition issue.

A strategy designed to deny something to an adversary is potentially costly for the same reason that a manipulative strategy designed to increase one's point total is: the additional points allocated to an issue out of spite have to come from another issue. Thereby the spiteful party runs the risk of losing on other issues. In this case, Egypt might risk losing part or all of, say, the Sinai in order to deny diplomatic recognition to Israel.

To convince a party that manipulation is hazardous when information is incomplete, one might have it go through the exercise of allocating insincere points for itself and then test (via AW) the outcome of such an assignment against various point assignments that its opponent might make. This exercise, in the absence of having complete information about the other side's point distribution, should convince that party that honest allocations are generally a sound strategy. I indicated earlier that honest allocations always guarantee a party at least 50 of its points—even if the other party has advance information on its allocation and follows an optimal manipulative strategy—making the outcome envy-free but not equitable. In fact, honesty is the *only* strategy that assures a party of an envy-free outcome.

2. Make Point Assignments Truthful

While honesty generally pays, it will not always be a simple matter to come up with point assignments that mirror one's valuations of the different issues. One way to facilitate this task is to have the parties begin by ranking the issues, from most to least important, in terms of their desire to get their way on each.

After the issues have been ranked, the parties face the problem of turning a ranking into point assignments that reflect their *intensities* of preferences for the different issues. This problem is discussed in considerable detail in Raiffa (1982), who concluded that a party must carefully weigh how much it would be willing to give up on one issue to obtain more on another. Thus for the Israelis in our example, West Bank and Gaza Strip (20 points) and Jerusalem (20 points) are each worth twice as much as Diplomatic Recognition (10 points) and Linkage (10 points), which in turn are each worth twice as much as Palestinian Rights (5 points).

To come up with such point assignments, one option for a party would be to begin by rating the importance of winning on its highest-ranked issue, compared with its next-highest ranked issue, by specifying a ratio. Continuing down the list, comparing the second-highest ranked issue with the third-highest ranked issue, and so on, a player would indicate, in relative terms, an "importance ratio" between adjacent issues.

For example, if there were three issues, and the importance ratios were 2:1 on the first issue relative to the second, and 3:2 on the second issue relative to the third, these would translate into a 6:3:2 proportion over the three issues. Rounding to the nearest integer, the point assignments would be 55, 27, and 18, respectively, on the three issues. A more systematic method for eliciting weightings, called "analytic hierarchy processing," is developed in Saaty (1995); other analytic approaches to negotiation are developed in Young (1991).

Another option for a party is to begin by intuitively assigning points to items. These assignments could be "tested" by asking whether various 50-point packages represent half the total value. To the extent that they do not, the initial point assignment for items would need to be modified. This process would continue until a party is satisfied that no further adjustments in its allocation of points to each item are necessary.

3. Render Issues Separable

There is also the problem of making the issues in a dispute as separable as possible in order to render the addition of points on different issues meaningful. If winning on, say, issue 1 affects the value of winning on issue 2, then the points a party receives on issue 2 cannot simply be added to the points it receives on other issues—this depends on what happens on issue 1. In this sense,

West Bank and Gaza Strip was probably best treated as a single issue—even though the West Bank and the Gaza Strip are two geographically separate territories—because it would have been difficult to make decisions on one independently of the other.

Although the 1993 Oslo accord between Israel and the PLO intricately linked the withdrawal of Israeli administrative and security personnel from the West Bank and from Gaza, the withdrawal of settlers is an entirely different story. In 2006, there was a unilateral withdrawal of Israeli settlers from Gaza. Although there has been a partial withdrawal of settlers from the West Bank, finding a settlement for this region is likely to be more drawn-out and contentious.

4. Optimize Timing

When is it most advantageous for disputants to use AW? According to former secretary of state Henry Kissinger, "Stalemate is the most propitious condition for settlement" (*New York Times*, October 12, 1974). Former president Jimmy Carter echoed this sentiment, saying that "parties must know they cannot win on the battlefield." Carter (1994, p. 390) added that "politicians have to see a significant difference between the costs of continuing with the status quo and the benefits of sitting down with the other side. A modest difference is not enough."

According to this view, it might be best to let the disputants try, on their own, to reach an agreement without AW. If, after repeated attempts, they fail, they may well become so frustrated and weary as to take seriously the adoption of a formal procedure like AW to break the impasse.

Of course, leaving the final shape of an agreement to any formal procedure is somewhat of a gamble, because one cannot predict the outcome with certainty. It becomes an acceptable risk to the degree that the disputants see AW as a procedure

- from which they can benefit equally, which equitability ensures;
- that provides a guarantee of getting at least 50 points (which is the same guarantee as provided by cut-and-choose), implying that it is envy-free;
- that is efficient, so the disputants can rest assured that there is no equitable agreement that can benefit both more.

In lieu of using AW, if one side thinks that it can frighten the other side into submission by threats, or that it can wear down the other side through endless haggling, then the equitability and efficiency of AW will not be so compelling. Indeed, it may take months or even years of impasse, as was the case in the negotiations leading up to Camp David, before the two sides are willing to contemplate certain compromises and then, perhaps with the help of a mediator, hammer out an agreement.

By comparison, AW allows the parties to reach closure immediately, at least once they agree on what the issues are and what winning and losing on each means. These, of course, are no small matters, but they are probably easier to reach agreement on—and then let AW find a settlement—than striking an overall agreement without AW.

5. Define Issues

Identifying the key issues, and rendering them as separable as possible, is likely to be time-consuming, requiring protracted negotiations before the parties can implement AW. But if the costs of delay are substantial, and the issues are quite narrowly defined, then the two sides should be able to reach agreement on these issues more quickly than they could reach a consensus without AW.

The determination of what is entailed by winning and losing for each side on the issues would have to be worked out beforehand. As with the definition of the issues, this determination will require good-faith negotiations, possibly aided by a mediator. Also, some way of monitoring and enforcing the agreements reached on each issue—whoever wins or loses—would have to be built into the agreement once it is implemented.

After AW is applied, the two sides will also have to decide what winning and losing in relative terms means on the one issue on which there is an equitability adjustment. In the case of Camp David, I suggested earlier that the demilitarization of the Sinai, and the stationing of U.S. forces to monitor the agreement, was tantamount to Israel's winning 17 percent on this issue.

Negotiations on what partially winning and partially losing mean on the equitability-adjustment issue can probably await the application of AW. Once the equitability adjustment is known, and on what issue, the parties can be told this information (for example, a 5:1 split on the Sinai issue is required), but not which party is the relative (5/6) winner and which the relative (1/6) loser.

At this point they would be told to negotiate two agreements, one in which Israel is the 5/6 winner and one in which Egypt is the 5/6 winner. This negotiation will be facilitated by the fact that either party could be the 1/6 loser. Thus, if one side asks for the moon—figuratively speaking—should it be the winner, so can the other side. This will chasten both sides to be fair-minded, lest the loser, which could be either side, ends up doing very badly. Thus, both sides will be motivated to reach agreement on what being the 5/6 winner means, whichever side this is.

None of the aforementioned practical considerations presents insuperable barriers to the use of AW. In order for the procedure to work best, the two sides would have to be educated as to the risks of trying to manipulate AW to their advantage or out of spite, including the likelihood that such manipulative strategies could backfire. They would also have to be advised on how best to define issues to make them as separable as possible. Thereby, they would ensure that the addition of points across different issues, once AW is applied, is sensible. Finally, they would have to reach agreement about what winning and losing on each issue means.

AW does not so much eliminate negotiations as require that they be structured in a certain way, which might help the disputants avoid minutae that otherwise might entangle them. Once this structuring is accomplished, AW finds a settlement that is envy-free, equitable, and efficient without further haggling.

This method of achieving closure is likely to save the two sides valuable time. Additionally, it should produce a better agreement than one reached after rancorous negotiations, which often leave both sides with a bitter taste that impedes future negotiations. Not only does AW diminish this problem, but it also offers a quick way of renegotiating agreements should priorities change due to a change, say, in government leaders or possibly fortuitous circumstances.

12.7. SUMMARY AND CONCLUSIONS

Was the Camp David agreement fair? If not, then it is difficult to make a case for AW, given that it would largely have duplicated the Camp David outcome.

Many Egyptians were disappointed with the results of the Camp David talks. A former foreign minister of Egypt, Ismail Fahmy (1983, p. 292), wrote, "The treaty gives all the advantages to Israel while Egypt pays the price. As a result, peace cannot last unless the treaty undergoes radical revision."

Quandt (1986, p. 255) also claimed that Israel did better in the negotiations. However, my reconstruction of the negotiations using AW suggests that the settlement was probably as fair as it could be. If Fahmy were correct in his belief that an unfair peace could not last, then the last three decades of peaceful relations (albeit a "cold" peace) between Israel and Egypt is testimony to the contrary.

Reinforcing this view is the fact that the negotiators, while undoubtedly desiring to "win," realized that this was not feasible because they were not in a total-conflict situation (i.e., a zero-sum game), wherein what one side wins the other side necessarily loses. Abetted by Jimmy Carter, they were driven to seek a settlement that, because it benefited both sides more or less equally, could be considered fair.

If it is surprising that a fair agreement was reached in the Middle East, it is probably more surprising that *any* agreement was concluded. In political disputes in general, and in international disputes in particular, players often expend much time and energy on procedural matters before substantive questions are even addressed.[1] The Egyptian-Israeli negotiations were no exception;

---

[1] For other applications of AW to international disputes, see Denoon and Brams (1997) and Brams and Taylor (1999) on the Spratly Islands conflict, and Massoud (2000) on the Israeli-Palestinian conflict. It is worth noting that in 1999 New York University patented the algorithm underlying AW; the U.S. patent number is 5,983,205. While the patent offers protection against

the two sides fought vigorously over procedural issues at several points in the negotiations (Quandt, 1986, p. 108).

Disputants have a strong incentive to do this, because procedures can be manipulated so as to bring about better or worse outcomes. By comparison, in guaranteeing a resolution that is fair according to several important criteria, AW affords disputants the opportunity to focus on substantive issues—while largely protecting them from procedural manipulation.

Another problem that plagues international disputes is the concern that one side will come out looking worse than the other. This sometimes pushes the more anxious side to abandon talks altogether rather than settle for a one-sided resolution, and then attempt to explain it back home. At Camp David, Sadat at one point expressed such a fear and packed his bags on the eleventh day with the intent of returning to Egypt. Only a strong personal appeal from Jimmy Carter, coupled with pointed threats, kept Sadat from breaking off the negotiations (Brzezinski, 1983, p. 272).

By guaranteeing an outcome with very appealing properties, AW can reduce the fears of the disputants and help keep negotiations on track. In all likelihood, it would have worked well in the Egyptian-Israeli conflict—producing a less crisis-driven atmosphere than was the case at Camp David, and possibly speeding up a settlement by two or three years—even if the outcome would not have differed much from that which actually was achieved.

This is not to say that fair-division procedures like AW are without shortcomings. For one thing, formal procedures do not have the flexibility of informal approaches, though "flexibility" can be a double-edged sword. In finding shortcuts, flexibility may produce arbitrary results. While the synergy of issues poses difficulties for rendering them separable, their adroit packaging can diminish this problem.

The benefits of a straightforward procedure that guarantees important properties of fairness should not be underestimated. Fairness, or the lack thereof, has long been a battle cry of people who feel either disadvantaged or exploited. To the extent that AW can help relieve their distress by resolving conflicts, it offers substantial promise for the future.

---

commercial use of AW by unauthorized parties, individual use of AW is encouraged. In fact, there is a free website at which AW calculations are made: http://www.nyu.edu/projects/adjustedwinner.

13

Allocating a Single Heterogeneous Good:
Cutting a Cake

13.1. INTRODUCTION

In chapter 11 I proposed several procedures (DD1, DD2, and DD3) for dividing a single homogeneous good (money), and in chapter 12 I analyzed a procedure (Adjusted Winner, or AW) for dividing multiple homogeneous goods. In this chapter I describe three new procedures for dividing a single heterogeneous good, such as a cake with different flavors or toppings whose parts players may value differently.

If there are n players, the procedures assume that the players make the minimal number of cuts (i.e., $n - 1$) required to divide the cake. In addition, they provide insight into the difficulties underlying the simultaneous satisfaction of certain properties of fair division, including *strategy-proofness*, or the incentive for players to be truthful about their valuations of a cake.

As is usual in the cake-cutting literature, I postulate that the goal of each player is to maximize the value of the minimum-size piece (*maximin piece*) that it can guarantee for itself, regardless of what the other players do. Thereby, I assume that each player is *risk-averse*—it will never choose a strategy that may yield a more valuable piece of cake if it entails the possibility of getting less than a maximin piece.

In section 13.2 I begin with the well-known 2-person cake-cutting procedure, cut-and-choose, that I discussed in section 12.2. It goes back at least to the Hebrew Bible (Brams and Taylor, 1999, p. 53) and, as I showed, satisfies two desirable properties that AW also does:

1. *Envy-freeness*: Each player thinks that it receives at least a tied-for-largest piece and so does not envy the other player.
2. *Efficiency (Pareto-optimality)*: There is no other allocation that is better for one player and at least as good for the other player.

But cut-and-choose does not satisfy the third desirable property that AW does:

Note: This chapter is adapted from Brams, Jones, and Klamler (2006a) with permission; see also Barbanel and Brams (2004).

3. *Equitability.* Each player's subjective valuation of the piece that it receives is the same as the other player's subjective valuation.

To bypass this problem, I propose in section 13.3 a new 2-player cake-cutting procedure that, while it does not satisfy equitability in an absolute sense, does satisfy it in a relative sense, which I call *proportional equitability.* After ensuring that each player receives exactly 50 percent of the cake, it gives each the same proportion of the cake that remains, called the *surplus,* as each values it. Thereby this procedure, which I call the *surplus procedure* (SP), gives each player at least 50 percent of the entire cake and generally more. By contrast, cut-and-choose limits the cutter to exactly 50 percent if it is ignorant of the chooser's preferences.

Remarkably, maximin strategies under SP require that each player be truthful about its preferences for different parts of the cake, rendering SP strategy-proof. This is because if a player is not truthful, it cannot guarantee at least a 50 percent share or, even if it does, may decrease the proportion of the surplus that a truthful strategy guarantees. By comparison, giving each player the same absolute amount of the surplus is *strategy-vulnerable,* because each player will have an incentive to lie about its preferences.

In section 13.4 I give a 3-person example that proves that if there are $n \geq 3$ players, envy-freeness and equitability may be incompatible. In section 13.5 I describe a simple 3-person, 2-cut envy-free procedure, called the *squeezing procedure,* which uses so-called moving knives. If $n > 3$, it is not known whether there exists an n-person, $(n - 1)$-cut envy-free procedure.

In section 13.6 I present an n-person *equitable procedure* (EP) that gives all players the maximal *equal* value that they all can achieve. Like SP and the squeezing procedure, it is strategy-proof.

I discuss trade-offs in cake division in section 13.7. Whereas SP does not limit one player to exactly 50 percent of the cake, as does cut-and-choose limit the divider, it is more information demanding, requiring that both players report their value functions over the entire cake, not just indicate their 50–50 points. On the other hand, the squeezing procedure does not require that the players indicate anything about their preferences.

Like SP, EP is information demanding; in addition, it may create envy, which SP never does. I conclude in section 13.7 by briefly discussing the applicability of SP, the surplus procedure, and EP to real-world problems.

13.2. CUT-AND-CHOOSE: AN EXAMPLE

Assume that two players, A and B, value a cake along a line that ranges from $x = 0$ to $x = 1$. More specifically, assume that the players have continuous value functions, $v_A(x)$ and $v_B(x)$, where $v_A(x) \geq 0$ and $v_B(x) \geq 0$ for all x over

[0, 1], and their measures are finitely additive, nonatomic probability measures.

Finite additivity ensures that the value of a finite number of disjoint pieces is equal to the value of their union. It follows that no subpieces have greater value than the larger piece that contains them. *Nonatomic measures* imply that a single cut, which defines the border of a piece, has no area and so contains no value. In addition, I assume that the measures of the players are *absolutely continuous*, so no portion of cake is of positive measure for one player and zero measure for another player.

Like *probability density functions* (pdfs), the total valuations of the players—the areas under $v_A(x)$ and $v_B(x)$—are 1. I assume that only parallel, vertical cuts, perpendicular to the horizontal x-axis, are made, which will be illustrated later.

Under cut-and-choose, recall that one player cuts the cake into two portions, and the other player chooses one. To illustrate, assume a cake is vanilla over [0, 1/2] and chocolate over (1/2, 1]. Suppose that the cutter, player A, values the left half (vanilla) twice as much as the right half (chocolate). This implies that $v_A(x) = 4/3$ on [0, 1/2], and $v_A(x) = 2/3$ on (1/2, 1].

To guarantee envy-freeness when the players have no information or beliefs about each other's preferences, A should cut the cake at some point x so that the value of the portion to the left of x is equal to the value of the portion to the right.[1] The two portions will be equal when A's valuation of the cake between 0 and x is equal to the sum of its valuations between x and 1/2 and between 1/2 and 1:

$$(4/3)(x - 0) = (4/3)(1/2 - x) + (2/3)(1 - 1/2),$$

which yields $x = 3/8$. In general, the only way that A, as the cutter, can ensure itself of getting half the cake is to give B the choice between two portions that A values at exactly one-half each.

To illustrate when cut-and-choose does not satisfy equitability, assume B values vanilla and chocolate equally. Thus, when A cuts the cake at $x = 3/8$, B will prefer the right portion, which it values at 5/8, and consequently will choose it. Leaving the left portion to A, B does better in its eyes (5/8) than A does in its eyes (1/2), rending cut-and-choose inequitable.

If the roles of A and B as cutter and chooser are reversed, the division remains inequitable. In this case, B will cut the cake at $x = 1/2$. A, by choosing the left half (all vanilla), will get 2/3 of its valuation, whereas B, getting the right half, will receive only 1/2 of its valuation. Because cut-and-choose selects the endpoints of the interval of envy-free cuts (3/8 for A, 1/2 for B), any cut strictly between 3/8 and 1/2 will be envy-free.

---

[1] When the cutter does have information or beliefs about the chooser's preferences, it may do better with a less conservative strategy (Brams and Taylor, 1996, 1999).

13.3. THE SURPLUS PROCEDURE (SP)

The rules of SP ensure that both A and B will obtain at least 50 percent of the cake, as they value it, and generally give each more:

1. Independently, A and B report their value functions, $f_A(x)$ and $f_B(x)$, over [0, 1] to a referee. These functions may be different from the players' true value functions, $v_A(x)$ and $v_B(x)$.
2. The referee determines the 50–50 points, a and b, of A and B—that is, the points on [0, 1] such that each player reports that half the cake, as it values it, lies to the left and half to the right (these points are analogous to the median points of pdfs).[2]
3. If a and b coincide, the cake is cut at $a = b$. One player is randomly assigned the piece to the left of this cutpoint and the other player the piece to the right. The procedure ends.
4. Assume that a is to the left of b, as illustrated below:

Then A receives the portion [0, a], and B the portion [b, 1], which each player values at one-half according to its reported value function.
5. Let c (for cutpoint) be the point in [a, b] at which the players receive the same *proportion p* of the cake in this interval, as each values it:

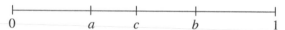

Then A receives the portion [a, c], and B the portion (c, b], so the players' combined portions are piece [0, c] for A and piece (c, 1] for B.

To determine c, we set the proportion p that A receives from subinterval [a, c] equal to the proportion that B receives from subinterval (c, b]:

$$p = \frac{\displaystyle\int_a^c f_A(x)dx}{\displaystyle\int_a^b f_A(x)dx} = \frac{\displaystyle\int_c^b f_B(x)dx}{\displaystyle\int_a^b f_B(x)dx}. \tag{13.1}$$

---

[2] One could assume that the referee asks the players first to indicate their 50–50 points and then to submit their pdfs for the half of the cake that includes the 50–50 point of the other player, which the referee would identify. This procedure would be somewhat less information-demanding than asking the players to submit their pdfs for the entire cake, but it would require the extra step of the referee's informing the players, after they have indicated their 50–50 points, of which half, as each player defines it, it needs to provide information on its value function.

In the earlier example, in which $a = 3/8$ and $b = 1/2$ and the pdfs are as given in section 13.2,

$$p = \frac{\displaystyle\int_{3/8}^{c} (4/3)dx}{\displaystyle\int_{3/8}^{1/2} (4/3)dx} = \frac{\displaystyle\int_{c}^{1/2} dx}{\displaystyle\int_{3/8}^{1/2} dx}$$

$$p = \frac{(4/3)(c - 3/8)}{(4/3)(1/2 - 3/8)} = \frac{(1/2 - c)}{(1/2 - 3/8)},$$

which yields $c = 7/16$, the midpoint of the interval $[3/8, 1/2]$ between the players' 50–50 points.

Whenever the players have uniform densities over the interval between their 50–50 points, as they do in this example, they will receive the same proportions of the interval at all points in it equidistant from a and b. In particular, at $c = 7/16$,

$$p = \frac{(1/2 - 7/16)}{(1/2 - 3/8)} = \frac{1}{2},$$

so each player obtains half the value it places on the entire interval, $[a, b]$.

Note that giving A and B the same proportion of the interval does not ensure equitability, because A and B value the interval differently. A values it at $(1/8)(4/3) = 1/6$ (and obtains $1/12$ at c), and B values it at $(1/8)(1) = 1/8$ (and obtains $1/16$ at c).

To ensure that A and B obtain exactly the same value from the interval rather than the same proportion of value, one can set equal the numerators of equation (13.1). Substituting e (for equitable point) for c in the limits of integration in the example gives

$$\int_{3/8}^{e} (4/3)dx = \int_{e}^{1/2} dx$$

$$(4/3)(e - 3/8) = (1/2 - e),$$

which yields $e = 3/7$. At this cutpoint, A and B each obtain $1/14$ from the interval.

There are conflicting arguments for cutting the cake at c (proportional equitability) and at e (equitability). An argument for cutting it at c is that the player that values the interval more (A in the example) should derive more value from it. The opposite argument reflects the egalitarian view that the players, in addition to the 50 percent portions they receive outside the interval, should get exactly the same value from the interval.

I will not try to resolve these conflicting claims for proportional equitability versus (absolute) equitability. Instead, I introduce a new property that only proportional equitability satisfies.

Define a procedure to be *strategy-vulnerable* if a maximin player can, by misrepresenting its value function, assuredly do better, whatever the value function of the other player (or, as I will discuss later, other players). A procedure that is not strategy-vulnerable is *strategy-proof*, giving maximin players always an incentive to let $f_A(x) = v_A(x)$ and $f_B(x) = v_B(x)$.

Proposition 13.1. *SP is strategy-proof, whereas any procedure that makes e the cut-point is strategy-vulnerable.*

Proof. To show that maximin players will be truthful when they submit their value functions to a referee, I show that A or B may do worse if they are not truthful in reporting:

1. *Their 50–50 points, a and b, based on their value functions.*
Assume B is truthful and A is not. If A misrepresents a and causes it to crisscross b, as illustrated by the location of a' below,

then A will obtain $[a', 1]$ and, in addition, get some less-than-complete portion of (b, a'). But this is less than 50 percent of the cake for A and, therefore, less than what A would obtain under SP if it were truthful.

2. *Their value functions over [a, b].*
Assume, again, B is truthful. Assume A considers misrepresenting its value function in a way that moves c rightward, as shown below:

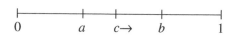

It can do this by (i) decreasing the value of $\int_a^c f_A(x)dx$, the numerator on the left side of equation (13.1), or (ii) increasing the value of the denominator, $\int_a^b f_A(x)dx$. But in order for A to misrepresent in this manner, it would have to know $f_B(x)$ and therefore b, which c depends on. But A does not know $f_B(x)$ and b and, consequently, cannot determine c. Hence, it cannot assuredly reduce its value of the interval $[a, c]$ relative to $[a, b]$ in order to make this proportion less than its true proportion and so move c rightward. Indeed, A's attempted misrepresentation could backfire by moving c leftward rather than rightward, which would give A a smaller proportion of $[a, b]$.

To be sure, if A knew the location of b, it could concentrate its value just to the left of b, which would move c rightward. But I assumed that A is ignorant of the location of b, and even which side of a (left or right) it is on. Hence, A cannot misrepresent its value function and assuredly do better, which makes SP strategy-proof.

On the other hand, assume the cake is cut at e, so its division is equitable rather than proportionally equitable (at c). When the players are truthful so $f_A(x) = v_A(x)$ and $f_B(x) = v_B(x)$, one player will receive [0, e] and the other player will receive (e, 1], which they will value equally and at least as much as 1/2; point e will be unique when the players' measures are absolutely continuous with respect to Lebesgue measure (Jones, 2002).

I next show that A can submit a value function different from $v_A(x)$ that moves e to a position favorable to it regardless of (i) player B's value function, and (ii) whether or not player A receives the left or the right piece of the cake. Because A is unaware of whether e is to the left or to the right of its 50–50 point, a, A should submit a value function that has the same 50–50 point as $v_A(x)$, as discussed in (1) above.

But to increase the value of its piece beyond 50 percent, A should submit a value function, $f_A(x)$, that decreases the value of $\int_a^e f_A(x)dx$ if a is to the left of b, and decreases the value of $\int_e^a f_A(x)dx$ if a is to the right of b. The former strategy will move e rightward, whereas the latter strategy will move e leftward, of its true value.

Clearly, A can effect *both* movements of e by decreasing the value of $\int_a^b f_A(x)dx$. Because A does not know the value of b, however, and even whether it is to the left or to the right of a, A can best minimize its value over [a, b] by concentrating almost 1/2 its value near 0 and almost 1/2 near 1—the endpoints of the cake—while ensuring that

$$\int_0^a f_A(x)dx = \int_a^1 f_A(x)dx = 1/2$$

so that its 50–50 point is truthful. Thereby, A decreases its value around its 50–50 point, which will move e toward B's 50–50 point—whichever side of a that b is on—and so help A. (Optimally, A should let the value strictly between its 50–50 point and the edges of the cake, where its value is concentrated, approach 0 in the limit.) Thereby any procedure that makes e the cutpoint is strategy-vulnerable. ∎

Because both A and B receive at least 50 percent of their valuations under SP, the resulting division is not only proportionally equitable but also envy-free. If there are more than two players, however, an envy-free allocation may be neither equitable nor proportionally equitable.

13.4. THREE OR MORE PLAYERS: EQUITABILITY AND ENVY-FREENESS MAY BE INCOMPATIBLE

To show that it is not always possible to divide a cake among three players into envy-free and equitable portions using two cuts, assume that A and B have the following (truthful) piecewise linear value functions that are symmetric and V-shaped,

$$v_A(x) = \begin{cases} -4x + 2 & \text{for } x \in [0, 1/2] \\ 4x - 2 & \text{for } x \in (1/2, 1] \end{cases}$$

$$v_B(x) = \begin{cases} -2x + 3/2 & \text{for } x \in [0, 1/2] \\ 2x - 1/2 & \text{for } x \in (1/2, 1] \end{cases}.$$

Whereas both functions have maxima at $x = 0$ and $x = 1$ and a minimum at $x = 1/2$, A's function is steeper (higher maximum, lower minimum) than B's, as illustrated in Figure 13.1. In addition, suppose that a third player, C, has a uniform value function, $v_C(x) = 1$, for $x \in [0, 1]$.

In this example, every envy-free allocation of the cake will be one in which A gets the portion to the left of x, B the portion to the right of $1 - x$

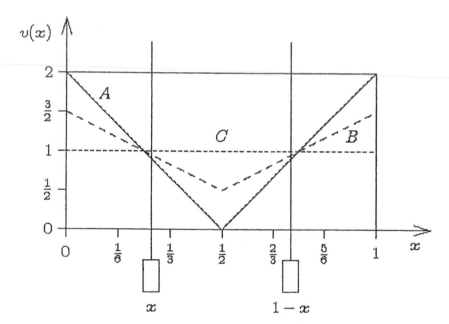

Figure 13.1. Impossibility of Envy-Free and Equitable Cuts for Three Players

(A and B could be interchanged), and C the portion in the middle. If the horizontal lengths of A's and B's portions are not the same (i.e., x), the player whose portion is shorter in length will envy the player whose portion is longer. But an envy-free allocation in which the lengths are the same will not be equitable, because A will receive a larger portion in its eyes than B receives in its eyes, violating equitability. Thus, an envy-free allocation is not equitable in this example, nor is an equitable allocation envy-free, although both these allocations will be efficient with respect to parallel, vertical cuts.[3]

Two envy-free procedures have been found for 3-person, 2-cut cake division. Whereas one of the envy-free procedures requires four simultaneously moving knives (Stromquist, 1980), the other requires only two simultaneously moving knives (Barbanel and Brams, 2004).

I describe the simpler procedure of Barbanel and Brams that involves "squeezing" pieces. It assumes that virtual cuts, or what Shishido and Zeng (1999) call "marks," can be made on the line segment defining the cake. These marks may subsequently be changed before real cuts are made.

Players are given "instructions." If they are followed, an envy-free allocation results; if they are not, then a player may do worse, violating the maximin goal assumed of players.

13.5. THE SQUEEZING PROCEDURE

A referee slowly moves a knife from left to right across a cake.[4] The players are instructed to call stop when the knife reaches the 1/3 point for each. Let the *first* player to call stop be player A. (If two players call stop at the same time, choose one randomly.)

Let A place a mark at the point where it calls stop (the right boundary of piece 1 in the diagram below), and a second mark to the right that bisects the

[3] Not all equitable divisions need be efficient. If C were given an end piece and A or B the middle piece in the example, cutpoints could be found such that all the players receive, in their own eyes, the same value. However, this value would be less than what another equitable allocation, in which C gets the middle piece and A and B the end pieces, yields. By contrast, an envy-free allocation that uses $n - 1$ parallel, vertical cuts is always efficient (Gale, 1993; Brams and Taylor, 1996, pp. 150–151).

[4] Moving-knife procedures are discussed in, among other places, Brams, Taylor, and Zwicker (1995), Brams and Taylor (1996), and Robertson and Webb (1998). In general, players are assumed not to know the value functions of other players and, therefore, have no basis for trying to manipulate them to their advantage, which makes them strategy-proof. For nonconstructive results on cake-cutting, which address the existence but not the construction of fair divisions that satisfy different properties, see Barbanel (2005).

remainder of the cake (the right boundary of piece 2 below). Thereby A indicates the two points that, for it, trisect the cake into pieces 1, 2, and 3:

Because neither player B nor player C called stop before A did, each of B and C thinks that piece 1 is at most 1/3. They are then asked whether they prefer piece 2 or piece 3. There are three cases to consider:

1. If B and C each prefers a different piece—one player prefers piece 2 and the other piece 3—the procedure ends: A, B, and C can each be assigned a piece that they consider to be at least tied for largest.

2. Assume B and C both prefer piece 2. A referee places a knife at the right boundary of piece 2 and moves it to the left. Meanwhile, A places a knife at the left boundary of piece 2 and moves it to the right in such a way that the amounts of cake traversed on the left and right are equal for A. Thereby pieces 1 and 3 increase equally in A's eyes. At some point, piece 2 will be diminished sufficiently to piece 2'—in either B's or C's eyes—to tie with either piece 1' or piece 3', the enlarged 1 and 3 pieces. Assume B is the first, or the tied-for-first, player to call stop when this happens; then give C piece 2', which it still thinks is the largest or the tied-for-largest piece. Give B the piece it thinks ties for largest with piece 2' (say, piece 1'), and give A the remaining piece (piece 3'), which it thinks ties for largest with the other enlarged piece (1'). Clearly, each player will think it gets at least a tied-for-largest piece.

3. Assume B and C both prefer piece 3. A referee places a knife at the right boundary of piece 2 and moves it to the right. Meanwhile, A places a knife at the left boundary of piece 2 and moves it to the right in such a way as to maintain the equality, in its view, of pieces 1 and 2 as they increase. At some point, piece 3 will be diminished sufficiently to piece 3'—in either B or C's eyes—to tie with either piece 1' or piece 2', the enlarged 1 and 2 pieces. Assume B is the first, or the tied-for-first, player to call stop when this happens; then give C piece 3', which it still thinks is the largest or the tied-for-largest piece. Give B the piece it thinks ties for largest with piece 3' (say, piece 1'), and give A the remaining piece (piece 2'), which it thinks ties for largest with the other enlarged piece (1'). Clearly, each player will think it got at least a tied-for-largest piece.

Note that which player moves a knife or knives varies, depending on what stage is reached in the procedure. In the beginning, I assume a referee moves a single knife, and the first player to call stop (A) then trisects the cake. But at the next stage of the procedure, in cases 2 and 3, it is the referee and A that move two knives simultaneously, "squeezing" what players B and C consider to be the largest piece until it eventually ties, for one of them, with one of the two other pieces. While Barbanel and Brams show that squeezing can also be used to produce an "almost" envy-free 4-person, 3-cut division (at most one player is envious), their extension of the squeezing procedure to obtain envy-freeness for 4 players requires up to 5 cuts, which may necessitate that some players receive disconnected pieces.

Earlier, Brams, Taylor, and Zwicker (1997) gave a 4-person, envy-free procedure that requires up to 11 cuts; chore division for 4 players requires even more (16 cuts) (Peterson and Su, 2002). Because the Brams-Taylor-Zwicker 4-person procedure involves fewer cases than the Barbanel-Brams procedure, it is arguably simpler, even though it requires more cuts (11 versus 5).

Beyond four players, no procedure is known that yields an envy-free division of a cake unless an unbounded number of cuts is allowed (Brams and Taylor, 1995, 1996; Robertson and Webb, 1998). While this number can be shown to be finite, it cannot be specified in advance—this will depend on the specific cake being divided. The complexity of what Brams and Taylor call the "trimming procedure" to obtain an n-person envy-free division makes it of dubious practical value.

I next show that it is always possible to find an equitable division of a cake among three or more players that is efficient (see note 3). In fact, the equitability procedure (EP) enables $n \geq 3$ players to achieve an equitable and efficient division of a cake that is also strategy-proof.

13.6. THE EQUITABILITY PROCEDURE (EP)

The rules of EP are as follows:

1. Independently, A, B, C, . . . report their value functions $f_A(x)$, $f_B(x)$, $f_C(x)$, . . . over [0, 1] to a referee. These functions may be different from the players' true value functions, $v_A(x)$, $v_B(x)$, $v_C(x)$ . . . .
2. The referee determines the cutpoints that equalize the common value that all players receive for each of the $n!$ possible assignments of pieces to the players from left to right.
3. The referee chooses the assignment that gives the players their maximum common value.

I next illustrate EP using the 3-person example in section 13.4. It is evident that the ordering of players that will maximize the common value to the players is to give the left piece to A (or B), the middle piece to C, and the right piece to B (or A).

Let the cutpoints be e_1 and e_2. Assume A receives the piece defined by the interval $[0, e_1]$, C the piece defined by the interval $(e_1, e_2]$, and B the piece defined by the interval $(e_2, 1]$. The players' values will be equal when

$$\int_0^{e1} (-4x + 2)dx = \int_{e1}^{e2} dx$$

$$\int_{e1}^{e2} dx = \int_{e2}^{1} (2x - 1/2)dx.$$

Integration and evaluation of these equations yields two quadratic equations in two unknowns:

$$-2e_1^2 + 2e_1 = e_2 - e_1$$
$$e_2 - e_1 = 1/2 - e_2^2 + e_2/2.$$

When solved simultaneously, these equations give $e_1 \approx 0.269$ and $e_2 \approx 0.662$. Players A, C, and B (from left to right, in that order) all value their pieces at 0.393, so each thinks it receives nearly 40 percent of the cake.

For n players, there will be $n - 1$ cutpoints e_i. For each assignment of pieces to the players from left to right, solving simultaneously the $n - 1$ equations that equalize the value functions of adjacent players will give the e_i's.

Choosing the assignment that gives the players a maximum common value yields a division that is efficient. This is because one player cannot get more value without another player's getting less, which would, of course, destroy equitability.

Proposition 13.2. *EP is strategy-proof.*

Proof. Assume that some player X is not truthful under EP but that all other players are truthful. For X to increase its allocation, it would have to know its borders in order to misrepresent its true value function and guarantee itself more. But because X is ignorant of the reported value functions of the other players, it will not be able to determine these borders, nor even where its piece lies in the left-right assignment of pieces to players if it did know these borders. Hence, X cannot ensure itself of a more valuable piece if it does not know the value functions of the other players. ∎

Assume that X, by misrepresenting its own value function, increases the value of its piece, as I showed was possible in the 2-person case of equitable division even with no information about the value functions of the other player. But then X will have no assurance that it will receive this more valuable piece, because its misrepresentation may change the left-right assignment of pieces to players. This was not possible in the 2-person case as long as X was truthful about its 50–50 point. By undervaluing the cake around its 50–50 point, X could increase its portion of the surplus while still retaining its 50 percent portion on the left or right side.

However, when there are additional players and there is no identifiable surplus to be divided among them—as in the 2-person case between A and B—X has no assurance that it will retain the piece that its misrepresentation might increase in size. Indeed, X may end up with a piece that it values less than $1/n$ of the cake.

Proposition 13.3. *If a player is truthful under EP, it will receive at least $1/n$ of the cake regardless of whether or not the other players are truthful; otherwise, it may not.*

Proof. Consider the moving-knife procedure, due to Dubins and Spanier (1961), in which a referee moves a knife slowly across a cake from left to right. A player that has not yet received a piece calls "stop," and makes a mark, when the knife reaches a point that gives it $1/n$ of the cake rightward of the last point at which the knife was stopped by a player (or from the left edge for the first player to call stop). It is easy to show that a truthful player will be able to get a $1/n$ piece, with some cake generally remaining near the right edge.[5] By moving all players' marks rightward (Shishido and Zeng, 1999), one can give each player an equal amount greater than $1/n$, exhausting the remainder, because the players' measures are nonatomic. If a player is not truthful, it will *appear* that it received a piece that is at least $1/n$ under EP, but its true value may be less than $1/n$. ∎

To illustrate a misrepresentation that may give a player less than $1/n$, assume player C in the example given in section 13.4 knows the value functions of players A and B, but A and B do not know C's value function. I first show how C can maximize its value function when it knows the value functions of A and B, which I assume are truthful.

Let c_1 be the cutpoint on the left that defines A's piece, starting from the left edge, and let c_2 be the cutpoint on the right that defines B's piece, starting from the right edge. Then C should undervalue the middle portion between c_1 and c_2 so that A and B receive exactly the same value from their pieces—as required by EP—

$$\int_0^{c1} (-4x + 2)dx = \int_{c2}^1 (2x - 1/2)dx,$$

while C receives as much of the middle portion of the cake as possible.

---

[5] Even though the Dubins-Spanier assignment gives each player at least $1/n$, it may not be the assignment from left to right that gives the players the maximal equitable division. Under EP, a different assignment of equal-valued pieces to players could give each more.

C can maximize the value of the middle portion by making B, which values the middle portion more than A does, indifferent between receiving this portion and obtaining the right portion:

$$\int_{c1}^{1/2} (-2x + 3/2)dx + \int_{1/2}^{c2} (2x - 1/2)dx = \int_{c2}^{1} (2x - 1/2)dx.$$

This "optimal" misrepresentation by C ensures that it obtains as physically large a middle piece as possible at the same time that it appears to receive the same-value pieces as A and B do on the left and right.[6]

Integration and evaluation of these equations yields two quadratic equations in two unknowns:

$$4c_1^2 - 4c_1 = 2c_2^2 - c_2 - 1$$
$$2c_1^2 - 3c_1 = -4c_2^2 + 2c_2.$$

Solving these equations simultaneously gives $c_1 \approx 0.230$ and $c_2 \approx 0.707$. A and B receive the same value of 0.354 for the left and right pieces, respectively, whereas C *appears* to receive this value for the middle piece.

But the true value for C of its now enlarged middle piece, $c_2 - c_1 \approx 0.477$, is 21 percent greater than its value when it is truthful (0.393), so C clearly benefits from this misrepresentation. But had C undervalued the middle portion more, and consequently overvalued the left or right portions by a greater amount, it would have received one of the latter under EP, which would have given it a true value of less than 0.393.

Thus, without information on the value functions of the other players, a player may misrepresent in a way that lowers its value over being truthful. Indeed, such misrepresentation may give it less than $1/n$ of the cake, making truthfulness not only a maximin strategy but also one that guarantees a player at least $1/n$ of the cake (Proposition 13.3).

The guarantee of at least $1/n$ to the players under EP generalizes the guarantee of at least one-half to the two players under SP.[7] The additional players

---

[6] Just as C must lower its value of $[c_1, c_2]$ to that which gives it the same value that B attaches to this middle portion, it must also raise its values of $[0, c_1)$ and $(c_2, 1]$ to those of B as well. Because this will allow the middle or the right pieces to be assigned to either B or C, C should slightly perturb its values so that it appears that it values the middle portion more, ensuring that it, rather than B, receives it.

[7] A minimal-cut envy-free procedure also gives this guarantee, because a player cannot receive less than $1/n$ without envying another player. However, EP maximizes the minimum amount *greater than* $1/n$ that a player receives, whereas an envy-free procedure may give one player less than this amount. In the 3-person example in section 13.4, for instance, an envy-free allocation will give A a larger proportion of the cake than B receives, though both players will

under EP create greater uncertainty about their allocations, making EP more difficult to exploit than SP. Consequently, EP is able to ensure a maximal equitable allocation that is strategy-proof, whereas SP can only ensure a proportionally equitable allocation that is strategy-proof.

13.7. SUMMARY AND CONCLUSIONS

I have described a new 2-person, 1-cut cake-cutting procedure, called the surplus procedure (SP). Like cut-and-choose, it is envy-free and efficient and also induces the players to be truthful when they have no information about each other's preferences, rendering it strategy-proof. But unlike cut-and-choose, SP produces a proportionally equitable division, whereas an analogous equitable procedure is strategy-vulnerable.

SP is more information demanding than cut-and-choose, requiring that the players submit to a referee their value functions over an entire cake, not just indicate a 50–50 point. Practically, players might sketch such functions, or choose from a variety of different-shaped functions, to indicate how they value a divisible good like land.

Thus, land bordering water might be more valuable to one player (A), whereas land bordering a forest might be more valuable to the other (B). Even if players know these basic preferences of each other, and hence that a will be closer to the water and b will be closer to the forest, uncertainty about the other player's 50–50 point makes it impossible for maximin players to exploit SP without knowledge of the other player's value function.

For three players, there may be no envy-free division that is also equitable, so a choice may have to be made between these two properties. The squeezing procedure gives an envy-free, efficient division, whereas the equitability procedure (EP) gives an equitable, efficient division; both are strategy-proof. EP also works for four or more players, but there is no known minimal-cut, envy-free procedure.

It is pleasing to have strategy-proof procedures that yield efficient, envy-free, and proportionally equitable divisions in the case of two players, and efficient and equitable divisions in the case of more than two players. If there are multiple divisible goods that must be divided, however, 2-person procedures like AW (chapter 12) seem more applicable than cake-cutting procedures, though Jones (2002) shows that AW can be viewed as a cake-cutting procedure.

---

value the two (equal) envy-free pieces each gets exactly the same. Under EP, by contrast, the players will value their pieces differently, causing A to envy B for getting a physically larger piece, though each player values its proportion of the cake exactly the same as the other player values its proportion.

The small literature on pie-cutting, in which radial cuts are made from the center of a disk instead of vertical cuts along a horizontal axis, raises new issues, including whether there always exists an envy-free and efficient division of a pie (Gale, 1993). Barbanel and Brams (2004) and Brams, Jones, and Klamler (2006a) provide some positive as well as negative answers, but suffice it to say that several questions remain open.

14

Allocating Divisible and Indivisible Goods

14.1. INTRODUCTION

So far I have analyzed the fair division of indivisible goods (chapters 9 and 10) and three kinds of divisible goods—homogeneous goods like money (chapter 11), multiple homogeneous goods (chapter 12), and a heterogeneous good like a cake (chapter 13). In this chapter I assume that there are multiple indivisible goods but also a single divisible homogeneous good—namely, money.

The money is a "bad" in the sense that it is used by the players to pay for the indivisible goods they receive. The fair-division procedure that I propose for allocating the indivisible goods and the divisible bad assumes that the players make bids for the goods.

As in a standard auction, the bids do double duty. They are the basis for (i) assigning the goods to the players, and (ii) determining the prices they pay for them. But unlike a standard auction, the player who bids the highest for a good does not necessarily receive it. Moreover, the price that a player pays for a good depends, in general, not just on its own bid but lower bids as well.[1]

I analyze the "one-good-to-a-player problem," or for convenience the *one-good problem*. Assume, for example, that the government wants to lease several goods (e.g., geographic sites) to multiple players for development, but it does not want any player to receive more than one good. This may be because it wants to ensure competition in the future by supporting multiple bidders in the present.

To simplify this problem, assume there are n players and n goods.[2] If the goods have different value for the different players, based on their location and potential for development, the players are likely to bid different amounts for the different goods.

Note: This chapter is adapted from Brams and Kilgour (2001a) with permission; see also Brams and Kilgour (1999).

[1] In this respect, the procedure resembles a Vickrey (1961), or second-price, auction, in which the highest bidder wins the item but pays only the second-highest bid for it. However, the procedure here does not equate the prices of goods and particular lower bids but, rather, makes prices dependent on possibly several lower bids.

[2] A generalization later allows for the division of m goods among n players. If $m < n$, some players will get no goods, which I illustrate in section 14.8. If $m > n$, some players may get more than one good, which I illustrate in section 14.9 wherein goods are distributed in successive rounds.

I assume that the government is willing to lease *all* the sites for some fixed amount, or *total price*.[3] Each player submits nonnegative bids for each good that sum to this price. On the basis of these bids, each bidder is assigned, and pays the price for, exactly one good.

A procedure, called the Gap Procedure, is proposed whereby the bidding competition is balanced by fairness constraints that produce a fair distribution of both the goods that the players receive and the bads, or prices, they must pay. Under this procedure, the prices the players pay for the goods reflect the bids of not only the winners but also those of players that bid less. This market-oriented approach to fair division (i) ensures nonnegative prices that never exceed a player's winning bid; (ii) is Pareto-optimal, precluding mutually beneficial trades, though not envy; (iii) is monotonic in bids, so higher bids never hurt in obtaining a good; (iv) encourages sincere bids; and (v) produces prices that are partially independent of the amounts bid (as in a Vickrey auction).[4]

I begin with definitions and assumptions in section 14.2 and then analyze certain difficulties that may arise with bidding in section 14.3. Next, I describe and illustrate the Gap Procedure and then discuss a variety of possible applications.

14.2. DEFINITIONS AND ASSUMPTIONS

A player's bids are *sincere* if they mirror its valuations of the goods, making it indifferent to receiving any good at its bid. I interpret the bids as the maximum amounts (possibly zero) that the bidders would be willing to pay for the

---

[3] How the total might be set I leave open, but presumably it would be based on a value appraisal of some kind as well as past experience. The alternative would be to hold a standard auction, with the highest bidder or bidders winning the goods. However, this approach could result in one predominant player obtaining all the goods. The *raison d'être* of the approach taken here is to prevent monopolistic control by one player that could eliminate competition and make a mockery of the fair division of goods.

[4] The Gap Procedure was inspired by the so-called housemates problem (Su, 1999), in which a procedure is proposed to determine how much rent each housemate must pay for the room he or she receives. This problem is different from the related problem of "house allocation," in which there are indivisible objects (houses) but no medium of exchange, as here. Different strategy-proof mechanisms that enable the players, who may or may not have property rights, iteratively to trade up to a Pareto-optimal core matching of themselves and the houses have recently been proposed in Abdulkadiroğlu and Sönmez (1998, 1999), Svensson (1999), and Pápai (2000); earlier work on this problem is discussed in these papers. Difficulties of allocating indivisible goods over which players have ordinal preferences are discussed in, among other places, Brams, Edelman, and Fishburn (2001, 2003). The so-called Knaster procedure (Brams and Taylor, 1996), first discussed in Steinhaus (1948), allows for bidding, as here, but it is a highest-bidder-wins procedure, with adjustments for fairness made via monetary side payments (either positive or negative) to the players; refinements of this procedure are analyzed in Raith (2000) and Chisholm (2000).

goods. In general, a player will never pay more than its bid for the good it receives under the Gap Procedure.

A problem that immediately arises is that of *feasibility*: Have the players, collectively, bid enough to cover the total price? I address this problem by finding the *maxsum* distribution of goods, which is the assignment of goods to players that maximizes the sum of the bids. If this sum is less than the total price, then the distribution problem is *infeasible*.

In order to focus on the distributional aspects of fair division, I assume in most of the examples that each player's bids for the goods sum to exactly the total price of all the goods. This condition is not essential; if some players' bids sum to less than the total price, a maxsum allocation may still be feasible if the other players' bids are sufficiently different.

The assumption that each player's bids sum to the total price *relativizes* the auction for goods: because each player bids the same total amount, only the relative bids matter. The one-good problem is somewhat akin to the apportionment problem, which concerns how many representatives should be allocated to different-size districts in a legislature (Balinski and Young, 1982, 2001).[5]

Unfortunately, some of the most obvious solutions to this problem run amok, even in the simple case in which there are only three goods and three players:

- One player might not make the highest bid for any good, which creates a problem of which good to assign that player.
- Even if each player bids highest on exactly one good, these winning bids will, in general, sum to more than the required total price.
- Reducing the highest bids equally or proportionally—so they sum to the total price—may lead to some players' (i) paying negative prices (i.e., being paid to receive a good) or (ii) paying more than their bids.

To eliminate these difficulties, the Gap Procedure has a marketlike flavor:

- It assigns goods to different players so as to maximize the *sum* of the bids.
- It makes prices for each good a function not just of the bid of the player receiving that good but also of lower bids (the more competitive these bids, the closer the winner's price is to its bid).

I show that the Gap Procedure assigns each player a good at a price no more, and generally less, than its bid. Moreover, there exist no trades of goods that can make all the traders better off, rendering the allocation efficient or Pareto-optimal.

---

[5] The apportionment problem, discussed in chapter 9, may be further constrained. For example, in the U.S. Senate, in which the districts are the states, all states have two senators, so size does not matter. Although size does matter in the U.S. House of Representatives, all states are entitled to at least one representative.

Two well-known procedures provide solutions that are related to the one-good problem:

1. In Hylland and Zeckhauser (1979), players are assumed to have equal endowments. They bid on goods, from which prices are derived that give each player a probability share in the goods and yield an ex-ante Pareto-optimal distribution. These prices are incentive-compatible, so the players are motivated to bid truthfully. But the Hylland-Zeckhauser procedure depends on a pseudomarket, whereby "money," because it is internal to the model, is not usable as a price.[6]

2. In Leonard (1983), which does use real money, goods are assigned and their prices determined so that each player's surplus—its valuation of its good minus the price it pays for it—is maximized. The procedure is also incentive-compatible, and the distribution of the goods and their prices is Pareto-optimal and envy-free. However, the sum of the prices cannot be controlled, so it is impossible to guarantee the extraction of a (prespecified) total price. Moreover, if the sum of the prices is less than the required total, the prices cannot be "corrected" to achieve that total without sacrificing other properties. In effect, the Leonard procedure makes the total price endogenous, whereas the total price is exogenous in the one-good problem.

Besides the Gap Procedure's marketlike properties, I investigate other properties, including its envy-freeness, monotonicity, sincerity, and propensity to make prices independent of winning bids. I also suggest the application of the Gap Procedure to choices over time, in which endowments change as players use up their bidding points. I conclude that competition and fairness can be coherently combined—while encouraging healthy competition, the Gap Procedure respects the rights of players to obtain certain minimum numbers of goods.

I begin with several examples that illustrate the clash between competitive bidding and fairness. I then show how the Gap Procedure lessens these difficulties by analyzing its properties and proving several propositions.

14.3. DIFFICULTIES WITH EQUAL AND PROPORTIONAL REDUCTIONS IN THE HIGH BIDS

I start with the simplest case. Two goods (G1 and G2) need to be allocated to two players (P1 and P2). Assume that the total price for the goods is 100, and

---

[6] Other random mechanisms have been proposed for achieving ex-post efficiency and the stronger notion of ordinal efficiency (Bogomolnaia and Moulin, 2001; Abdulkainoğlu and Sönmez, 2003). The *realization* of these mechanisms, however, is a deterministic assignment, which will not generally possess certain properties—to be discussed subsequently—even if the lotteries on which they are based give positive weight only to efficient assignments or are restricted to those that are not stochastically dominated.

each player's bids for the goods are nonnegative. This gives rise to three possibilities:

1. P1 bids more than P2 for G1, and P2 bids more than P1 for G2.
2. P2 bids more than P1 for G1, and P1 bids more than P2 for G2.
3. One player, P1 or P2, bids more for both goods.

For now, assume that P1 and P2 bid sincerely. Then the obvious solution in cases 1 and 2 is to give the high bidders their preferred goods. In case 3, assign different goods to the two players so as to maximize the bid total.

In all three cases, there may be a feasibility problem: the sum of the two bids for the goods may or may not be greater than 100. If it is not, the one-good problem is infeasible; henceforth I will assume that this is not the case.

Even when the one-good problem is feasible, how much players should pay for the goods is by no means evident. To encourage competitive pricing, I begin by considering two different ways of reducing the maximum bids for goods to prices that sum to 100, which Example 14.1 illustrates.

Example 14.1. Bids by Two Players for Two Goods

	G1	G2	Total
P1	75	45	120
P2	30	60	90

The high bids for each good (75 and 60), which are underscored, sum to 135. Consider two different ways in which the *total surplus* of 35 (above the price of 100 for both goods) might be subtracted from the two players' high bids:

1. *Equal reductions.* Subtract half the total surplus ($35/2 = 17.5$) from each player's high bid, so that P1 pays 57.5 for G1 and P2 pays for 42.5 for G2.
2. *Proportional reductions.* Divide the total surplus of 35 between P1 and P2 in proportion to their winning bids ($75/60 = 5/4$), so P1's high bid is reduced by $(5/9)(35) \approx 19.4$, and P2's high bid by $(4/9)(35) \approx 15.6$. Thus, P1 will pay 55.6 for G1, and P2 will pay 44.4 for G2.

Clearly, the player whose winning bid is less (P2 in Example 14.1) will prefer equal reductions, whereas the player whose winning bid is more (P1) will prefer proportional reductions.

While arguments based on different equity considerations could be adduced for each kind of reduction, both are flawed according to other criteria: (i) equal reductions can lead to a player's paying a negative price; and (ii) both equal and proportional reductions can lead to a player's paying more than its bid for a good. These difficulties arise when one adds one more good and one more player to the auction. To highlight distributional issues, suppose that each player's bids sum to exactly 100, which guarantees feasibility.

Negative Price

Example 14.2. Bids by Three Players for Three Goods

	G1	G2	G3	*Total*
P1	97	2	1	100
P2	1	97	2	100
P3	32	33	35	100

As in Example 14.1, each player most prefers a different good, again assuming the bids are sincere. The high bids for each good (underscored) sum to 229, giving a total surplus of 129, which can be reduced in the two different ways described earlier:

1. If the reductions are *equal*, 129/3 = 43 is subtracted from the high bids, so P1 and P2 pay 54 each for G1 and G2, respectively, but P3 pays a negative price of − 8 for G3. That is, P3 will have to be paid 8 in addition to receiving G3. This seems absurd, especially in light of the fact that P3 prefers G3 to the other two goods.[7] Why should P1 and P2 compensate P3 to obtain the good it most desires?

2. If the reductions are *proportional* to the high bids of 97, 97, and 35, the portions of 129 that are subtracted from these bids are 54.6, 54.6, and 19.7, respectively. Consequently, P1 pays 42.4 for G1, P2 pays 42.4 for G2, and P3 pays 15.3 for G3.[8]

Proportional reductions, which can never produce negative prices because the amounts subtracted from the bids are always less than the bids, are not without their difficulties, as I next show.

---

[7] This would not be absurd if the players were bidding for bads, like chores or burdens, in which case each would bid highest for the chores it considers least bad. In this situation, it would be reasonable for P3 to be paid 8 for G3 (if G3 were interpreted as a chore) since P1 and P2 consider it so onerous. I will return to the issue of negative prices in section 14.6, where I analyze the Pareto-optimality of the Gap Procedure when negative prices are both allowed and disallowed.

[8] A consequence of rounding is that the reductions sum to 128.9 (rather than 129.0), and the prices sum to 100.1 (rather than 100.0). Henceforth, I ignore the effects of rounding in this and other numerical examples.

Loss to a Player

Example 14.3. Bids by Three Players for Three Goods

	G1	G2	G3	Total
P1	50	1	49	100
P2	29	40	31	100
P3	31	38	31	100

This example differs from Examples 14.1 and 14.2 in that the high bidder on each good is not always different: P1 is the high bidder on G1 and G3, with bids of 50 and 49, respectively (underscored). By comparison, P2 is the high bidder on G2 (with an underscored bid of 40), and P3 is the high bidder on no good. Because the high bids sum to 139 (50 + 40 + 49), the total surplus is 39:

1. If the reductions are *equal*, 39/3 = 13 is subtracted from each of the high bids, giving prices of 37, 27, and 36 for G1, G2, and G3, respectively. Notice that G1 and G3 cannot be assigned to any player other than P1, because only P1's bids for these goods are at least equal to their prices. Thus, whichever of G1 or G3 is assigned to P1 (say, G1), neither of the other players, which both bid 31 for G3, will be willing to pay G3's price of 36. Hence, there is no assignment of different goods to the three players such that all pay no more than their bids and, therefore, do not suffer a loss.
2. If the reductions are *proportional* to the high bids of 50, 40, and 49, the portions of 39 that are subtracted from these bids are 14.0, 11.2, and 13.7, giving prices of 36.0 for G1, 28.8 for G2, and 35.3 for G3. Once again, G1 and G3 cannot be given to any player other than P1, precluding an allocation of the three goods to the three different players such that they all pay no more than their bids.

The lesson of Example 14.3 is that under the no-loss constraint—that each player's price be no greater than its bid—pricing must take account of the good assignments; one cannot simply reduce the highest bids for each good. The Gap Procedure, to be described in section 14.3, starts by assigning goods so as to maximize the total surplus, which sometimes entails giving a good to a player that is not the highest bidder.[9] But instead of making equal reductions in the total surplus, or reductions proportional to the high bids—in order to find prices that sum to 100—the Gap Procedure takes into account the lower bids for each good, setting prices competitively.

---

[9] In this manner, it promotes the *collective* welfare of players rather than that of just the highest bidders.

14.4. THE GAP PROCEDURE

I begin by describing the Gap Procedure, which proceeds in three steps, and then illustrate its application to the three preceding examples.

Gap Procedure

1. *Maxsum Assignment.* Assign the goods to different players so as to maximize the sum of player bids. Call this sum "maxsum."[10]
2. *Feasibility.* If maxsum < 100, the one-good problem is infeasible, and the procedure stops. Otherwise, the goods prices are based on the maxsum bids, as described next.
3. *Prices for Goods.* Descend from the maxsum assignment to the next-lower bids for each good, the next-lower bids to these, and so on until the sum of the current set of bids is less than or equal to 100.[11] Stop the descent on a good when the lowest bid on that good is reached; if necessary, continue the descent on the other goods until the sum of the current set of bids is less than or equal to 100.[12] If this last sum is exactly 100, then these bids are the goods prices, and the procedure stops. Otherwise, return to the next-higher sum (which is necessarily greater than 100) and reduce the bids it comprises in proportion to the gaps between them and the next-lower bids so that the sum of the reduced bids equals 100. These are the goods prices, and the procedure stops.

Example 14.1. Maxsum is 135 (75 + 60) along the main diagonal, and the next-lower bids (30 and 45) along the off-diagonal sum to 75. Reduce the maxsum bids in proportion to the gaps—$75 - 30 = 45$ and $60 - 45 = 15$, which sum to 60—between the maxsum and the next-lower bids. Because the reductions are in the ratios $45:15 = 3:1$, P1's price is $75 - (3/4)(35) = 49.75$, and P2's is $60 - (1/4)(35) = 51.25$. Note that P2 pays more, but not because it bid more for the good it receives (G2). Rather, bidding for G2 is more competitive, with the second-highest bid for G2 (45) being closer to the highest bid (60) than for G1 (30 and 75, respectively).

---

[10] If there is more than one maxsum assignment, one can be chosen randomly.

[11] "Next-lower bids" may not, in fact, be lower but tied with those from which they descend, as Example 14.3 illustrates.

[12] After sufficiently many descents, the sum of the current set of bids will be less than or equal to 100, except in one extreme case—when the sum of the minimum bids for the goods exceeds 100. This case is rather unlikely: How often would it occur that the lowest bids for the goods are sufficient to pay the entire price? Nonetheless, should this be the case, I assume one further descent—to bids of 0 for each good. Prices could then be determined according to step 3, based on the gap between 0 and the minimum bids.

313

Example 14.2. Maxsum is 229 (97 + 97 + 35) along the main diagonal, and the next-lower bids (32, 33, and 2) sum to 67. Reduce the maxsum bids in proportion to the gaps—97 − 32 = 65, 97 − 33 = 64, and 35 − 2 = 33, which sum to 162—between the maxsum and the next-lower bids. Thus, the total surplus of 229 − 100 = 129 is subtracted from the maxsum bids for G1, G2, and G3, respectively, in the amounts of (65/162)(129) ≈ 51.8, (64/162)(129) ≈ 51.0, and (33/162)(129) ≈ 26.3. This gives goods prices of 45.2, 46.0, and 8.7 for G1, G2, and G3, respectively.[13]

Example 14.3. Maxsum is 121 (50 + 40 + 31) along the main diagonal. The next-lower bids for G1 and G2—and the tied-for-next-lower bid in the case of G3—are 31, 38, and 31, respectively. Because these bids sum to exactly 100, no further descent is necessary—these bids *are* the goods prices under the Gap Procedure. Unlike Examples 14.1 and 14.2, these prices do not depend directly on the maxsum bids but, rather, on the next-lower bids, which is a matter I will return to in section 14.8.

With these examples in mind, I now formally state two properties of the Gap Procedure. Property 14.1 eliminates the first difficulty identified in section 14.3.

Property 14.1. *Under the Gap Procedure, no player ever pays a negative price.*

Proof. The lowest price that a player can pay for a good is the low bid for that good (or, in the case described in note 12, some number between zero and this low bid). ∎

Note in Example 14.3 that P3 pays the low bid of 31 for G3.

Define Pi's *surplus* for Gj to be

$$s_{ij} = b_{ij} - p_j,$$

or the difference between i's bid for Gj, b_{ij}, and the price, p_j, for Gj.[14] In Example 14.3, the surpluses of P1, P2, and P3 for the goods they are assigned are $s_{11} = 50 − 31 = 19$, $s_{22} = 40 − 38 = 2$, and $s_{33} = 31 − 31 = 0$. Clearly, P1 is greatly

---

[13] These prices would change if—reflecting the competition of *all* players in the marketplace—they were based on the *average* bids below those of the winners. To illustrate, the average of the second-highest *and* third-highest bids in Example 14.2 is 16.5 for G1, 17.5 for G2, and 1.5 for G3. Consequently, the maxsum bids would be reduced to goods prices of 43.3, 44.0, and 12.7. Compared with the Gap prices of 45.2, 46.0, and 8.7, P1 and P2 would pay somewhat less, and P3 substantially more, if the averages, rather than the second-highest bids, were used to set prices. Because it is the second-highest bids, not the averages, that *directly* compete with the bids of the maxsum bidders, however, I think they are the proper basis for pricing goods in the marketplace.

[14] If players are sincere in their bidding, their bids, b_{ij}, will reflect their (true) valuations, v_{ij}, of the goods. In this case, s_{ij} would be a plausible measure of player utilities, $u_{ij} = v_{ij} - p_j$. I consider later the incentives of players to be sincere under the Gap Procedure and argue that, because of the procedure's relative invulnerability to manipulation, especially when information is incomplete, s_{ij} may well be a good approximation of u_{ij}.

helped by the lack of competitive bids for G1, whereas P2 and P3 fare much worse, but no player receives a negative surplus. This result is general, eliminating the second difficulty identified in section 14.3.

Property 14.2. *Under the Gap Procedure, no player ever pays more than its bid: if Pi is assigned Gj, then $s_{ij} \geq 0$.*

Proof. Under the Gap Procedure, a player pays either (i) its bid for the good it receives (if maxsum = 100) or (ii) a nonnegative price that is less than or equal to its bid (if maxsum > 100). Since $p_j \leq b_{ij}$, s_{ij} is nonnegative. ∎

Property 14.2 establishes that the Gap Procedure satisfies the no-loss constraint.

So far I have shown that the Gap Procedure avoids the problem of negative prices (Property 14.1), which can occur if the high bids are reduced by equal amounts (Example 14.2). It also ensures that no player ever suffers a negative surplus (Property 14.2), which I showed can happen if the high bids are reduced either equally or proportionally (Example 14.3).

Because there are $n!$ possible assignments of the goods to different players, one might think that finding the maxsum assignment for large n will be computationally infeasible. It turns out, however, that this problem is equivalent to the problem of assigning n workers to n jobs so as to maximize the value of their work (Bondy and Murty, 1976, pp. 86–90) and to the weighted-matching problem for bipartite graphs (Lawler, 1976, pp. 201–207). For these problems, there are algorithms computable in polynomial time that render the determination of maxsum assignments feasible in almost any conceivable situation (more on practical applications later).

14.5. PARETO-OPTIMALITY

I next show that the Gap Procedure, because of its use of maxsum assignments, always leads to Pareto-optimal surpluses. Nonmaxsum assignments can also yield Pareto-optimal surpluses, as I will illustrate, but this cannot happen if negative prices are allowed.

Property 14.3. *The Gap Procedure maximizes the total surplus. There is no other assignment of goods, or prices that the players pay for them, that yields individual surpluses Pareto-superior to the Gap surpluses, rendering the Gap surpluses Pareto-optimal. In particular, no trade involving two or more players can improve the surpluses of all the traders without hurting nontraders.*

Proof. Because the assignment of goods under the Gap Procedure maximizes the sum of bids, it maximizes the total surplus, which equals the bid sum minus 100. In particular, the Gap prices, in addition to ensuring that each player's surplus is nonnegative, maximize the total surplus. Consequently, in order for one or more players to obtain a greater surplus from a different assignment or different prices, one or more other players would have to do worse. Hence, the Gap surpluses are Pareto-optimal, precluding trades that benefit some players without hurting others. ■

But the existence of Pareto-optimal surpluses is not exclusive to the Gap Procedure. Pareto-optimality is also possible if nonmaxsum assignments are used, as shown by the next example.

Example 14.4. Bids by Three Players for Three Goods

	G1	G2	G3	Total
P1	<u>40</u>	27	33	100
P2	27	<u>15</u>	58	100
P3	5	20	<u>75</u>	100

Maxsum is 130 (40 + 15 + 75) along the main diagonal. Applying the Gap Procedure, the descent to the next-lower bids goes to 27 for G1 and 58 for G3, but it stops at 15 for G2 since 15 is the lowest bid for G2. These three bids coincide with P2's bids in the second row of the bid matrix, which sum to 100, so they are the prices of the three goods under the Gap Procedure. The surpluses are $s_{11} = 40 - 27 = 13$, $s_{22} = 15 - 15 = 0$, and $s_{33} = 75 - 58 = 17$ for P1, P2, and P3, respectively.

P2 does particularly badly, obtaining a surplus of zero. Theoretically, the greatest surplus it could obtain under *any* maxsum assignment, given that negative prices are disallowed, is 15 (this is feasible, for example, if G2 is priced at 0, G1 at 35, and G3 at 65, giving surpluses of 5 and 10, respectively, to P1 and P3 for G1 and G3).

On the other hand, consider a nonmaxsum assignment of G1 to P2, G2 to P1, and G3 to P3. If G1 were priced at 0 (and, say, G2 were priced at 26 and G3 at 74, giving surpluses of 1 each to P1 and P3), then P2 would obtain a surplus of 27, well above the maximum surplus of 15 it could obtain under the maxsum assignment if prices are nonnegative.

So far I have ruled out negative prices in the one-good problem. In the context of chore assignments, however, it is perfectly reasonable to interpret negative prices as (positive) side payments to the players that get assigned chores that other players abhor. It turns out that allowing negative prices renders maxsum assignments the *only* ones that give Pareto-optimal surpluses, as demonstrated by the first proposition.

Proposition 14.1. *If an assignment of goods to players is not maxsum, then there are no prices that yield the players Pareto-optimal surpluses if negative prices are allowed.*

Proof. If the assignment is not maxsum, then there is a maxsum assignment in which at least two players receive different goods (since all players receive one good, one player cannot be assigned a different good unless its original good is assigned to someone else). If there are exactly two players with different goods under maxsum, then the sum of their bids for these two goods is higher than the sum of their bids for the two goods they get under the nonmaxsum assignment. Thus, it is possible to find prices, possibly negative, under the maxsum assignment that give *both* of these players greater surpluses than under the nonmaxsum assignment. A similar argument can be made if the number of players that receive different goods under the maxsum assignment is greater than two. The sum of their bids under the maxsum assignment will be greater, so prices can be found that give these players greater surpluses. ∎

Thus, nonmaxsum assignments are never Pareto-optimal when negative prices are allowed, making maxsum assignments a good starting point when there is the possibility of negative prices. To be sure, the Gap Procedure precludes such prices, which renders other Pareto-optimal allocations possible, as Example 14.4 illustrated.

What, then, is special about the Gap Procedure? In addition to its prices being competitive, it has other attractive features that I explore in section 14.7. First, however, I show that there is no procedure, including the Gap Procedure, that can preclude envy if there are four or more players, prices are nonnegative, and the total price is exogenous.

14.6. ENVY-FREENESS: AN IMPOSSIBLE DREAM

Consider again Example 14.3, wherein P3, especially, appears to get a "raw deal" under the Gap Procedure. It has to pay exactly its bid, giving it a surplus of 0 ($s_{33} = 31 - 31$), whereas P1 and P2 get surpluses of 19 ($s_{11} = 50 - 31$) and 2 ($s_{22} = 40 - 38$), respectively. However, P3 would *not* be better off getting G1 or G2 at the Gap prices of 38 or 31, respectively, because these prices are exactly P3's bids for these goods as well, yielding P3 the same surplus of 0.

Can either of the other players benefit from a different assignment of goods at the Gap prices? P1's surplus of 19 would not increase if it were assigned either G2 ($s_{12} = 1 - 38 = -37$) or G3 ($s_{13} = 49 - 31 = 18$). Neither can P2 do better than its surplus of 2 if it were assigned G1 ($s_{21} = 29 - 31 = -2$) or G3 ($s_{23} = 31 - 31 = 0$).

Call a player *envious* if its allocation of a different good would give it a greater surplus. If this is not the case for any player, the assignments and prices are *envy-free*. Not only is there envy-freeness in the Example 14.3 assignments, but this is also true in the Example 14.1, 14.2, and 14.4 assignments.

Unfortunately, this may not always be the case.

Property 14.4. *Under the Gap Procedure, one player can be envious.*

Proof. Consider the following example:

Example 14.5. Bids by Three Players for Three Goods

	G1	G2	G3	*Total*
P1	57	28	15	100
P2	24	60	16	100
P3	48	47	5	100

Maxsum is 123 (48 + 60 + 15) along the off-diagonal, and the next-lower bids (24, 47, and 5) sum to 76. Under the Gap Procedure, the goods prices are 36.3, 53.6, and 10.1 for G1, G2, and G3, respectively.

Now P1, which gets G3 ($s_{13} = 15 - 10.1 = 4.9$), envies P3, which gets G1; if P1 were assigned G1, its surplus would be $s_{11} = 57 - 36.3 = 20.7$, far exceeding its Gap surplus of 4.9. Consequently, maxsum assignments at the prices given by the Gap Procedure are *not* envy-free in Example 14.5: one player (P1) would get a greater surplus from being assigned a different good at the Gap price. ∎

In Example 14.5, it turns out, neither P2 nor P3 can benefit from being assigned different goods at the Gap prices. While P1 will desire to trade goods with P3, it will not be in P3's interest to agree to such a trade: it would receive a negative surplus of $s_{33} = 5 - 10.1 = -5.1$ if it got G3 at the Gap price of 10.1, compared with its Gap surplus of $s_{31} = 11.7$, violating the no-loss constraint.

Despite the Pareto-optimal surpluses under the Gap Procedure, a player may still prefer the good of another player, at the price the other player pays for it, to its assigned good at the Gap price. However, envy can never be two-way—if one player envies another, then the other player cannot envy the first—because if this were possible, the two players could both benefit by trading, which contradicts Pareto-optimality. Likewise, it is impossible for P1 to envy P2, P2 to envy P3, and P3 to envy P1, which would create the possibility of a three-way trade that would make all three players better off. In effect, Property 14.4 precludes mutually beneficial trades of any kind.

While the maxsum assignment along the off-diagonal in Example 14.5 at Gap prices of 36.3, 53.6, and 10.1 for G1, G2, and G3, respectively, is not envy-free, it is not difficult to check that P3's bids of 48, 47, and 5 (third row of bid matrix) yield envy-free prices, as do other goods prices close to these.[15] The

---

[15] Using geometric methods, it has been determined that the set of *all* envy-free prices in Example 14.4 can be expressed in (p_1, p_2)-space, where $p_3 = 100 - (p_1 + p_2)$. These prices are bounded by a trapezoid with vertices (p_1, p_2) = ($46\frac{2}{3}$, $48\frac{2}{3}$), ($47\frac{1}{3}$, $48\frac{1}{3}$), ($48, 47$), and ($47\frac{2}{3}$, $46\frac{2}{3}$). The fact that these coordinates are nearly equal says that the envy-free price range is quite narrow. The center of the figure that encompasses the three prices is the point (p_1, p_2, p_3) = ($47\frac{5}{12}$, $47\frac{2}{3}$, $4\frac{11}{12}$). Note that

aforementioned prices give surpluses of $s_{13} = 48 - 48 = 0$, $s_{22} = 60 - 47 = 13$, and $s_{31} = 15 - 5 = 10$, compared with the surpluses that the Gap prices give of 11.7, 6.4, and 4.9 for the maxsum assignments.

The next example illustrates the need to descend below the second-highest bids if there are four or more players, and each player's bids sum to the total rent. It also demonstrates that the envy created by the Gap prices cannot be eliminated by *any* nonnegative prices, making the problem of envy ineradicable.

Example 14.6. Bids by Four Players for Four Goods

	G1	G2	G3	G4	Total
P1	<u>36</u>	34	30	0	100
P2	31	<u>36</u>	33	0	100
P3	34	30	<u>36</u>	0	100
P4	32	33	35	<u>0</u>	100

Applying the Gap Procedure, maxsum is 108 $(36 + 36 + 36 + 0)$ along the main diagonal, the next-lower bids (34, 34, 35, and 0) sum to 103, and the next-lower bids to these (32, 33, 33, and 0) sum to 98.[16] Reducing the total surplus of 3 for the second set of bids in proportion to the gaps between the second set and the third set—2 $(34 - 32)$, 1 $(34 - 33)$, 2 $(35 - 33)$, and 0 $(0 - 0)$, which sum to 5—leads to reductions of 1.2, .6, 1.2, and 0 in the second-set bids, giving prices of 32.8 for G1, 33.4 for G2, 33.8 for G3, and 0 for G4. The surpluses of P1, P2, P3, and P4 for these four goods are

$$s_{11} = 36 - 32.8 = 3.2; \; s_{22} = 36 - 33.4 = 2.6;$$
$$s_{33} = 36 - 33.8 = 2.2; \; s_{44} = 0 - 0 = 0.$$

But P4 will envy P3 because $s_{43} = 35 - 33.8 = 1.2$, which is greater than $s_{44} = 0$.

Are there *any* feasible prices and assignments that can dispel envy for all four players?

Proposition 14.2. *If $n \geq 4$, there may be no envy-free assignment of goods at nonnegative prices.*

---

envy-free prices must be based on maxsum assignments; otherwise, trades could be found that imply two-way envy. While maxsum assignments at Gap prices allow for one-way envy, the fact that there exist situations in which such envy cannot be ruled out (see Proposition 14.2 below) means that the search for an algorithm, like the Gap Procedure, that guarantees envy-freeness is futile. On the other hand, there are other ways around the envy problem for dividing indivisible goods, with associated costs, that I will discuss shortly.

[16] Since the bids for G4 are all the same, the "next-lower" bids are all 0, or equivalently, the descent stops at 0.

Proof. In Example 14.6, because G4—the good nobody wants at any positive price—must be assigned to some player Pi, these conditions imply the following:

- *Nonnegativity*: G4's price must be 0 to ensure no loss to Pi.
- *Envy-freeness*: The prices of all other goods must at least equal Pi's bids to guarantee that Pi's surplus for these goods is not more than 0, which it receives from R4.

Since the sum of Pi's bids for all the goods is 100, the only prices that do not make Pi envious are *exactly* its bids for the four goods. Suppose $i = 1$. Then there is no assignment of G1 to any player other than P1 that is feasible— the other players' bids are all less than 36. Similarly for $i = 2$ or 3: the assignment of G2 or G3, also at a price of 36, to any player other than Pi is not feasible.

The only remaining possibility is $i = 4$, at prices of 32 for G1, 33 for G2, and 35 for G3. At these prices, however, G3 must be assigned to P3, because the bids of P1 and P2 are less than 35; likewise, G1 must be assigned to P1, because the bid of P2 is less than 32 (and P3 has already been assigned). This leaves G2 to be assigned to P2. But now P3's surplus from G3 ($s_{33} = 36 - 35 = 1$) is less than its surplus were it assigned G1 ($s_{31} = 34 - 32 = 2$), so P3 will envy P1. Hence, there is no feasible assignment of goods at envy-free prices. Examples like this can readily be embedded in situations in which $n > 4$. ∎

The impossibility of envy-freeness in Example 14.6 does not depend on one good's being worthless to all players. For example, if each player makes a sufficiently small positive bid for G4, it can be shown that there is no envy-free assignment of players to goods at nonnegative prices.

Proposition 14.2 applies whether the bids of all players sum to 100, as in Example 14.6, or not. If they do not, then it is possible to find smaller examples ($n = 2$) for which there is no envy-free assignment at nonnegative prices.

Example 14.7. Bids by Two Players for Two Goods

	G1	G2	*Total*
P1	150	0	150
P2	160	10	170

It is not difficult to show that the *only* envy-free prices that sum to 100 are $p_1 = 125$ and $p_2 = -25$; either of the two goods assignments work at these prices. On the other hand, if bid prices for each player sum to the total price, then the following holds:

Proposition 14.3. *If $n \leq 3$, there is always an envy-free assignment of goods at nonnegative prices if the bids of all players sum to the total price (e.g., 100).*

Proof (Partial). If $n = 2$, either (i) each player bids more than the other for a different good, or (ii) they bid the same for both goods. In (i), assign each player the good for which it bids more, with equal reductions in each player's bid so that their prices sum to 100. Then neither player envies the other, because the good a player does not get gives it less surplus (see Example 14.1 for an illustration). In (ii), the assignment of either good to either player, with equal reductions in the bids so they sum to 100 if they do not do so already, produces a tie in surpluses and is, therefore, envy-free.

If $n = 3$, a situation analogous to Example 14.6, but for three players rather than four, is given by Example 14.8.

Example 14.8. Bids by Three Players for Three Goods

	G1	G2	G3	*Total*
P1	x	$100 - x$	0	100
P2	y	$100 - y$	0	100
P3	z	$100 - z$	0	100

Without loss of generality, assume $x > y > z$. Then assigning G1 to P1 at price y, G2 to P3 at price $100 - y$, and G3 to P2 at price 0 is feasible and envy-free (note that all these prices are P2's bids in the second row of the bid matrix). Coincidentally, these are also the Gap assignments and prices.

This is a "worst-case" example for finding a feasible assignment at envy-free prices, because the zero bids for G3 by all players put the strongest possible constraint on prices for the three goods. (The fact that all bids sum to 100 does not tighten or loosen the envy-freeness constraint.) In fact, *whatever* P2 bids for G3 (not necessarily 0), the only envy-free prices for the aforementioned assignments are exactly P2's bids, provided $x > y > z$. Because of its length, I do not give the remainder of the proof here (for any possible bids), which involves showing how envy-free prices can be constructed geometrically for three players (see note 15 for an example).[17] ∎

Su (1999) offers a constructive proof, using Sperner's lemma, that if each player prefers a free good, or one whose price is 0, to a nonfree good—his "miserly tenants" condition in the housemates problem—there is always a feasible assignment of goods at envy-free prices. In Example 14.6, it is easy to show that this condition does not reflect player preferences when defined in terms of surplus. For instance, the Gap solution gives P4 a surplus of $s_{44} = 0 - 0 = 0$, but P4

---

[17] Specific solutions can also be found algebraically, using linear programming and related techniques (see note 18 for citations). Typically, there will be an infinity of envy-free prices—unlike the Gap prices—which are unique. Much of the work on envy-freeness cited in note 18 focuses on *which* envy-free prices are "best."

prefers the nonfree good G3, which P3 receives, to the free good G4 because $s_{43} = 35 - 33.8 = 1.2$. Consequently, P4 envies P3, as noted earlier. Although there is no solution that eradicates envy in Example 14.6 (Proposition 14.2), Su's miserly tenants condition makes an envy-free solution possible.[18]

In effect, Su's condition gives special status to a free good (priced at 0 but valued at, say, 1) compared to the most expensive good (e.g., priced at 40 but valued at, say, 50). But would a player really prefer the free good, giving it a surplus of 1, to the most expensive good, giving it a surplus of 10?

In fairness, Su admits that his miserly tenants condition—and even its relaxation, that tenants never choose the most expensive good if a free one is available—is not plausible in all situations. I concur; but unlike Su, I favor grounding preferences in surplus ($s_{ij} = b_{ij} - p_j$), wherein player valuations (v_{ij}) are traded off against prices (p_j) to give utilities ($u_{ij} = v_{ij} - p_j$), at least insofar as players are sincere.[19] In section 14.7, I explore the question of sincerity further.

To summarize, envy-freeness cannot be guaranteed if there are four or more players, at least for the surplus function I have postulated. On the other hand, envy-freeness can be guaranteed, in a nonbidding framework, by invoking Su's condition of the overriding desirability of a free good. But I find this condition implausible, because it does not trade off value against price in the construction of utility.

The Gap Procedure does not ensure envy-freeness, even when it is possible. With nonnegative prices, this is always the case for two or three players whose bids sum to the total price. However, Gap prices do take into account the competitiveness of bidding for goods, which makes the pricing mechanism market oriented. Although envy-freeness is a desirable property, I prefer a marketlike mechanism when there is a conflict between these two properties; players *should* pay more when bids are competitive, even at the sacrifice of causing envy.[20] An additional advantage of the Gap Procedure is that its prices, except

[18] Allowing negative prices also admits envy-free prices (Haake, Raith, and Su, 2002; Klign, 2000; Potthoff, 2002; Sung and Vlach, 2004; Abdulkadiroğlu, Sönmez, and Ünver, 2004). Thus in Example 14.6, prices (34, 34, 34, – 2) for (G1, G2, G3, G4) are envy-free, giving each of P1, P2, P3, and P4 a surplus of 2 for their maxsum assignments. These prices do not reflect the fact that there is more competitive bidding for G3 than for G1 or G2, which is why the Gap solution makes G3 more expensive than G1 or G2. Potthoff (2002) proposes that the envy-free prices closest to the Gap prices, which he demonstrates can be implemented by a linear program, be used if the Gap solution is not envy-free. In Example 14.6, these prices are (32.8, 33.4, 34.4, –0.6); Sung and Vlach (2004) also suggest the use of linear programming. Leonard (1983) prices, which are (33, 33, 35, 0), are envy-free, too, but they sum to 101, not the total price of 100.

[19] Su's (1999) algorithm, which is based on successive approximations, produces one solution but does not identify all envy-free solutions, as I did for Example 14.5 (see note 15). Likewise, the Gap Procedure, which is not iterative, finds a unique solution, except when there is more than one maxsum assignment (in which case a coin toss could be used to select one).

[20] Moulin (1995, p. 178) expresses a similar sentiment: "While no envy [i.e., envy-freeness] has a valid claim to the preeminence among other tests of justice, one should not forget that alternative, conflicting tests are worthy of our attention, too."

in the case of maxsum ties, are specific, whereas there may be a range of envy-free prices, rendering the choice of a particular price problematic.

14.7. SINCERITY AND INDEPENDENCE

I consider next the potential manipulability of the Gap Procedure, first by noting the monotonicity of assignments with respect to bidding.

Property 14.5. *By raising its bid for a good under the Gap Procedure, a player never hurts, and may help, its chances of receiving that good.*

Proof. If Pi raises its bid for G$_j$, then its bid for G$_j$ is at least as likely to be included in the maxsum assignment, holding the bids of other players constant. Hence, the likelihood of Pi's being assigned Gj under the Gap Procedure cannot be less and may be greater. ■

This is not to say that Pi will necessarily receive a greater surplus after it raises its bid for Gj and, as a consequence, receives it. For example, its surplus from its old maxsum assignment of Gk ($k \neq j$)—before it raised its bid—might exceed its surplus from its new maxsum assignment of Gj.

I think it would be very hard for a player to predict, not knowing the bids of the other players, whether raising its bid for a good would increase its surplus. Consequently, players will have good reason to be sincere, making bids that reflect their valuation, v_{ij}.

There are other good reasons for players to be sincere, as shown by the next two properties of the Gap Procedure.

Property 14.6. *Sincere bids are the only bids that preclude negative utility under the Gap Procedure.*

Proof. If Pi's bids are not sincere, then its bid b_{ij} for some Gj can exceed v_{ij}. In this case, Pi may be assigned Gj at a price greater than v_{ij} but not greater than b_{ij}, which gives it negative utility. ■

Define the price of a good to be *independent* if it does not depend directly on the winner's bid.[21]

---

[21] I distinguish here between the maxsum assignment of a good and its Gap price. A player's good assignment depends on the bids of all players, but the price it pays for its assigned good may depend only on *other* players' bids, in which case it is independent. Such independence implies that a small change in the winner's bid—small enough that its good assignment is not affected—does not change the Gap price. (In fact, only a sufficiently large decrease in a winner's bid can affect its good assignment and therefore the Gap price; no increase in its winning bid can ever change its good assignment and, hence, the price it pays for this good if there is independence.)

Property 14.7. *Under the Gap Procedure, good prices are independent if and only if the sum of the bids next-lower to the maxsum bids is equal to or greater than the total price.*

Proof. Assume the contrary. Then the Gap prices are greater than the bids next-lower to the maxsum bids. Consequently, the gap used in the determination of the price depends on the player's maxsum bid (as well as the next-lower bid for the good). Otherwise, the price is solely a function of lower bids and, hence, is independent. ∎

Good prices are not independent in Example 14.1. (When there are only two players, good prices are independent if and only if the sum of the two nonmaxsum bids is equal to or greater than the total price.) Goods prices also are not independent in Example 14.2 and Example 14.4, whereas they are independent in Example 14.3, Example 14.6 (four players), and Example 14.7. Generally speaking, the more players, or the more competitive they are in their bidding, the more likely their bids that are next-lower to the maxsum bids will sum to 100 or more. This makes the prices independent of the winning player's bids (as in a Vickrey auction).

This result suggests that players can "afford" to be sincere and bid their valuations v_{ij}, because what they pay will depend only on other players' bids. Unlike a Vickrey auction, however, sincere bidding is not a dominant strategy, even when there is independence, because maxsum assignments do depend on player bids. Thus, I call pricing under the Gap Procedure *partially independent*; in effect, it squeezes as much independence as is possible out of the bids in setting prices.

14.8. EXTENDING THE GAP PROCEDURE

The Gap Procedure, as specified in section 14.4, uses as input the bids of n players for n goods. If the maxsum bids are sufficient to pay the total price, the procedure assigns goods and sets prices for each player. The conditions on the goods assignments and prices are that (i) each player gets exactly one good, (ii) a player's bid for the good it receives is not less than the price of that good, and (iii) the prices sum to the total price. The Gap Procedure also requires that the bids be nonnegative, but the sum of the bids of each player is unconstrained.

The Gap Procedure can be extended to the distribution of n goods among $m \geq n$ players, based on each player's nonnegative bids for each good. In fact, the procedure, as described in section 14.4, is essentially unchanged. Example 14.9, with four players bidding for three goods, illustrates this extension of the Gap Procedure.

Example 14.9. Bids by Four Players for Three Goods

	G1	G2	G3	Total
P1	40	27	20	87
P2	27	15	58	100
P3	0	16	75	91
P4	20	10	20	50

Maxsum is $40 + 15 + 75 = 130$ along the main diagonal, so P1 gets G1, P2 gets G2, and P3 gets G3. P4 does not receive any good but, consistent with the approach for determining good prices as competitively as possible, P4's bids for goods are included in the price calculation. In particular, note that the next-lower bids are (27, 10, 58), which includes P4's bid of 10 for G2; these bids sum to 95. Thus, the Gap prices will equal these next-lower bids for each good plus 5/35 of the gap between this bid and the maxsum bid for each good. The prices are therefore $p_1 = 27 + (5/35)(13) = 28.9$, $p_2 = 10 + (5/35)(5) = 10.7$, and $p_3 = 58 + (5/35)(17) = 60.4$.[22]

A variant of this procedure would be to drop from the calculation all players that do not get a good by considering the competition only among bidders that are successful in getting some good. This revised procedure might seem appropriate on the grounds that unsuccessful bidders might not have been "serious."

But I do not favor this revision, because it ignores information about the competition. Applied to Example 14.9, it would produce prices of $p_1' = 27$, $p_2' = 15$, and $p_3' = 58$, exactly P2's bids that do not take into account P4's bid of 10 for G2. Moreover, it seems unfair to exclude P4, which already suffers from being the player cut out of getting a good, in the calculation. Note that including P4's bids raises the price of the two goods (G1 and G3) on which P4 is most competitive.

14.9. OTHER APPLICATIONS

I turn next to other possible applications of the Gap Procedure. Distribution problems are often constrained by fairness considerations that dictate that players receive some minimum number of items. In many business schools, for instance, MBA students bid points for courses or for interviews with companies recruiting on campus.

There is no assurance that these items will be equitably distributed if they go only to the highest bidders. Thus in Examples 14.3, 14.4, and 14.5, one of the three bidders is not highest, or tied for highest, on any good, so under a

---

[22] Even though there are more bidders than goods, the problem of determining the maxsum assignment can still be accomplished in polynomial time. The extreme case mentioned in note 12—the sum of the minimum bids exceeds 100—can also occur when there are more bidders than goods.

highest-bidder-wins auction, one bidder would get two goods, one would get one good, and one would get none.

The highest-bidder-wins auction creates incentives for players to overbid on the items they most desire, cutting out players who place middling bids on these items. By contrast, under an extension of the Gap Procedure, all players would get one item before any player gets two items, two items before any player gets three, and so on, if the procedure is repeated round by round.

To illustrate this extension of the procedure, consider Example 14.10, in which I assume that the items being allocated are MBA job interviews (I's) in four different business fields (consulting, finance, management, and marketing), and each MBA begins with 100 bidding points.[23]

Example 14.10. Bids by Four Players for Four Interviews (First Round)

	Consulting (I1)	Finance (I2)	Management (I3)	Marketing (I4)	Total
P1	<u>45</u>	30	10	15	100
P2	50	<u>36</u>	14	0	100
P3	30	0	<u>30</u>	40	100
P4	20	9	15	<u>56</u>	100

If the four interviews were sold in a standard highest-bidder-wins auction, then one player (P2) would get two interviews (I1 and I2), two players (P3 and P4) would get one interview each (I3 and I4, respectively), and one player (P1)—not being the highest bidder on any field—would get none. This seems grossly unfair.

Under the Gap Procedure, the maxsum assignment is along the main diagonal at prices of 27.5, 23.9, 14.8, and 33.8 for I1, I2, I3, and I4, respectively. This allocation is not envy-free because P2 envies P1 (P2 would obtain a surplus of 22.5 from I1, compared with its surplus of 12.1 from I2). Note that the highest bid for I1 (50 by P2) does not figure into the Gap price calculation, because this bid is higher, not lower, than the maxsum bid of 45 for I1.[24]

After the Gap Procedure assigns one interview to each player, assume the players can rebid their remaining points—the 100 points they started out with minus the prices they paid for the interviews they received on the first round—

---

[23] Hylland and Zeckhauser (1979) proposed a different method of allocating items (dormitory rooms) to students, but their scheme, as discussed in section 14.2, is quite different from the present one and, in addition, requires that the number of students be large relative to the capacity of the rooms.

[24] In this example, the Leonard prices for I1, I2, I3, and I4 are 14, 0, 0, and 10, respectively, which sum to 24; these envy-free, incentive-compatible prices can be adjusted upward to sum to 100 by adding 19 to each price, which does not change their property of envy-freeness, yielding new envy-free prices of 33, 19, 19, and 29. However, other properties of Leonard prices, such as incentive-compatibility, may be lost by such an adjustment.

in a second round. Suppose that the two players, P2 and P3, that did not receive their most-preferred interviews (i.e., those on which they placed the most points, presuming sincere bidding) on the first round put all their remaining points on these two interviews (I1 and I4, respectively) in the second round. In the case of P1 and P4 in Example 14.11, I suppose that P1 continues to put more points on its favorite interview from the first round (I1), whereas P4 now places its largest bid on its next-favorite interview from the first round (I1).

Example 14.11. Bids by Four Players for Four Interviews (Second Round)

	Consulting (I1)	Finance (I2)	Management (I3)	Marketing (I4)	Total
P1	30	<u>22.5</u>	10	10	72.5
P2	<u>76.1</u>	0	0	0	76.1
P3	0	0	0	<u>85.2</u>	85.2
P4	43.2	5	<u>18</u>	0	66.2

Under the Gap Procedure, the maxsum assignments underscored in Example 14.11 are at prices of 51.0, 9.2, 11.9, and 27.9 for I1, I2, I3, and I4. Notice on this round that all interviews happen to go to the highest bidders.

Putting the two rounds together, each player receives the following two interviews at the following total prices:

P1: I1 & I2 at 36.7;   **P2:** I1 & I2 at 74.9;
P3: I3 & I4 at 42.7;   **P4:** I3 & I4 at 45.7.

Observe that each player receives its two most-valued interviews, based on its first-round bids, although the prices they pay for them vary considerably. These variations mirror the competitiveness of bidding on each round, which seems appropriate.

Bidding under the Gap Procedure could be extended to more than two rounds, with players' bidding strategies presumably shifting as they learn the results on each round. The fact that each player always wins one item on each round ensures that the total number each wins is the same. More important, round-by-round bidding gives the players the ability to adjust their bids to make achieving a preferred *package* of items more likely.

Fair-division problems frequently arise at the international level. Thus, in the division of Germany into four zones after World War II, there was debate among the four Allies not only over how Germany would be divided but also about which zones each would control.[25] Since the Allies were considered to have more or less equal rights, they might have bid for the zones, with those

---

[25] See Brams and Taylor (1996, 1999) for a discussion of this case and citations of the literature on Germany's zonal division.

paying higher prices contributing more to the overall administration of the country. Of course, cost-sharing is more difficult to implement if the costs of administration are uncertain.

As another example, consider the division of an estate among children, and suppose there is bidding for the indivisible goods in rounds. Then it is reasonable to suppose, as with the allocation of interviews, that the price that an heir pays after each round will be deducted from its allocation. Alternatively, an heir might be allowed to bid for as many items as it desires in a single round, but there would be a stipulation that each heir is entitled to a certain minimum number of items (not necessarily one). Similarly, members of the U.S. Congress are generally assigned to a minimum number of committees and subcommittees.[26]

The Gap Procedure could be modified to allow for different minima as well as different endowments. For example, it could be used to assign ministerial posts in a parliamentary government to parties in the governing coalition as a function of their size (see chapter 9 for a different procedure).

If entitlements are equal, then one might presume that the players who pay less for the indivisible goods that they win will be entitled to receive more of the divisible goods (e.g., liquid assets) not included in the auction. Some formula, however, would probably have to be agreed to in advance to value the indivisible goods in terms of the divisible goods, which would be allocated to the players in proportion to their unspent points at the end of the auction.

I will not pursue further other applications or extensions of the Gap Procedure. Clearly, there is a variety of allocation problems in which market competition needs to be tempered by considerations of fairness.

14.10. SUMMARY AND CONCLUSIONS

I have focused on the one-good problem because it encapsulates the trade-offs that players face when the goods they want come at a price (the bad). Fairness comes to the fore when players, either with similar or different preferences, are entitled to certain minimum numbers of goods. Even in the simple case in which two players have a choice between two goods, it is important to assign and price the goods according to how much each player is willing to pay.

---

[26] In the U.S. House of Representatives, each member is usually entitled to at most one so-called exclusive or major committee assignment. In the Democratic Party, freshmen members, and continuing members who seek committee transfers, submit rankings of the committees to which they would like to be assigned, but both the rankings and the selection process are conditioned by political and strategic factors (Shepsle, 1978; Munger, 1988). Allowing bidding on successive rounds under the Gap Procedure, starting with exclusive and major committees and proceeding to lesser committees and subcommittees, would presumably make the selection process more objective, but whether this is desirable is, of course, debatable.

The Gap Procedure, which starts with maxsum assignments that maximize the total surplus, lets prices descend to the next-highest bids if their sum is greater than or equal to the total price of 100—and still lower until the sum of these bids is 100 or less. Once this level is reached, and provided the sum is not exactly 100, reductions in the next-higher bids are made in proportion to the differences, or gaps, between these bids for each good and the next-lower bids. Thereby the market helps to set prices by incorporating the most competitive lower bids into the pricing mechanism.

The Gap Procedure precludes the possibility of negative prices. In addition, it assigns goods to players at prices that never exceed their bids, which ensures that the allocation is feasible and that the total price equals a predetermined amount. Because no other goods assignments and prices can yield surpluses Pareto-superior to the Gap surpluses, the Gap surpluses are Pareto-optimal, though not exclusively so (if negative prices are disallowed). While this Pareto-optimality rules out mutually beneficial trades, it does not rule out envy: one player may prefer another's good at the price it pays. Indeed, there may be no feasible assignments at envy-free prices if there are four or more players.

While the Gap Procedure does not always give an envy-free allocation, even when this is possible (which is always the case when there are only two or three players), the Gap prices take into account market competition. It seems only fair that prices should be higher when there is greater demand for goods. Additional advantages of the Gap Procedure are that it is monotonic in bids, encourages sincere bidding, and generates prices that are partially independent of the bids.

The Gap Procedure is applicable not only to situations in which each player is entitled to receive one item but also more than one if the procedure is applied round by round. Besides MBA interviews, it might be applied in equal-entitlement auctions ranging from estate division to chore division, including burden sharing at the international level.

In addition, the Gap Procedure can be used when players do not have the equal endowments or entitlements—for example, when some players are larger or more senior than others. Suffice it to say that the Gap Procedure offers a promising approach to balancing competition and fairness, though neither it nor any other procedure can eliminate envy entirely.

15

Summary and Conclusions

In this brief chapter, I offer a capsule summary of findings in each chapter and then draw some more general conclusions.

CHAPTER 1. ELECTING A SINGLE WINNER: APPROVAL VOTING IN PRACTICE

Since the mid-1980s, approval voting (AV) has been successfully used to choose presidents and other officers in several major professional societies, some with tens of thousands of members. AV almost always elects Condorcet winners if they exist and, with only one exception (the Institute of Electrical and Electronic Engineers, or IEEE), has proved uncontroversial. Contrary to the fears of some, the winners in these societies are not "lowest common denominators" but tend to be supported by all classes of voters—both those that vote for one candidate and those that vote for more than one. But AV has not yet been adopted in public elections, in part, it seems, because there has not yet been a concerted and well-financed effort to implement it in such elections.

CHAPTER 2. ELECTING A SINGLE WINNER: APPROVAL VOTING IN THEORY

AV has several attractive features, including the sovereignty it gives voters to express themselves by voting for as many candidates as they like. It allows for the election of candidates that other voting systems do not, including even Condorcet losers, while remaining responsive to where voters draw the line between approved and disapproved candidates. Condorcet winners are always strong Nash equilibrium outcomes at their critical strategy profiles, which Condorcet voting systems do not ensure in equilibrium. However, other outcomes may be in equilibrium under AV, so AV does not necessarily single out a particular equilibrium (or nonequilibrium) outcome.

CHAPTER 3. ELECTING A SINGLE WINNER: COMBINING APPROVAL AND PREFERENCE

AV may be combined with preference rankings to expand the ways in which voters can express themselves. Preference approval voting (PAV) uses approval as a criterion for election if at most one candidate is majority approved, whereas if more than one is majority approved, preferences are used to determine a winner among the majority-approved candidates. Fallback voting (FV) requires that voters rank only their approved candidates and uses a descent process to determine a winner. FV and especially PAV, which may select winners different from AV and other voting systems, induce candidates to make broad-gauged appeals—in order to win majority approval—but not promise the moon.

CHAPTER 4. ELECTING MULTIPLE WINNERS: CONSTRAINED APPROVAL VOTING

Constrained approval voting (CAV) assumes candidates for a board or council can be categorized by different criteria, such as specialty and region. Although voters can approve of as many candidates as they like, independent of their category, those who can be elected are constrained. The election of the most popular candidates in the largest categories are mandated, whereas the election of candidates in the other categories depends on their popularity across all categories. CAV, which is based on "controlled roundings," ensures that those elected mirror to some degree the size of the categories, so no category is severely underrepresented or overrepresented.

CHAPTER 5. ELECTING MULTIPLE WINNERS: THE MINIMAX PROCEDURE

Like CAV, the minimax procedure starts with an approval ballot but does not categorize candidates. Instead, it aggregates approval votes in a way so as to favor candidates who receive approval from voters who cast similar ballots, as measured by their weighted Hamming distance. These candidates, who are well connected to other voters and not too far away from the outcome, may not receive the most votes; the latter candidates are those elected by the minisum procedure. In the 2003 election of twelve new members to the council of the Game Theory Society, the minimax procedure would, arguably, have provided a more representative council than that provided by the minisum procedure.

CHAPTER 6. ELECTING MULTIPLE WINNERS: MINIMIZING MISREPRESENTATION

I analyzed several procedures, all of which use integer programming, that minimize the misrepresentation of voters on a council or in a legislature. This methodology ensures, insofar as possible, the proportional representation (PR) of different interests in a voting body. But the precise meaning of "proportional" depends on the objective function that is being minimized. Should, for example, it be based on AV or a ranking system like the Borda count? I considered different alternatives and technical issues associated with each, but I did not make one recommendation. "Hierarchical PR" provided a way to identify the leader of a voting body without the necessity of a separate election for this position.

CHAPTER 7. SELECTING WINNERS IN MULTIPLE ELECTIONS

The paradox of multiple elections, highlighted in this chapter, raises the question of whether voting for multiple alternatives, such as propositions in a referendum, should aggregate votes separately for propositions (proposition voting) or for combinations of propositions (combination voting). These outcomes may be diametrically opposed; in addition, proposition voting may give a combination that receives the fewest votes, which occurred in a 1990 referendum in California. This paradox implies the Condorcet paradox of cyclical majorities and occurs in other settings, such as voting for bills in a legislature. I suggested that yes-no voting might be an efficient way for voters to express their approval of combinations without the need to make judgments about each and every one.

CHAPTER 8. SELECTING A GOVERNING COALITION IN A PARLIAMENT

Governing coalitions in parliamentary systems are usually put together by the leader of the largest party and often require costly and time-consuming bargaining to achieve. To circumvent this problem, I suggested that compatible coalitions might be formed by having parliamentary members (or parties) rank each other as coalition partners. When players can be ordered from left to right (single-peakedness), two processes of coalition formation, fallback (FB) and buildup (BU), were analyzed. FB yields majority coalitions whose members rank each other highest for the first time, whereas BU yields the first majority coalition whose members rank only each other highest. BU majority coalitions tend to be larger than FB majority coalitions and are always connected, whereas

FB coalitions may be disconnected, illustrating the phenomenon of "strange bedfellows." As predicted by the models, data from the U.S. Congress and the Supreme Court show that minimal and maximal majority coalitions form more frequently than intermediate-size majority coalitions.

CHAPTER 9. ALLOCATING CABINET MINISTRIES IN A PARLIAMENT

Which governing coalition forms in a parliament often depends on whether the parties that it includes are able to obtain the cabinet ministries they desire. I analyzed a procedure that has been used in Northern Ireland and Denmark wherein, once a governing coalition forms, individual parties choose cabinet ministries in a sequence determined by their size and a divisor method of apportionment. Using this procedure, however, parties may have good reason to be strategic in their choices—specifically, by not making sincere choices of their favorite ministries when their turn comes up, because they anticipate being able to obtain them later. This may result in inefficient outcomes and create monotonicity problems, whereby a party does better by choosing later rather than earlier. If there are only two parties that must reach an agreement, there is a mechanism that makes sincere choices rational, lest the parties have to trade ministries later to recover the sincere outcome.

CHAPTER 10. ALLOCATING INDIVISIBLE GOODS: HELP THE WORST-OFF OR AVOID ENVY?

Cabinet ministries are one kind of indivisible good, but there are many others that are not well suited to sequential choice, especially when entitlements (e.g., seat shares in a parliament) are not well defined. Even in the case of a 50-50 division between two players, one player may envy the allocation of the other. This problem carries over to three or more players that seek to divide a set of goods efficiently. In particular, satisfying the Rawlsian criterion of helping the worst-off player may conflict with eliminating envy. I discussed related conflicts, such as that between efficient and envy-free allocations, showing that a difficult choice may have to be made about which desiderata, when they are incompatible, to satisfy.

CHAPTER 11. ALLOCATING A SINGLE HOMOGENEOUS DIVISIBLE GOOD: DIVIDE-THE-DOLLAR

Money is certainly the most common homogeneous divisible good that people must divide. I showed that the two-person game of divide-the-dollar (DD) has severe consequences if the players together bid more than a dollar, which can be mitigated by a reasonable payoff scheme. Three procedures to implement this scheme were analyzed:

1. DD1 allows two players to make initial bids in a first stage, and to affirm their bids or usurp the other player's bid in a second stage.
2. DD2 allows two or more players to make bids such that those bidding the least amounts are paid off first.
3. DD3 combines DD1 and DD2, which greatly reduces the calculations required to induce the DD2 outcome.

All three procedures induce the players to make almost egalitarian bids, giving equal payoffs to each, but they vary in their rationales. The analysis suggests how salaries of members of a team might be determined so that each member would have an incentive to evaluate honestly his or her contribution to the team's effort.

CHAPTER 12. ALLOCATING MULTIPLE HOMOGENEOUS DIVISIBLE GOODS: ADJUSTED WINNER

Adjusted winner (AW) assumes two players make bids for two or more homogeneous divisible goods. If the players are truthful, the resulting allocation, which requires the division of at most one good, is efficient, envy-free, and equitable, giving each player the same total value above 50 percent. Any attempt to manipulate AW to obtain even more is highly risky; moreover, even if it is successful, the manipulated player still is guaranteed a minimum of 50 percent. The application of AW to the main issues that Israel and Egypt negotiated, under the auspices of Jimmy Carter at Camp David in 1978, showed that both sides, hypothetically, could have obtained 66 percent of what they wanted. This outcome matches closely the actual agreement that the two sides achieved, which I suggested AW might have expedited and facilitated.

CHAPTER 13. ALLOCATING A SINGLE HETEROGENEOUS GOOD: CUTTING A CAKE

A cake with different flavors or toppings is a metaphor for a single heterogeneous good. Assuming that players value its different parts differently, then cut-and-choose is the best-known procedure for dividing a cake fairly with one cut. But cut-and-choose limits the cutter to exactly 50 percent of the cake, as he or she values it, if the cutter does not know the chooser's preferences. By contrast, the proportionally equitable division of the surplus procedure (SP) gives, in general, both players more than 50 percent. Analogously, the equitable procedure (EP) gives $n > 2$ players more than $1/n$ shares, but this allocation may not be envy-free (the "squeezing procedure" gives an envy-free allocation, but only for $n = 3$). Both SP and EP are strategy-proof if the players are risk-averse; but unlike cut-and-choose, each procedure requires that the players report their value functions over the entire cake, not just indicate their 50-50 cutpoints. These procedures seem most applicable to dividing a divisible heterogeneous good such as land.

CHAPTER 14. ALLOCATING DIVISIBLE AND INDIVISIBLE GOODS

Often players must divide several indivisible goods and a single divisible good such as money. Assume, for example, that the government wants to auction off rights to develop n geographic sites. Suppose, in addition, that the government values all the sites at some specific total amount and wants to award one site to each of n bidders so that all remain viable in the future. When the players submit bids for these sites that sum to the total amount sought by the government, the Gap Procedure awards one site to each bidder at prices that sum to the total amount. These prices induce the players to bid truthfully by making their bids partially independent of the prices they pay, which are higher the more competitive the bids of other players are for each site. I illustrated different applications of the Gap Procedure, some of which assume bidding over stages.

CONCLUSIONS

Because there is not one voting or fair-division procedure that fits all circumstances, I have proposed and analyzed several of each. Although most of these procedures are not well known, each meets important needs in the diverse situations described.

What may surprise theorists of democracy is the degree to which the justifications of the procedures depend on mathematical analysis. Although commonsensical arguments can be made for most of them, precisely what properties each satisfies (or does not satisfy) requires careful and rigorous analysis that is fundamentally mathematical in nature.[1]

But as I showed in some cases, there may be no procedure that guarantees all the properties one might want in a procedure, especially when one is allocating indivisible goods. In this situation, I think it is important to understand

---

[1] The need for mathematical literacy and quantitative skills to promote responsible citizenship and to understand the contemporary world is made in Steen (2001), which has the same title as this book; so does Simeone and Pukelsheim (2006).

the trade-offs that must be made. Thereby, one can make a better-informed choice about which properties should be given priority when they all cannot, even in principle, be satisfied.

The fair-division procedures, especially, may seem quite remote from what is needed to make a democracy run well. But recall that once candidates are elected, their policies are often ill-formed, leading to arbitrary or inchoate choices that seem unfair to many citizens. In such a situation, all the goodwill in the world may not be enough to repair the damage caused.

Politicians' decisions about how to allocate goods need to be informed by more than rhetoric and vague notions of fairness. If they are not, the ad hoc decisions they make will be plagued by inconsistency and are more likely to cause anger and strife than satisfy the electorate. Worse, if the procedures are highly manipulable, they are likely to be undermined or corrupted by self-seeking (yet rational) politicians.

This is why I emphasized the stability of the outcomes the procedures produce and the strategy-proofness of some of the procedures. If it is not rational to be manipulative, outcomes are more likely to be responsive to all players' preferences and, consequently, be accepted as legitimate.

Despite the plethora of procedures I have discussed, there are surely new ones to be discovered, and old ones to be resurrected and rehabilitated, that will foster more robust democratic institutions. I urge political scientists, mathematicians, and other scholars to continue the search for these.

But I also exhort those who have a serious practical interest in making democracy work better to lend aid and encouragement, especially in helping to implement the theoretically most compelling procedures in order to test whether they work well in practice. The outcomes of these tests, especially if they are problematic, will inspire new theoretical advances, bringing the scientific enterprise full circle in helping to perfect democratic institutions.

Glossary

This glossary provides definitions, in relatively nontechnical language, of the most important concepts in this book. More rigorous definitions of technical concepts can be found in the text, as can descriptions of the more complicated voting and fair-division procedures. Italicized terms are defined elsewhere in the glossary.

Additional-member voting system. — A *proportional-representation (PR)* voting system in which some legislators are elected from districts and other members may be added to the legislature to ensure, insofar as possible, that the parties underrepresented on the basis of their national-vote proportions gain additional seats so as to reflect their size.

Adjusted district voting (ADV). — An *additional-member voting system* that results in a variable-size legislature, depending on the number of seats that must be added to satisfy *proportional representation (PR).*

Adjusted winner (AW). — A two-person point-allocation procedure that is efficient, envy-free, and equitable if the players are truthful in their point allocations.

Admissibility. — A strategy is admissible if it is not dominated—there is no other strategy that gives outcomes at least as good as, and in at least one situation better than, another strategy.

Apportionment method. — A procedure for allocating seats to states, or seats to parties in a parliament, that reflects their size. Among the methods used are ones attributed to Hamilton, Jefferson, and Webster.

Approval voting (AV). — A voting system in which voters can vote for as many candidates as they like or find acceptable, and the candidate with the most votes wins. See also *minisum procedure.*

AV dominance. — A candidate is AV-dominant if and only if, whatever sincere strategies voters choose, he or she is the winner under *approval voting (AV).*

Backward induction. — A reasoning process in which players, working backward from their last possible choices in a game, anticipate one another's earlier choices in order to make their own optimal choices.

Balanced alternation. — When taking turns choosing items, players choosing later may get two turns in a row.

Bandwagon strategy. — Occurs when a player, by misrepresenting its preferences, can benefit from being a member of a majority coalition sooner than it would be if it were truthful.

Borda count (BC). — A voting system that awards points to candidates according to their ranking of them. If there are n candidates, a voter's top-ranked candidate receives $n-1$ points, the second-ranked $n-2$ points, . . . , to 0 points for the lowest-ranked candidate. The candidate with the most points wins. See also *modified Borda score* and *Borda maxsum allocation*.

Borda maxsum allocation.—In the fair division of indivisible goods, an allocation that maximizes the maximum *modified Borda score*.

Build-Up coalition process (BU).—Players progressively descend in their preference rankings to form majority coalitions whose members mutually prefer only one another, so no players outside the coalition are more preferred.

Condorcet loser.—A candidate who loses to all other candidates in separate pairwise contests.

Condorcet paradox.—Occurs if there is no *Condorcet winner*. Every candidate can be defeated by at least one other candidate in separate pairwise contests, resulting in cyclical majorities whereby who wins in a contest depends on who is pitted against whom. See also *cyclical preferences*.

Condorcet voting system.—A voting system that elects a Condorcet winner, if there is one, when voters are *sincere*.

Condorcet winner.—A candidate who defeats all other candidates in separate pairwise contests.

Constrained approval voting (CAV).—A *proportional-representation (PR)* voting system in which candidates are categorized by different criteria; the number of candidates elected in each category depends on both the sizes of the categories and the popularity of the candidates.

Critical strategy profile.—Under *approval voting (AV)*, a critical strategy profile of candidate i is one in which every voter who ranks i as his or her worst candidate votes only for the candidate that he or she ranks top; the remaining voters vote for i and all candidates they prefer to i.

Cumulative voting.—A *proportional-representation (PR)* voting system in which each voter can distribute a fixed number of votes among one or more candidates, with the candidates with the most votes winning.

Cut-and-choose.—A two-person fair-division procedure in which one player (the cutter) cuts a cake into two pieces, and the other player (the chooser) selects one of the pieces. It is *efficient* and *envy-free* but not *equitable*.

Cyclical preferences.—Voters' preferences are cyclical when there is a *Condorcet paradox*.

DD1, DD2, DD3.—Variations of *divide-the-dollar (DD)*—DD1 has a reasonable payoff scheme; DD2 involves a second stage; and DD3 has both a reasonable payoff scheme and a second stage.

Disconnected coalition.—If players' preferences are single-peaked, there is a "hole" in a coalition due to the absence of a player in the left-right ordering.

Divide-the-dollar (DD).—A two-person auction procedure in which players make bids for a dollar, which each receives if the sum of their bids does not exceed a dollar; otherwise, they receive nothing.

Dominance.—An outcome is dominant if there is no other outcome that is as good for all players and better for at least one. A strategy is dominant if it leads to outcomes that are as good as, and in at least one situation better than, all other strategies.

Dominance inducibility.—An outcome is dominance-inducible if, given a prior choice by a player, it is *dominance-solvable*.

Dominance solvability.—An outcome is dominance-solvable if it is the result of the successive elimination of dominated strategies by the players.

Domination.—An outcome is dominated if there is another outcome that is as at least as good for all players and better for at least one. A strategy is dominated if it leads to outcomes that are no better than, and in at least one situation worse than, all other strategies.

Efficiency.—An outcome is efficient if there is no other outcome that is better for one player and at least as good for all the other players.

Egalitarian behavior.—Occurs when all n players in *divide-the-dollar (DD)* or its variations bid $1/n$ of a dollar.

Egalitarian outcome.—An allocation in *divide-the-dollar (DD)* or its variations that gives to each of n players the payoff of $100/n$ cents.

Envy-freeness.—Each player thinks it receives at least a tied-for-largest portion, so it does not envy the allocation of another player. If an allocation is not envy-free, it may be envy-possible (envy is possible but not guaranteed to occur) or envy-ensuring (envy is sure to occur).

Equitability.—Each player's valuation of the portion that it receives is the same as every other player's valuation, based on either their utilities or their having the same rankings.

Equitability adjustment.—Under *adjusted winner (AW)*, the equitability adjustment equalizes the point totals of the two players.

Equitability procedure (EP).—An n-person cake-cutting procedure ($n > 2$) that is equitable and strategy-proof. See also *surplus procedure (SP)*.

Fairness.—A procedure is fair to the degree that it satisfies certain properties, such as proportionality, envy-freeness, efficiency, equitability, or invulnerability to manipulation.

Fallback coalition process (FB).—Players progressively descend in their preference rankings to form majority coalitions whose members mutually prefer one another.

Fallback voting (FV).—A voting system in which voters rank only candidates of whom they approve. For these candidates, the winner is determined according to the *majoritarian compromise (MC)*, except that the descent process stops for a voter when he or she ranks no candidates lower.

Feasibility.—Under the *Gap Procedure*, an allocation is feasible if the *maxsum* bids equal or exceed the total price.

Fixed rule.—A voting system in which voters vote for a predetermined number of candidates.

Gap Procedure.—A procedure for assigning a single indivisible good to each player, and determining the price each pays for it, so that the prices sum to the total price.

Grand coalition.—A coalition of all players.

Hamming distance.—The number of components on which two approval ballots differ.

Hare system of single transferable vote (STV).—A *proportional representation (PR)* voting system in which candidates who receive the fewest first-choice votes are progressively eliminated and their votes transferred to second choices—and lower choices if necessary—until one candidate obtains a quota (simple majority if only one candidate is to be elected).

Heterogeneous good.—A good whose parts may be valued differently by the players.

Hierarchical PR.—A *proportional-representation (PR)* voting system in which candidate(s) who minimize misrepresentation values are elected to a higher office, whereas candidates who score less well fill lower positions.

Homogeneous good.—A good whose parts are valued the same by all players.

Integer programming.—A mathematical technique to minimize or maximize an objective function subject to constraints, including that the solution be in integers.

Majoritarian compromise (MC).—A voting system in which first-choice votes, then second-choice votes, and then lower-choice votes are counted until at least one candidate receives a majority of votes; if more than one candidate receives a majority, the candidate with the most votes wins. This system is also called Bucklin voting.

Manipulability.—The degree to which an outcome can be changed to a player's advantage when it acts strategically.

Manipulable coalition.—Occurs when a player, by reporting a preference ranking different from its true preference ranking, can induce a majority coalition to form that it prefers.

Maximin allocation.—In the fair division of indivisible goods, an allocation that maximizes the minimum rank of items any player receives. A Borda maximin allocation maximizes the minimum *modified Borda score* that any player receives.

Maximin player.—A player that maximizes the minimum value that it can ensure for itself. See also *risk-aversity.*

Maxsum.—Under the Gap Procedure, the assignment of goods to players that maximizes the sum of their bids.

Minimax procedure.—A voting system based on approval balloting that chooses a committee that minimizes the maximum weighted Hamming distance to all voters. It uses proximity weights, which helps to ensure that the committee elected represents different interests in a voting body, based in part on their connectedness to other voters.

Minisum procedure.—A voting system based on approval balloting that chooses a committee that minimizes the weighted sum of Hamming distances to all voters. It uses count weights, which elects the same committee—the most popular candidate(s)—that *approval voting (AV)* does.

Misrepresentation value.—Measures the extent to which a voter is misrepresented by a candidate.

Modified Borda score.—In fair division, point scoring according to the *Borda count (BC),* except that the scoring begins with 1 instead of 0 for a worst-ranked item.

Monotonicity.—A voting system is approval-monotonic if a class of voters, by approving of a new candidate—without changing their approval of other candidates—never hurts and may help this candidate get elected. It is rank-monotonic, or just monotonic, if a class of voters, by raising a candidate in their ranking—without changing their ranking of other candidates—never hurts and may help this candidate get elected. Sequential choices are monotonic if early choices never give outcomes worse than later choices do when players choose *sophisticated strategies.*

Nash equilibrium.—The strategies of players that yield an outcome such that no player has an incentive to depart unilaterally, because its departure would lead to a worse, or at least not a better, outcome. At a *strong* Nash equilibrium, no set of players would have an incentive to depart unilaterally from the strategies that produce this outcome.

Negative price. — A payment to a player for receiving a good (rather than a payment by the player for the good).

Paradox of multiple elections. — Occurs when the set of winners in multiple elections is a combination that receives the fewest votes.

Pareto candidate. — A candidate is a Pareto candidate if there is no other candidate that some voters rank higher and no voters rank lower.

Pareto-dominance/domination. — In the fair division of indivisible items, allocation X Pareto-dominates allocation Y for a player if, for every item in X that is not in Y, there is a different item in Y that the player ranks lower. In this case, X is said to Pareto-dominate Y.

Pareto-optimality. — See *efficiency.*

Partial independence. — Under the *Gap Procedure,* the price a player pays for the good it receives does not depend on its bid in some but not necessarily all situations.

Party-list voting. — A *proportional representation (PR)* voting system in which voters vote for parties, which receive seats in parliament proportional to the number of votes they receive.

Plurality voting (PV). — A voting system in which voters can vote for exactly one candidate, and the candidate with the most votes wins.

Preference approval voting (PAV). — A voting system in which voters rank candidates and draw a line in their rankings, approving of all candidates above the line and disapproving of all candidates below the line; the winner is determined by both approvals and preferences.

Preference profile. — A complete listing of all players' preferences.

Procedure. — Rules of play that produce outcomes.

Proportional equitability. — Under the *surplus procedure (SP),* each player receives exactly the same proportion of the surplus.

Proportionality. — An allocation is proportional if each of n players thinks it received a portion that is at least $1/n$ of its total value.

Proportional representation (PR). — The representation of different parties or interests in a legislature according to their size in the electorate, which is usually implemented by an *apportionment method.*

Random society. — One in which all preference rankings of the players are equally likely.

Reasonable payoff scheme. — In *divide-the-dollar (DD)* and its variations, a scheme that satisfies five conditions that do not entail harsh punishment for overbidding.

Representativeness. — Degree to which those elected to an office reflect the views of their constituents. See also *Proportional representation (PR).*

Risk-aversity. — A player is risk-averse if it never chooses a strategy that may yield a better outcome if it entails the possibility of getting a worse outcome. See also *maximin player.*

Scoring rule. — A voting system that gives at least as many points to more preferred candidates as to less preferred candidates. See also *Borda count (BC).*

Separable preferences. — Occurs when voters' preferences in one election are not affected by outcomes in other elections, so the elections may be thought of as independent of each other.

Sincerity.—To vote or choose sincerely is to do so truthfully, in accordance with one's preferences. Under *approval voting (AV)*, a strategy is sincere if, given the lowest-ranked candidate that a voter approves of, he or she also approves of all candidates ranked higher.

Single-peakedness.—Players can be ordered along a line from left to right, with their preferences declining monotonically to the left and right of their positions in the ordering.

Single transferable vote (STV).—See *Hare system of single transferable vote (STV).*

Social-choice rule.—Aggregates preferences or approvals to produce an outcome.

Sophisticated strategy.—A strategy in which a player anticipates choices of other players, based on *backward induction*, in making its own optimal choices.

Squeezing procedure.—A 3-person minimal-cut cake-cutting procedure that is envy-free and strategy-proof.

Stability.—An outcome is stable if it is the product of strategies chosen at a *Nash equilibrium;* it is strongly stable if it is the product of strategies chosen at a strong Nash equilibrium. A coalition is stable if no member would prefer to be in another coalition of the same size; otherwise it is semi-stable.

Strategic voting.—Occurs when a voter does not vote sincerely or truthfully because it is advantageous not to do so.

Strategy profile.—A complete listing of all players' strategies.

Strategy-proofness.—A player cannot ensure a preferred outcome other than by being truthful. If a procedure is not strategy-proof, it is strategy-vulnerable.

Strict alternation.—Each player gets one turn in a sequence but not two turns in a row.

Surplus.—Under the *Gap Procedure,* the difference between a player's bid for the good it receives and the price that it pays for it.

Surplus procedure (SP).—A 2-person cake-cutting procedure that is *proportionally equitable* and *strategy-proof.*

Top cycle.—A voting cycle in which there are no majority-approved candidates preferred to candidates in the cycle.

Transitivity.—If a voter prefers *a* to *b* and *b* to *c*, then he or she prefers *a* to *c*.

Two-party mechanism.—A procedure for making sequential choices that induces rational players to be sincere.

Voting cycle.—Occurs when there is a *Condorcet paradox.*

Weighted voting.—Voting in which members of a voting body cast different numbers of votes; their voting power, or ability to change outcomes, may not be directly related to the number of votes they cast.

Yes-no voting.—A voting system in which voters can vote for subsets of alternatives by indicating all those that must be included in the subset (Y), those that must be excluded (N), and those that may either be included or excluded.

Zero-sum game.—A game in which payoffs at each outcome sum to zero. This is not the case in nonzero-sum games like Prisoners' Dilemma or *divide-the-dollar (DD)*, in which all players may win (i.e., all receive positive payoffs) or lose simultaneously.

References

Abdulkadiroğlu, Atila, and Tayfun Sönmez (1998). "Random Serial Dictatorship and the Core from Random Endowment in House Allocation Problems." *Econometrica* 66, no. 3 (May): 689–701.

Abdulkadiroğlu, Atila, and Tayfun Sönmez (1999). "House Allocation with Existing Tenants." *Journal of Economic Theory* 88, no. 2 (October): 233–259.

Abdulkadiroğlu, Atila, and Tayfun Sönmez (2003). "Ordinal Efficiency and Dominated Sets of Assignments." *Journal of Economic Theory* 112, no. 1 (September): 157–172.

Abdulkadiroğlu, Attila, Tayfun Sönmez, and Utku Ünver (2004). "Room Assignment—Rent Division: A Market Approach." *Social Choice and Welfare* 22, no. 3 (June): 515–538.

Abreu, Dilip, and H. Matsushima (1994). "Exact Implementation." *Journal of Economic Theory* 64, no. 1 (October): 1–19.

Adger, W. Neil, Jouni Paavola, Saleemul Huq, and M. J. Mace (2006). *Fairness in Adaptation to Climate Change.* Cambridge, MA: MIT Press.

Alesina, Alberto, and Howard Rosenthal (1995). *Partisan Politics, Divided Government, and the Economy.* Cambridge, UK: Cambridge University Press.

Alger, Dan (2006). "Voting by Proxy." *Public Choice* 126, nos. 1–3 (January): 1–26.

Amar, Akhil Reed (1984). "Choosing Representatives by Lottery Voting." *Yale Law Journal* 93, no. 7: 1283–1308.

"Amendment to AMS By-Laws" (1987). *Amstat News* 135 (May): 1.

Anbarci, Nejat (2001). "Divide-the-Dollar Game Revisited." *Theory and Decision* 50, no. 4 (June): 295–304.

Anscombe, G.E.M. (1976). "On Frustration of the Majority by Fulfillment of the Majority's Will." *Analysis* 36, no. 4 (June): 161–168.

Arrington, Theodore S., and Saul Brenner (1984). "Another Look at Approval Voting" and "Arrington and Brenner to Brams and Fishburn." *Polity,* 17, no. 1 (Fall): 118–134, 144.

Arrow, Kenneth J. (1951, 1963). *Social Choice and Individual Values,* 1st ed., 2nd ed. New Haven, CT: Yale University Press.

Arrow, Kenneth J., Amartya K. Sen, and Kotaro Suzumura (eds.) (2002). *Handbook of Social Choice and Welfare,* vol. 1. Amsterdam: North-Holland.

Aumann, Robert J., and Michael Maschler (1985). "Game-Theoretic Analysis of a Bankruptcy Problem from the Talmud." *Journal of Economic Theory* 36, no. 2 (August): 195–213.

Austen-Smith, David, and Jeffrey S. Banks (1999). *Positive Political Theory I: Collective Preference.* Ann Arbor: University of Michigan Press.

Austen-Smith, David, and Jeffrey S. Banks (2005). *Positive Political Theory II: Strategy and Structure.* Ann Arbor: University of Michigan Press.

Bacharach, Michael (1993). "Variable Universe Games." In Ken Binmore, Alan Kirman, and Piero Tani (eds.), *Frontiers of Game Theory*, pp. 255–275. Cambridge, MA: MIT Press.

Bag, Parimal Kanti, and Hamid Sabourian (2005). "Distributing Awards Efficiently: More on King Solomon's Problem." *Games and Economic Behavior* 53, no. 1 (October): 43–58.

Baharad, Eyal, and Shmuel Nitzan (2002). "Ameliorating Majority Decisiveness through Expression of Preference Intensity." *American Political Science Review* 96, no. 4 (December): 745–754.

Baharad, Eyal, and Shmuel Nitzan (2005). "Approval Voting Reconsidered." *Economic Theory* 26, no. 3: 619–628.

Balinski, Michel (2006). "Apportionment: Uni- and Bi-Proportional." In Bruno Simeone and Friedrich Pukelsheim (eds.), *Mathematics and Democracy: Recent Advances in Voting Systems and Collective Choice*, pp. 43–53. Berlin, Germany: Springer.

Balinski, Michel L., and Gabrielle Demange (1989). "An Axiomatic Approach to Proportionality between Matrices." *Mathematics of Operations Research* 14, no. 4 (November): 700–719.

Balinski, Michel L., and Rida Laraki (2007). "A Theory of Measuring, Electing and Ranking." *Proceedings of the National Academy of Sciences* 104, no. 21 (May 22): 8720–8725.

Balinski, Michel L., and H. Peyton Young (1982, 2001). *Fair Representation: Meeting the Ideal of One Man, One Vote*, 1st ed., 2nd ed. New Haven, CT: Yale University Press.

Banzhaf, John F., III (1965). "Weighted Voting Doesn't Work: A Mathematical Analysis." *Rutgers Law Review* 19, no. 2: 317–343.

Banzhaf, John F., III (1966). "Multi-Member Electoral Districts—Do They Violate the 'One Man, One Vote' Principle?" *Yale Law Journal* 75, no. 8: 1309–1338.

Banzhaf, John F., III (1968). "One Man, 3.312 Votes: A Mathematical Analysis of the Electoral College." *Villanova Law Review* 13, no. 2: 304–332.

Barbanel, Julius B. (2005). *The Geometry of Efficient Fair Division.* New York: Cambridge University Press.

Barbanel, Julius B., and Steven J. Brams (2004). "Cake Division with Minimal Cuts: Envy-Free Procedures for 3 Person, 4 Persons, and Beyond." *Mathematical Social Sciences* 48, no. 4 (November): 251–269.

Barberà, Salvador, Walter Bossert, and Prasanta K. Pattanaik (1998). "Ranking Sets of Objects." In Salvador Barberà, Peter J. Hammond, and Christian Seidl (eds.), *Handbook of Utility Theory*, vol. 2. Boston: Kluwer Academic Publishers.

Barberà, S., M. Maschler, and J. Shalev (2001). "Voting for Voters: A Model of Electoral Evolution." *Games and Economic Behavior* 37, no. 1 (October): 40–78.

Barberà, Salvador, Hugo Sonnenschein, and Lin Zhou (1991). "Voting by Committees." *Econometrica* 59, no. 3 (May): 595–609.

Baron, Jonathan, Nicole Y. Altman, and Stephen Kroll (2005). "Approval Voting and Parochialism." *Journal of Conflict Resolution* 49, no. 6 (December): 895–907.

Begley, Sharon (2003). "Why We Sometimes Get Tofu for President When We Want Beef." *Wall Street Journal* (March 14): B1.

Benoit, Jean-Pierre, and Lewis A. Kornhauser (1994). "Social Choice in a Representative Democracy." *American Political Science Review* 88, no. 1 (March): 185–192.

Beviá, Carmen (1998). "Fair Allocation in a General Model with Indivisible Goods." *Review of Economic Design* 3, no. 3 (June): 195–213.

Binmore, Ken (2005). *Natural Justice.* Oxford, UK: Oxford University Press.

Black, Duncan (1958). *Theory of Committees and Elections.* Cambridge, UK: Cambridge University Press.

Blais, André, and Louis Massicotte (2002). "Electoral Systems." In Lawrence LeDuc, Richard S. Niemi, and Pippa Norris (eds.), *Comparing Democracies 2: New Challenges in the Study of Elections and Voting*, pp. 40–69. London: Sage.

Bloch, Francis, and Stéphana Rottier (2002). "Agenda Control in Coalition Formation." *Social Choice and Welfare* 19, no. 4 (October): 769–788.

Blondel, Jean, and Maurizio Cotta (eds.) (1996). *Party and Government: An Inquiry into the Relationship between Governments and Supporting Parties in Liberal Democracies.* Houndmills, UK: Macmillan Press.

Blydenburgh, John C. (1971). "The Closed Rule and the Paradox of Voting." *Journal of Politics* 33, no. 1 (February): 57–71.

Boehm, George A. W. (1976). "One Fervent Vote against Wintergreen." Preprint.

Bogomolnaia, Anna, and Hervé Moulin (2001). "A New Solution to the Random Assignment Problem." *Journal of Economic Theory* 100, no. 2 (October): 295–328.

Bondy, J. A., and U.S.R. Murty (1976). *Graph Theory with Applications.* London: Macmillan.

Bossert, Walter (1993). "An Alternative Solution to Bargaining Problems with Claims." *Mathematical Social Sciences* 25, no. 3 (May): 205–220.

Bowler, Shaun, Todd Donovan, and David Brockington (2003). *Electoral Reform and Minority Representation: Local Experiments with Alternative Elections.* Columbus: Ohio State University Press.

Brady, David W. (1993). "The Causes and Consequences of Divided Government: Toward a New Theory of American Politics?" *American Political Science Review* 87, no. 1 (March): 189–194.

Brams, Steven J. (1975, 2004). *Game Theory and Politics*, 1st ed., 2nd ed. New York: Free Press; Mineola, NY: Dover.

Brams, Steven J. (1976). *Paradoxes in Politics: An Introduction to the Nonobvious in Political Science.* New York: Free Press.

Brams, Steven J. (1978). *The Presidential Election Game.* New Haven, CT: Yale University Press.

Brams, Steven J. (1982a). "The AMS Nomination Procedure Is Vulnerable to 'Truncation of Preferences'" and "Rejoinder [to Chandler Davis]." *Notices of the American Mathematical Society* 29, no. 2 (February): 136–138.

Brams, Steven J. (1982b). "Strategic Information and Voting Behavior." *Society* 19, no. 6 (September/October): 4–11.

Brams, Steven J. (1988). "MAA Elections Produce Decisive Winners." *Focus: The Newsletter of the Mathematical Association of America* 8, no. 3 (May–June): 1–2.

Brams, Steven J. (1990). "Constrained Approval Voting: A Voting System to Elect a Governing Board." *Interfaces* 20, no. 5 (September–October): 67–79.

Brams, Steven J. (2002). "Approval Voting: A Better Way to Select a Winner." http://alum.mit.edu/ne/whatmatters/200211.

Brams, Steven J. (1980, 2003). *Biblical Games: Game Theory and the Hebrew Bible*, 1st ed., 2nd ed. Cambridge, MA: MIT Press.

Brams, Steven J. (2006a). "Fair Division." In Barry Weingast and Donald Wittman (eds.), *Handbook of Political Economy*, pp. 425–437. New York: Oxford University Press.

Brams, Steven J. (2006b). "The Normative Turn in Public Choice." *Public Choice* 127, nos. 3–4 (June): 245–250.

Brams, Steven J., and Morton D. Davis (1978). "Optimal Jury Selection: A Game-Theoretic Model for the Exercise of Peremptory Challenges." *Operations Research* 26, no. 6 (November–December): 966–991.

Brams, Steven J., Paul H. Edelman, and Peter C. Fishburn (2001). "Paradoxes of Fair Division." *Journal of Philosophy* 98, no. 6 (June): 300–314.

Brams, Steven J., Paul H. Edelman, and Peter C. Fishburn (2003). "Fair Division of Indivisible Goods." *Theory and Decision* 55, no. 2 (September): 147–180.

Brams, Steven J., and Peter C. Fishburn (1978). "Approval Voting." *American Political Science Review* 72, no. 3 (September): 831–847.

Brams, Steven J., and Peter C. Fishburn (1979). "Reply [to Gordon Tullock]." *American Political Science Review* 73, no. 2 (June): 552.

Brams, Steven J., and Peter C. Fishburn (1983, 2007). *Approval Voting* 1st ed., 2nd ed. Cambridge, MA: Birkhäuser Boston; New York: Springer.

Brams, Steven J., and Peter C. Fishburn (1984a). "A Careful Look at 'Another Look at Approval Voting.' " *Polity* 17, no. 1 (Fall 1984): 135–143.

Brams, Steven J., and Peter C. Fishburn (1984b). "A Note on Variable-Size Legislatures to Achieve Proportional Representation." In Arend Lijphart and Bernard Grofman (eds.), *Choosing an Electoral System: Issues and Alternatives*, pp. 175–177. New York: Praeger.

Brams, Steven J., and Peter C. Fishburn (1984c). "Proportional Representation in Variable-Size Legislatures." *Social Choice and Welfare* 1, no. 3 (October): 211–229.

Brams, Steven J., and Peter C. Fishburn (1985). "Comment on 'The Problem of Strategic Voting under Approval Voting' " and "Rejoinder to Niemi." *American Political Science Review* 79, no. 3 (September): 816–819.

Brams, Steven J., and Peter C. Fishburn (1988). "Does Approval Voting Elect the Lowest Common Denominator?" *PS: Political Science and Politics* 21, no. 2 (Spring): 277–284.

Brams, Steven J., and Peter C. Fishburn (1991). "Alternative Voting Systems." In L. Sandy Maisel (ed.), *Political Parties and Elections in the United Sates: An Encyclopedia*, vol. 1, pp. 23–31. New York: Garland.

Brams, Steven J., and Peter C. Fishburn (1992a). "Approval Voting in Scientific and Engineering Societies." *Group Decision and Negotiation* 1 (April): 41–55.

Brams, Steven J., and Peter C. Fishburn (1992b). "Coalition Voting." *Mathematical and Computer Modelling* (Paul E. Johnson (ed.), *Formal Theory of Politics II: Mathematical Modelling in Political Science*) 10 (August/September): 15–26.

Brams, Steven J., and Peter C. Fishburn (2000). "Fair Division of Indivisible Items between Two People with Identical Preferences: Envy-Freeness, Pareto-Optimality, and Equity." *Social Choice and Welfare* 17, no. 2 (February): 247–267.

Brams, Steven J., and Peter C. Fishburn (2001). "A Nail-Biting Election." *Social Choice and Welfare* 18, no. 3: 409–414.

Brams, Steven J., and Peter C. Fishburn (2002). "Voting Procedures." In Kenneth Arrow, Amartya Sen, and Kotaro Suzumura (eds.), *Handbook of Social Choice and Welfare*, pp. 175–236. Amsterdam: Elsevier Science.

Brams, Steven J., and Peter C. Fishburn (2005). "Going from Theory to Practice: The Mixed Success of Approval Voting." *Social Choice and Welfare* 25, no. 2: 457–474.

Brams, Steven J., Peter C. Fishburn, and Samuel Merrill III (1988a). "The Responsiveness of Approval Voting: Comments on Saari and Van Newenhizen." *Public Choice* 59, no. 2 (November): 121–131.

Brams, Steven J., Peter C. Fishburn, and Samuel Merrill (1988b). "Rejoinder to Saari and Van Newenhizen." *Public Choice* 59, no. 2 (November): 149.

Brams, Steven J., Michael W. Hansen, and Michael E. Orrison (2006). "Dead Heat: The 2006 Public Choice Society Election." *Public Choice* 128, nos. 3–4 (September): 1314–1321.

Brams, Steven J., and Dudley R. Herschbach (2001a). "Response to Richie, Bouricius, and Macklin." *Science* 294, no. 5541 (12 October): 305–306.

Brams, Steven J., and Dudley R. Herschbach (2001b). "The Science of Elections." *Science* 292, no. 5521 (25 May): 1449.

Brams, Steven J., Michael A. Jones, and D. Marc Kilgour (2002). "Single-Peakedness and Disconnected Coalitions." *Journal of Theoretical Politics* 14, no. 3 (July): 359–383.

Brams, Steven J., Michael A. Jones, and D. Marc Kilgour (2005). "Forming Stable Coalitions: The Process Matters." *Public Choice* 125, nos. 1–2 (October): 67–94.

Brams, Steven J., Michael A. Jones, and Christian Klamler (2006a). "Better Ways to Cut a Cake." *Notices of the AMS [American Mathematical Society]* 53, no. 11 (December): 1314–1321.

Brams, Steven J., Michael A. Jones, and Christian Klamler (2006b). "Proportional Pie-Cutting." *International Journal of Game Theory*, forthcoming.

Brams, Steven J., and Todd R. Kaplan (2004). "Dividing the Indivisible: Procedures for Allocating Cabinet Ministries in a Parliamentary System." *Journal of Theoretical Politics* 16, no. 2 (April): 143–173.

Brams, Steven J., and D. Marc Kilgour (1999). "Competitive Fair Division." In Harrie de Swart (ed.), *Logic, Game Theory and Social Choice: Proceedings of the International Conference, LGS '99, May 13–16, 1999*, pp. 104–124. Tilburg, The Netherlands: Tilburg University Press.

Brams, Steven J., and D. Marc Kilgour (2001a). "Competitive Fair Division." *Journal of Political Economy* 109, no. 2 (April): 418–443.

Brams, Steven J., and D. Marc Kilgour (2001b). "Fallback Bargaining." *Group Decision and Negotiation* 10, no. 4 (July): 287–316.

Brams, Steven J., D. Marc Kilgour, and M. Remzi Sanver (2004, 2007). "A Minimax Procedure for Negotiating Multilateral Treaties." In Matti Wiberg (ed.), *Reasoned Choices: Essays in Honor of Hannu Nurmi*. Turku, Finland: Finnish Political Science Association (2004). Reprinted in Rudolf Avenhaus and I. William Zartman (eds.) (2007), *Diplomacy Games: Formal Models and International Negotiations*, pp. 265–282. Heidelberg: Springer (2007).

Brams, Steven J., D. Marc Kilgour, and M. Remzi Sanver (2007). "A Minimax Procedure for Electing Committees." *Public Choice*, forthcoming.

Brams, Steven J., D. Marc Kilgour, and William S. Zwicker (1997). "Voting on Referenda: The Separability Problem and Possible Solutions." *Electoral Studies* 16, no. 3 (September): 359–377.

Brams, Steven J., D. Marc Kilgour, and William S. Zwicker (1998). "The Paradox of Multiple Elections." *Social Choice and Welfare* 15, no. 2 (February): 211–236.

Brams, Steven J., and Daniel L. King (2005). "Efficient Fair Division: Help the Worst Off or Avoid Envy?" *Rationality and Society* 17, no. 4 (November): 387–421.

Brams, Steven J., and Samuel Merrill (1994). "Would Ross Perot Have Won the 1992 Presidential Election under Approval Voting?" *PS: Politics and Political Science* 27, no. 1 (March): 39–44.

Brams, Steven J., and Jack H. Nagel (1991). "Approval Voting in Practice." *Public Choice* 71, nos. 1–2 (August): 1–17.

Brams, Steven J., and M. Remzi Sanver (2006). "Critical Strategies under Approval Voting: Who Gets Ruled In and Ruled Out." *Electoral Studies* 25: 287–305.

Brams, Steven J., and M. Remzi Sanver (2008). "Voting Systems That Combine Approval and Preference." In Steven J. Brams, William V. Gehrlein, and Fred S. Roberts (eds.), *The Mathematics of Preference and Choice: Essays in Honor of Peter C. Fishburn*. Heidelberg: Springer, forthcoming.

Brams, Steven J., and Philip D. Straffin, Jr. (1979). "Prisoners' Dilemma and Professional Sports Drafts." *American Mathematical Monthly* 86, no. 2 (February): 80–88.

Brams, Steven J., and Philip D. Straffin, Jr. (1982). "The Apportionment Problem." *Science* 217, no. 4558 (30 July): 437–438.

Brams, Steven J., and Alan D. Taylor (1994). "Divide the Dollar: Three Solutions and Extensions." *Theory and Decision* 37, no. 2 (September): 211–231.

Brams, Steven J., and Alan D. Taylor (1995). "An Envy-Free Cake Division Protocol." *American Mathematical Monthly* 102, no. 1 (January 1995): 9–18.

Brams, Steven J., and Alan D. Taylor (1996). *Fair Division: From Cake-Cutting to Dispute Resolution*. New York: Cambridge University Press.

Brams, Steven J., and Alan D. Taylor (1999). *The Win-Win Solution: Guaranteeing Fair Shares to Everybody*. New York: W.W. Norton.

Brams, Steven J., Alan D. Taylor, and William S. Zwicker (1995). "Old and New Moving-Knife Schemes." *Mathematical Intelligencer* 17, no. 4 (Fall): 30–35.

Brams, Steven J., Alan D. Taylor, and William S. Zwicker (1997). "A Moving-Knife Solution to the Four-Person Envy-Free Cake Division Problem." *Proceedings of the American Mathematical Society* 125, no. 2 (February): 547–554.

Brams, Steven J., and Jeffrey M. Togman (1996). "Camp David: Was the Agreement Fair?" *Conflict Management and Peace Science* 13, no. 3: 99–112. Reprinted in Paul F. Diehl (ed.) (1999), *A Road Map to War: Territorial Dimensions of International Conflict*, pp. 238–253. Nashville, TN: Vanderbilt University Press.

Broome, John (1991). *Weighing Goods*. Oxford, UK: Basil Blackwell.

Browne, Eric C., and John Dreijmanis (eds.) (1982). *Government Coalitions in Western Democracies*. New York: Longman.

Brualdi, Richard A. (1999). *Introductory Combinatorics*. Upper Saddle River, NJ: Prentice Hall.

Brzezinski, Zbigniew (1983). *Power and Principle: Memoirs of the National Security Adviser, 1977–81*. New York: Farrar, Straus and Giroux.

Budge, Ian, and Hans Pieter Keman (1990). *Parties and Democracy: Coalition Formation and Government Functioning in Twenty States.* Oxford, UK: Oxford University Press.

Carmignani, Fabrizio (2001). "Cabinet Formation in Coalition Systems." *Scottish Journal of Political Economy* 48, no. 3 (August): 313–329.

Carr, Edward H. (1964). *The Twenty-Years' Crisis, 1919–1939: An Introduction to the Study of International Relations.* New York: Harper.

Carter, Jimmy (1982). *Keeping Faith: Memoirs of a President.* New York: Bantam.

Carter, Jimmy (1994). "The Power of Moral Suasion in International Mediation." In Deborah M. Kolb and Associates (eds.), *When Talk Works: Profiles of Mediators,* pp. 375–391. San Francisco: Jossey Bass.

Cechlárová, Katarina, and Antonio Romero-Medina (2000). "Stability in Coalition Formation Games." *International Journal of Game Theory* 29, no. 4 (December): 487–494.

Center for Range Voting (2007). http://rangevoting.org.

Chamberlin, John R., and Paul N. Courant (1983). "Representative Deliberations and Representative Decisions: Proportional Representation and the Borda Rule." *American Political Science Review* 77, no. 3 (September): 718–733.

Chappell, Henry W., Jr., Rob Roy McGregor, and Todd Vermilyea (2005). *Committee Decisions on Monetary Policy: Evidence from Historical Records of the Federal Open Market Committee.* Cambridge, MA: MIT Press.

Chatterjee, K., and L. Samuelson (1990). "Perfect Equilibria in Simultaneous Offers Bargaining." *International Journal of Game Theory* 19, no. 3: 237–267.

Chisholm, John (2000). "A Modification of Knaster's 'Sealed Bids' Method of Fair Division Yielding Envy-Free Distributions." Preprint, Department of Mathematics, Western Illinois University.

Chun, Youngsub (1988). "The Proportional Solution for Rights Problems." *Mathematical Social Sciences* 15, no. 3 (June): 231–246.

Chun, Youngsub, and William Thomson (1994). "Bargaining Problems with Claims." *Mathematical Social Sciences* 25, no. 3 (August): 19–33.

Colomer, Josep M. (2001). *Political Institutions: Democracy and Social Choice.* Oxford, UK: Oxford University Press.

Colomer, Josep M. (ed.) (2004). *Handbook of Electoral System Choice.* New York: Palgrave.

Colomer, Josep M., and Iain McLean (1998). "Electing Popes: Approval Balloting and Qualified-Majority Rule." *Journal of Interdisciplinary History* 29, no. 1 (Summer): 1–22.

Conlan, Timothy J. (1991). "Competitive Government in the United States: Policy Promotion and Divided Party Control." *Governance* 4, no. 4 (October): 403–419.

Cox, Gary W. (1987). *The Cabinet and the Development of Political Parties in Victorian England.* New York: Cambridge University Press.

Cox, Gary W. (1997). *Making Votes Count: Strategic Coordination in the World's Electoral Systems.* Cambridge, UK: Cambridge University Press.

Cox, Lawrence H. (1987). "A Constructive Procedure for Unbiased Controlled Rounding." *Journal of the American Statistical Association* 82, no. 398 (June): 520–524.

Cox, Lawrence H., and Lawrence R. Ernst (1982). "Controlled Rounding." *INFOR* 20 no. 4 (November): 423–432.

Curiel, Imma J., Michael Maschler, and Stef H. Tijs (1988). "Bankruptcy Games." *Zeitschrift für Operations Research* 31, no. 1: A143–159.

Dagan, Nir, and Oscar Volij (1993). "The Bankruptcy Problem: A Cooperative Bargaining Approach." *Mathematical Social Sciences* 26, no. 3 (November): 287–297.

Dantzig, George B. (1963). *Linear Programming and Extensions.* Princeton, NJ: Princeton University Press.

Daverman, Robert J. (2002). Private communication (September 17).

Davis, Chandler (1982). "Comment." *Notices of the American Mathematical Society* 29, no. 2 (February): 138.

Davis, K. Roscoe, and Patrick G. McKeown (1984). *Quantitative Models for Management,* 2nd ed. Boston: Kent.

Dayan, Moshe (1981). *Breakthrough: A Personal Account of the Egypt-Israel Peace Negotiations.* New York: Alfred A. Knopf.

Deb, Rajat, and David Kelsey (1987). "On Constructing a Generalized Ostrogorski Paradox: Necessary and Sufficient Conditions." *Mathematical Social Sciences* 14, no. 2: 161–174.

de Clippel, Geoffroy, Hervé Moulin, Florenz Plassmann, and Nicholas Tideman (2006). "Impartial Division of a Dollar." Preprint, Department of Economics, Rice University.

Dellis, Arnaud, and Mandar P. Oak (2006). "Approval Voting with Endogenous Candidates." *Games and Economic Behavior* 54, no. 1 (January): 47–76.

Demange, Gabrielle (1984). "Implementing Efficient Egalitarian Equivalent Allocations." *Econometrica* 52, no. 5 (September): 1167–1177.

Denoon, David B. H., and Steven J. Brams (1997). "Fair Division: A New Approach to the Spratly Islands Controversy." *International Negotiation* 2, no. 2 (December): 303–329.

De Sinopoli, Francesco, Bhaskar Dutta, and Jean-François Laslier (2006). "Approval Voting: Three Examples." *International Journal of Game Theory* 35, no. 1: 27–38.

Diamond, Larry, and Marc F. Plattner (eds.) (2006). *Electoral Systems and Democracy.* Baltimore, MD: Johns Hopkins University Press.

Dietrich, Franz, and Christian List (2007). "Arrow's Theorem in Preference Aggregation." *Social Choice and Welfare* 29, no. 1 (July): 19–33.

Dubin, Jeffrey A., and Elizabeth R. Gerber (1992). "Patterns of Voting on Ballot Propositions: A Mixture Model of Voter Types." Social Science Working Paper 795. California Institute of Technology (May).

Dubins, Lester E., and E. H. Spanier (1961). "How to Cut a Cake Fairly." *American Mathematical Monthly* 84, no. 5 (January): 1–17.

Dummett, Michael (1984). *Voting Procedures.* Oxford, UK: Oxford University Press.

Edelman, Paul H. (2006a). "Getting the Math Right: Why California Has Too Many Seats in the House of Representatives." *Vanderbilt Law Review* 59, no. 2 (March): 296–346.

Edelman, Paul H. (2006b). "Minimum Total Deviation Apportionments." In Bruno Simeone and Friedrich Pukelsheim (eds.), *Mathematics and Democracy: Recent Advances in Voting Systems and Social Choice,* pp. 55–64. Berlin, Germany: Springer.

Edelman, Paul H., and Peter C. Fishburn (2001). "Fair Division of Indivisible Items among People with Similar Preferences." *Mathematical Social Sciences* 41, no. 3 (May): 327–347.

Edelman, Paul H., and Suzanna Sherry (2000). "All or Nothing: Explaining the Size of Supreme Court Majorities." *North Carolina Law Review* 78, no. 5 (June): 1225–1252.

Ernst, Lawrence R. (1994). "Apportionment Methods of the House of Representatives and the Court Challenges." *Management Science* 40, no. 10: 1207–1227.

Fagan, J. T., B. V. Greenberg, and B. Hemmig (1988). "Controlled Rounding of Three-Dimensional Tables." Statistical Research Division Report Series, Census/SRD/RR-88/02, Bureau of Census, U.S. Department of Commerce.

Fahmy, Ismail (1983). *Negotiating for Peace in the Middle East.* Baltimore, MD: Johns Hopkins University Press.

Federal Election Commission (1989). "Report on the Visit by the Federal Election Commission to the Soviet Union, June 1989." Washington, DC: Federal Election Commission.

Felsenthal, Dan S. (1989). "On Combining Approval with Disapproval Voting." *Behavioral Science* 34 (1989): 53–60.

Felsenthal, Dan S., and Moshé Machover (1998). "Postulates and Paradoxes of Relative Voting Power—A Critical Re-Appraisal." *Theory and Decision* 38, no. 2: 195–229.

Fiorina, Morris P. (1992). *Divided Government.* New York: Macmillan.

Fishburn, Peter C. (1982). "Monotonicity Paradoxes in the Theory of Elections." *Discrete Applied Mathematics* 4, no. 2 (April): 119–134.

Fishburn, Peter C. (2004). Personal communication to Steven J. Brams (January 24).

Fishburn, Peter C., and Steven J. Brams (1983). "Paradoxes of Preferential Voting." *Mathematics Magazine* 56, no. 4 (September): 207–214.

Fishburn, Peter C., and Steven J. Brams (1993). "Yes-No Voting." *Social Choice and Welfare* 10: 35–50.

Fishburn, Peter C., and John D. C. Little (1988). "An Experiment in Approval Voting." *Management Science* 34, no. 5 (May): 555–568.

Fleischacker, Samuel (2004). *A Short History of Distributive Justice.* Cambridge, MA: Harvard University Press.

Fleurbaey, Marc (1992). "L'absence d'envie: un critère de justice?" Preprint, Départment Recherche, INSEE, Malakoff, France.

Forsythe, Robert, Joel L. Horowitz, N. E. Savin, and Marc Sefton (1994). "Fairness in Simple Bargaining Experiments." *Games and Economic Behavior* 6, no. 3 (May): 347–369.

Fossum, Robert M. (2002). Private communication (September 17).

Gaertner, Wulf (2006). *A Primer in Social Choice Theory.* New York: Oxford University Press.

Gale, David (1993). "Mathematical Entertainments." *Mathematical Intelligencer* 15, no. 1 (Winter): 48–52.

García-LaPresta, José, and Miguel Martínez-Panero (2002). "Borda Count versus Approval Voting: A Fuzzy Approach." *Public Choice* 112, nos. 1–2 (July): 167–184.

Gardner, Martin (1976). "Mathematical Games." *Scientific American* 234 (March): 119–124.

Gardner, Martin (written by Lynn Arthur Steen) (1980). "Mathematical Games (From Counting Votes to Making Votes Count: The Mathematics of Elections)." *Scientific American* 243, no. 4 (October): 16ff.

Gassner, Marjorie B. (1991). "Biproportional Delegations: A Solution for Two-Dimensional Proportional Representation." *Journal of Theoretical Politics* 3, no. 3 (July): 321–342.

Gersbach, Hans (2005). *Designing Democracy: Ideas for Better Rules*. Berlin, Germany: Springer-Verlag.

Gilbert, Margaret (1989). "Rationality and Salience." *Philosophical Studies* 57, no. 1 (September): 61–77.

Gilbert, Margaret (1990). "Rationality, Coordination, and Convention." *Synthese* 84, no. 1 (July): 1–21.

Gillman, Leonard (1987). "Approval Voting and the Coming MAA Elections." *Focus: The Newsletter of the Mathematical Association of America* 7, no. 2 (March–April): 2, 5.

Glazer, Jacob, and Ching-To Albert Ma (1989). "Efficient Allocation of a 'Prize'—King Solomon's Dilemma." *Games and Economic Behavior* 1, no. 3 (September): 222–233.

Gordon, Julie P. (1981). "Report of the Secretary." *Econometrica* 48, no. 1 (January): 229–233.

Gottron, Martha V. (ed.) (1983). *Congressional Districts in the 1980s*. Washington, DC: Congressional Quarterly.

Gould, F. J., G. D. Eppen, and C. P. Schmidt (1993). *Introductory Management Science*, 4th ed. Englewood Cliffs, NJ: Prentice Hall.

Greenberg, Joseph, and Shlomo Weber (1993). "Stable Coalition Structures with a Unidimensional Set of Alternatives." *Journal of Economic Theory* 60, no. 1: 62–82.

Grillidi Cortono, Pietro, Cecilia Manzi, Aline Pennisi, Federica Ricca, and Bruno Simeone (1999). *Evaluation and Optimization of Electoral Systems*. Philadelphia: Society for Industrial and Applied Mathematics (SIAM).

Grofman, Bernard (1982). "A Dynamic Model of Protocoalition Formation in Ideological N-Space." *Behavioral Science* 27: 77–99.

Grofman, Bernard, and Peter van Roozendal (1997). "Modeling Cabinet Durability/Cabinet Termination: A Synthetic Literature Review and Critique." *British Journal of Political Science* 27, no. 4 (October): 419–451.

Gusfield, Dan, and Robert W. Irving (1989). *The Stable Marriage Problem: Structure and Algorithms*. Cambridge, MA: MIT Press.

Guterman, Lila (2002). "When Votes Don't Add Up: Mathematical Theory Reveals Problems in Election Procedures." *Chronicle of Higher Education* (November 3): A18–A19.

Güth, Werner, and Reinhard Tietz (1990). "Ultimatum Bargaining Behavior: A Survey and Comparison of Experimental Results." *Journal of Economic Psychology* 11, no. 3 (September): 417–449.

Haake, Claus-Jochen, Matthias G. Raith, and Francis Edward Su (2002). "Bidding for Envy-Freeness: A Procedural Approach to n-Player Fair-Division Problems." *Social Choice and Welfare* 19, no. 4 (October): 723–749.

Herlihy, Patrick (1981). "Does Operations Research Have a Role in the Revision of Irish Parliamentary Constituencies?" In J. P. Brans (ed.), *Operational Research '81*, pp. 257–268. Amsterdam: North-Holland.

Herreiner, Dorothea, and Clemens Puppe (2002). "A Simple Procedure for Finding Equitable Allocations of Indivisible Goods." *Social Choice and Welfare* 19, no. 2 (February): 415–430.

Hillinger, Claude (2005). "The Case for Utilitarian Voting." *Homo Oeconomicus* 22, no. 3: 295–321.

Hoag, Clarence Gilbert, and George Hervey Hallet, Jr. (1926). *Proportional Representation*. New York: Macmillan.

Hodge, Jonathan K., and Richard E. Klima (2005). *The Mathematics of Voting and Elections*. Providence, RI: American Mathematical Society (AMS).

Hodge, Jonathan K., and Peter Schwaller (2006). "How Does Separability Affect the Desirability of Referendum Election Outcomes?" *Theory and Decision* 101, no. 3 (November): 205–249.

Hurwicz, Leonid, and Murat R. Sertel (1999). "Designing Mechanisms, in Particular for Electoral Systems: The Majoritarian Compromise." In Murat R. Sertel (ed.), *Economic Design and Behaviour*. London: Macmillan.

Hylland, Aanund, and Richard Zeckhauser (1979). "The Efficient Allocation of Individuals to Positions." *Journal of Political Economy* 87, no. 2 (April): 293–314.

Iizuka, Hiroyuki, Masahito Yamamoto, Keji Suzuki, and Azuma Ohuchi (2002). "Bottom-up Consensus Formation in Voting Games." *Nonlinear Dynamics, Psychology, and Life Sciences* 6, no. 2 (April): 185–195.

Jackson, Matthew O., Thomas R. Palfrey, and Sanjay Srivastava (1994). "Undominated Nash Implementation in Bounded Mechanisms." *Games and Economic Behavior* 6, no. 3 (May): 474–501.

Jacobson, Gary C. (1991). "The Persistence of Democratic House Majorities." In Gary W. Cox and Samuel Kernell (eds.), *The Politics of Divided Government*, pp. 57–84. Boulder, CO: Westview.

Jarvis, John J. (1984). "Council—Actions and Issues: Approval Voting." *OR/MS Today* 11, no. 4 (August): 16.

Jones, Michael A. (2002). "Equitable, Envy-Free, and Efficient Cake Cutting for Two People and Its Application to Divisible Goods." *Mathematics Magazine* 75, no. 4 (October): 275–283.

Kacowicz, Arie Marcelo (1994). *Peaceful Territorial Change*. Columbia, SC: University of South Carolina Press.

Keller, Bill (1987). "In Southern Russia, a Glimpse of Democracy." *New York Times* (June 20): 1, 4.

Keller, Bill (1988). "Moscow Says Changes in Voting Usher in Many New Local Leaders." *New York Times* (September 21): A1, A7.

Kelly, James P., Bruce L. Golden, and Arjang A. Assad (1989). "The Controlled Rounding Problem: A Review." Working Paper Series MS/S 89-016, College of Business and Management, University of Maryland (July).

Kelly, J. S. (1989). "The Ostrogorski Paradox." *Social Choice and Welfare* 6 no. 1 (January): 71–76.

Kiely, Tom (1991). "A Choice, Not an Echo?" *Technology Review* 94, no. 6 (August/September): 19–20.

Kiewiet, D. Roderick (1979). "Approval Voting: The Case of the 1968 Election." *Polity* 12, no. 1 (Fall): 528–537.

Kilgour, D. Marc, Steven J. Brams, and M. Remzi Sanver (2006). "How to Elect a Representative Committee Using Approval Balloting." In Bruno Simeone and Friedrich Pukelsheim (eds.), *Mathematics and Democracy: Recent Advances in Voting Systems and Collective Choice*, pp. 83–95. Berlin: Springer.

King, Gary (1996). Personal communication.

Klarreich, Erica (2002). "Election Selection: Are We Using the Worst Voting Procedure?" *Science News* 162, no. 18 (November 2): 280–282.

Klijn, Flip (2000). "An Algorithm for Envy-Free Allocations in an Economy with Indivisible Objects and Money." *Social Choice and Welfare* 17, no. 2 (February): 201–215.

Kohler, David A., and R. Chandrasekaran (1970). "A Class of Sequential Games." *Operations Research* 19, no. 2 (March–April): 270–277.

Kolm, Serge-Christophe (1996). *Modern Theories of Justice*. Cambridge, MA: MIT Press.

Lacy, Dean, and Emerson M. S. Niou (2000). "A Problem with Referendums." *Journal of Theoretical Politics* 12, no. 1: 5–31.

Laffond, G., and J. Laine (2006). "Singe-Switch Preferences and the Ostrogorski Paradox." *Mathematical Social Sciences* 52, no. 1 (July): 49–66.

Lagerspetz, Eerik (1996). "Paradoxes and Representation." *Electoral Studies* 15, no. 1 (February): 83–92.

Laslier, Jean-François (2003). "Analysing a Preference and Approval Profile." *Social Choice and Welfare* 20, no. 2 (April): 229–242.

Laslier, Jean-François (2006). "Spatial Approval Voting." *Political Analysis* 14, no. 2: 160–185.

Laslier, Jean-François, and Karine Vander Straeten (2003). "Approval Voting: An Experiment during the French 2002 Presidential Election." Laboratoire d'Econométrie, Ecole Polytechnique, Paris.

Laver, Michael, and W. Ben Hunt (1992). *Policy and Party Competition*. New York: Routledge.

Laver, Michael, and Norman Schofield (1998). *Multiparty Government: The Politics of Coalition in Europe* (reprint, with a new preface). Ann Arbor: University of Michigan Press.

Laver, Michael, and Kenneth A. Shepsle (eds.) (1994). *Cabinet Ministers and Parliamentary Government*. Cambridge, UK: Cambridge University Press.

Laver, Michael, and Kenneth A. Shepsle (1996). *Making and Breaking Governments: Cabinets and Legislatures in Parliamentary Democracies*. Cambridge, UK: Cambridge University Press.

Lawler, Eugene L. (1976). *Combinatorial Optimization: Networks and Matroids*. New York: Holt, Rinehart, and Winston.

LeGrand, Julian (1991). *Equity and Choice: An Essay in Economics and Applied Philosophy*. London: HarperCollins Academic.

LeGrand, Rob, Evangelos Markakis, and Aranyak Mehta (2006). "Approval Voting: Local Search Heuristics and Approximation Algorithms for the Minimax Solution." Preprint, Department of Computer Science, Washington University.

Leonard, Herman B. (1983). "Elicitation of Honest Preferences for the Assignment of Individuals to Positions." *Journal of Political Economy* 91, no. 3 (June): 461–479.

Lewis, David (1969). *Convention: A Philosophical Study*. Cambridge, MA: Harvard University Press.

Lijphart, Arend, and Bernard Grofman (eds.) (1984). *Choosing an Electoral System: Issues and Alternatives*. New York: Praeger.

Lines, Marjorie (1986). "Approval Voting and Strategy Analysis: A Venetian Example." *Theory and Decision* 20: 155–172.

Little, John, and Peter Fishburn (1986). "TIMS Tests Voting Method." *OR/MS Today* 13, no. 5 (October): 14–15.

MacKenzie, Dana (2000). "May the Best Man Lose." *Discover* 21, no. 11 (November): 84–91.

Mackie, Gerry (2003). *Democracy Defended*. Cambridge, UK: Cambridge University Press.

Maier, Sebastian (2006). "Algorithms for Biproportional Apportionment." In Bruno Simeone and Friedrich Pukelsheim (eds.), *Mathematics and Democracy: Recent Advances in Voting Systems and Collective Choice*, pp. 105–116. Berlin: Springer.

Maltzman, Forrest, James F. Spriggs II, and Paul J. Wahlbeck (2000). *Crafting Law on the Supreme Court: The Collegial Game*. New York: Cambridge University Press.

Maoz, Zeev (1990). *Paradoxes of War: On the Art of National Self-Entrapment*. Boston: Unwin Hyman.

Marshall, Albert W., Ingram Olkin, and Friedrich Pukelsheim (2002). "A Majorization Comparison of Apportionment Methods in Proportional Representation." *Social Choice and Welfare* 19, no. 4 (October): 885–900.

Massoud, Tansa George (2000). "Fair Division, Adjusted Winner Procedure (AW), and the Israeli-Palestinian Conflict." *Journal of Conflict Resolution* 44, no. 3 (June): 333–358.

Mayhew, David R. (1991). *Divided We Govern: Party Control, Lawmaking, and Investigations, 1946–1990*. New Haven, CT: Yale University Press.

McGarry, John, and Brendan O'Leary (2006a). "Consociational Theory, Northern Ireland's Conflict, and Its Agreement. Part 1: What Critics Can Learn from Northern Ireland." *Government and Opposition* 41, no. 1 (Winter): 43–63.

McGarry, John, and Brendan O'Leary (2006b). "Consociational Theory, Northern Ireland's Conflict, and Its Agreement. Part 2: What Consociationalists Can Learn from Northern Ireland." *Government and Opposition* 41, no. 2 (Spring): 249–277.

McKay, David (1994). "Review Article: Divided and Governed? Recent Research on Divided Government in the United States." *British Journal of Political Science* 24, Part 4 (October): 517–534.

McLean, Ian, and Arnold B. Urken (eds.) (1995). *Classics of Social Choice*. Ann Arbor: University of Michigan Press.

Meirowitz, Adam (2004). "Polling Games and Information Revelation in the Downsian Framework." *Games and Economic Behavior* 51 no. 2 (May): 464–489.

Merrill, Samuel, III (1988). *Making Multicandidate Elections More Democratic*. Princeton, NJ: Princeton University Press.

Merrill, Samuel, III, and Jack H. Nagel (1987). "The Effect of Approval Balloting on Strategic Voting under Alternative Decision Rules." *American Political Science Review* 81, no. 2 (June): 509–524.

Miller, Nicholas R. (1983). "Pluralism and Social Choice." *American Political Science Review* 77, no. 3 (September 1983): 734–747.

Monroe, Burt L. (1995). "Fully Proportional Representation," *American Political Science Review* 89, no. 4 (December): 925–940.

Moulin, Hervé (1979). "Dominance Solvable Voting Schemes." *Econometrica* 47, no. 6 (November): 1337–1351.

Moulin, Hervé (1994). "Serial Cost Sharing of Excludable Public Goods." *Review of Economic Studies* 61, no. 2 (April): 305–325.

Moulin, Hervé (1995). *Cooperative Microeconomics: A Game-Theoretic Introduction.* Princeton, NJ: Princeton University Press.

Moulin, Hervé (2003). *Fair Division and Collective Welfare.* Cambridge, MA: MIT Press.

Moulin, Hervé, and Scott Shenker (1992). "Serial Cost Sharing." *Econometrica* 60, no. 5 (September): 1009–1037.

Mueller, John E. (1969). "Voting on the Propositions: Ballot Pattern and Historical Trends in California." *American Political Science Review* 63, no. 4 (December): 1197–1212.

Müller, Wolfgang C., and Kaare Strøm (eds.) (2000). *Coalition Governments in Western Europe.* Oxford, UK: Oxford University Press.

Munger, Michael C. (1988). "Allocation of Desirable Committee Assignments: Extended Queues versus Committee Expansion." *American Journal of Political Science* 32 (May): 317–344.

Myerson, Roger (1991). *Game Theory: Analysis of Conflict.* Cambridge, MA: Harvard University Press.

Myerson, Roger B. (2002). "Comparison of Scoring Rules in Poisson Voting Games." *Journal of Economic Theory* 103, no. 1 (March): 219–251.

Nagel, Jack (1984). "A Debut for Approval Voting." *PS: Political Science & Politics* 17, no. 1 (Winter): 62–65.

Nagel, Jack (2006). "A Strategic Problem in Approval Voting." In Bruno Simeone and Friedrich Pukelsheim (eds.), *Mathematics and Democracy: Recent Advances in Voting Systems and Collective Choice*, pp. 133–150. Berlin: Springer.

Nagel, Jack H. (2007). "The Burr Dilemma in Approval Voting." *Journal of Politics* 69, no. 1 (February): 43–58.

National Academy of Sciences (1981). *Constitution and Bylaws* (April 28).

Nicholson, Stephen P. (2005). *Voting the Agenda: Candidates, Elections, and Ballot Propositions.* Princeton, NJ: Princeton University Press.

Niemi, Richard G. (1984). "The Problem of Strategic Voting under Approval Voting." *American Political Science Review* 78, no. 4 (December): 952–958.

Niemi, Richard G. (1985). "Reply to Brams and Fishburn." *American Political Science Review* 79, no. 3 (September): 818–819.

Niou, Emerson M. S., and Guofu Tan (1994). "An Analysis of Dr. Sun Yat-sen's Self-Assessment Scheme for Land Taxation." *Public Choice* 78, no. 1: 103–114.

Norris, Pippa (2004). *Electoral Engineering: Voting Rules and Political Behavior.* Cambridge, UK: Cambridge University Press.

Nurmi, Hannu (1987). *Comparing Voting Systems.* Dordrecht, Holland: D. Reidel.

Nurmi, Hannu (1999). *Voting Procedures and How to Deal with Them.* Berlin: Springer-Verlag.

Nurmi, Hannu (2002). *Voting Procedures under Uncertainty.* Berlin: Springer-Verlag.

Nurmi, Hannu (2006). *Models of Political Economy.* London: Routledge.

O'Leary, Brendan, Bernard Grofman, and Jørgen Elklit (2005). "Divisor Methods for Sequential Portfolio Allocation in Multi-Party Executive Bodies: Evidence from Northern Ireland and Denmark." *American Journal of Political Science* 49, no. 1 (January): 198–211.

Olszewski, Wojciech (2003). "A Simple and General Solution to King Solomon's Problem." *Games and Economic Behavior* 42, no. 2 (February): 315–318.

O'Neill, Barry (1982). "A Problem of Rights Arbitration from the Talmud." *Mathematical Social Sciences* 2, no. 4 (June): 345–371.

Ossipoff, Mike, and Warren D. Smith (2005). "Survey of Voting Methods That Avoid Favorite-Betrayal." Preprint available at http://rangevoting.org/FBCsurvey.html.

Özkal-Sanver, Ipek, and M. Remzi Sanver (2006). "Ensuring Pareto Optimality by Referendum Voting." *Social Choice and Welfare* 27, no. 1 (August): 211–219.

Pápai, Szilvia (2000). "Strategyproof Assignment by Hierarchical Exchange." *Econometrica* 68, no. 6 (November): 1403–1434.

Perry, Motty, and Philip J. Reny (1999). "A General Solution to King Solomon's Problem." *Games and Economic Behavior* 26, no. 2 (January): 279–285.

Peterson, Elisha, and Francis Edward Su (2002). "Four-Person Envy-Free Chore Division." *Mathematics Magazine* 75, no. 2 (April): 117–122.

Potthoff, Richard F. (1990). "Comment: Use of Integer Programming for Constrained Approval Voting." *Interfaces* 20, no. 5: 79–80.

Potthoff, Richard F. (2002). "Use of Linear Programming to Find an Envy-Free Solution Closest to the Brams-Kilgour Gap Solution for the Housemates Problem." *Group Decision and Negotiation* 11, no. 5 (September): 405–414.

Potthoff, Richard F., and Steven J. Brams (1998). "Proportional Representation: Broadening the Options." *Journal of Theoretical Politics* 10, no. 2 (April): 147–178.

Procaccia, Ariel D., Jeffrey S. Rosenschein, and Aviv Zohar (2007). "On the Complexity of Achieving Proportional Representation." *Social Choice and Welfare*, forthcoming.

Pukelsheim, Friedrich (2006). "Current Issues of Apportionment Methods." In Bruno Simeone and Friedrich Pukelsheim (eds.), *Mathematics and Democracy: Recent Advances in Voting Systems and Collective Choice*, pp. 167–176. Berlin: Springer.

Quandt, William B. (1986). *Camp David: Peacemaking and Politics*. Washington, DC: Brookings Institution.

Rae, Douglas W., and Hans Daudt (1976). "The Ostrogorski Paradox: A Peculiarity of Compound Majority Decision." *European Journal of Political Research* 4, no. 4 (December): 391–398.

Rae, Douglas W., and Michael Taylor (1970). *The Analysis of Political Cleavages*. New Haven, CT: Yale University Press.

Raiffa, Howard (1982). *The Art and Science of Negotiation*. Cambridge, MA: Harvard University Press.

Raith, Matthias G. (2000). "Fair-Negotiation Procedures." *Mathematical Social Sciences* 39, no. 3 (May): 303–322.

Rapoport, Amnon, Dan S. Felsenthal, and Zeev Maoz (1988a). "Microcosms and Macrocosms: Seat Allocations in Proportional Representation Systems." *Theory and Decision* 24, no. 1 (January): 11–33.

Rapoport, Amnon, Dan S. Felsenthal, and Zeev Maoz (1988b). "Proportional Representation of Single-Staged, Non-Ranked Voting Procedures." *Public Choice* 59, no. 2 (November): 151–165.

Rapoport, Amnon, and Ramzi Suleiman (1992). "Equilibrium Solutions for Resource Dilemmas." *Group Decision and Negotiation* 1, no. 3 (November): 269–274.

Ratliff, Thomas C. (2003). "Some Startling Inconsistencies When Electing Committees." *Social Choice and Welfare* 21: 433–454.

Ratliff, Thomas C. (2006). "Selecting Committees." *Public Choice* 126, nos. 3–4 (January): 343–355.

Rawls, John (1971). *A Theory of Justice*. Cambridge, MA: Harvard University Press.

Regenwetter, Michel, and Bernard Grofman (1998). "Approval Voting, Borda Winners, and Condorcet Winners: Evidence from Seven Elections." *Management Science* 44, no. 4 (April): 520–533.

Regenwetter, Michel, Bernard Grofman, A.A.J. Marley, and Ilia Tsetlin (2006). *Behavioral Social Choice: Probabilistic Models, Statistical Inference, and Applications*. Cambridge, UK: Cambridge University Press.

Reilly, Benjamin (2001). *Democracy in Divided Societies: Electoral Engineering for Conflict Management*. Cambridge, UK: Cambridge University Press.

Reilly, Benjamin (2002). "Social Choice in the South Seas: Electoral Innovation and the Borda Count in the Pacific Island Countries." *International Political Science Review* 23, no. 4 (October): 355–372.

Richie, Rob, Terrill Bouricius, and Philip Macklin (2001). "Candidate Number 1: Instant Runoff Voting." *Science* 294, no. 5541 (12 October): 303–304.

Riker, William H. (1962). *The Theory of Political Coalitions*. New Haven, CT: Yale University Press.

Riker, William H. (1982). *Liberalism against Populism: A Confrontation between the Theory of Democracy and the Theory of Social Choice*. New York: Freeman.

Riker, William H. (1986). *The Art of Political Manipulation*. New Haven, CT: Yale University Press.

Riker, William H. (1996). *The Strategy of Rhetoric: Campaigning for the American Constitution*. New Haven, CT: Yale University Press.

Riker, William H., and Lloyd S. Shapley (1968). "Weighted Voting: A Mathematical Analysis for Instrumental Judgments," pp. 199–216. In J. Roland Pennock and John W. Chapman (eds.), *Representation*. New York: Atherton.

Risse, Matthias (2005). "Why the Count de Borda Cannot Beat the Marquis de Condorcet." *Social Choice and Welfare* 25, no. 1 (September): 95–113.

Robertson, Jack, and William Webb (1998). *Cake-Cutting Algorithms: Be Fair If You Can*. Natick, MA: A K Peters.

Roemer, John E. (1996). *Theories of Distributive Justice*. Cambridge, MA: Harvard University Press.

Roemer, John E. (2000). *Equality of Opportunity*. Cambridge, MA: Harvard University Press.

Roth, Alvin, Jay B. Kadane, and Morris DeGroot (1977). "Optimal Peremptory Challenges in Trials by Juries: A Bilateral Sequential Process." *Operations Research* 25, no. 6 (November–December): 901–919.

Roth, Alvin E., and Marilda A. Oliveira Sotomayor (1990). *Two-Sided Matching: A Study in Game-Theoretic Modeling and Analysis.* Cambridge, UK: Cambridge University Press.

Saari, Donald G. (1994). *Geometry of Voting.* New York: Springer-Verlag.

Saari, Donald G. (1995). "A Chaotic Exploration of Aggregation Paradoxes." *SIAM Review* 37, no. 1 (March): 37–52.

Saari, Donald G. (2001a). "Analyzing a Nail-Biting Election." *Social Choice and Welfare* 18, no. 3: 415–430.

Saari, Donald G. (2001b). *Chaotic Elections! A Mathematician Looks at Voting.* Providence, RI: American Mathematical Society.

Saari, Donald G. (2001c). *Decisions and Elections: Explaining the Unexpected.* New York: Cambridge University Press.

Saari, Donald G. (2006). "Which Is Better: The Condorcet or Borda Winner?" *Social Choice and Welfare* 26, no. 1 (January): 107–129.

Saari, Donald G., and Jill Van Newenhizen (1988a). "The Problem of Indeterminacy in Approval, Multiple, and Truncated Voting Systems." *Public Choice* 59, no. 2 (November): 101–120.

Saari, Donald G., and Jill Van Newenhizen (1988b). "Is Approval Voting an 'Unmitigated Evil': A Response to Brams, Fishburn, and Merrill." *Public Choice* 59, no. 2 (November): 133–147.

Saaty, Thomas L. (1995). *The Fundamentals of Decision Making and Priority with the Analytic Hierarchy Process.* Pittsburgh: RWS Publications.

Scarsini, Marco (1998). "A Strong Paradox of Multiple Elections." *Social Choice and Welfare* 15, no. 2 (February): 237–238.

Schelling, Thomas C. (1960). *The Strategy of Conflict.* Cambridge, MA: Harvard University Press.

Schofield, Norman (2006). *Architects of Political Change: Constitutional Quandries and Social Choice Theory.* Cambridge, UK: Cambridge University Press.

Schofield, Norman, and Itai Sened (2006). *Multiparty Democracy: Elections and Legislative Politics.* New York: Cambridge University Press.

Schwartz, Bernard (1996). *Decision: How the Supreme Court Decides Cases.* New York: Oxford University Press.

Sertel, Murat R., and M. Remzi Sanver (1999). "Designing Public Choice Mechanisms," in Imed Limam (ed.), *Institutional Reform and Development in the MENA Region,* pp. 129–148. Cairo: Arab Planning Institute.

Sertel, Murat R., and M. Remzi Sanver (2004). "Strong Equilibrium Outcomes of Voting Games Are the Generalized Condorcet Winners." *Social Choice and Welfare,* 22, no. 2 (February): 331–347.

Sertel, Murat R., and Bilge Yilmaz (1999). "The Majoritarian Compromise Is Majoritarian-Optimal and Subgame-Perfect Implementable." *Social Choice and Welfare* 16, no. 4 (August): 615–627.

Shabad, Theodore (1987). "Soviets to Begin Multi-Candidate Election Experiment in June." *New York Times* (April 15): A6.

Shapley, L. S., and Martin Shubik (1954). "A Method for Evaluating the Distribution of Power in a Committee System." *American Political Science Review* 48, no. 3: 787–792.

Shepsle, Kenneth A. (1978). *The Giant Jigsaw Puzzle: Democratic Committee Assignments in the Modern House.* Chicago: University of Chicago Press.

Shishido, Harunori, and Dao-Zhi Zeng (1999). "Mark-Choose-Cut Algorithms for Fair and Strongly Fair Division." *Group Decision and Negotiation* 8, no. 2 (March 1999): 125–137.

Simeone, Bruno, and Friedrich Pukelsheim (eds.) (2006). *Mathematics and Democracy: Recent Advances in Voting Systems and Collective Choice.* Berlin: Springer.

Simpson, E. H. (1951). "The Interpretation of Interaction in Contingency Tables." *Journal of the Royal Statistical Society, Series B* 13: 238–241.

Sjöström, Tomas (1994). "Implementaton in Undominated Nash Equilibria without Integer Games." *Games and Economic Behavior* 6, no. 3 (May): 502–511.

Slinko, Arkadii (2002). "The Majoritarian Compromise in Large Societies." *Review of Economic Design* 7, no. 3 (November): 341–347.

Stearns, Maxwell L. (2000). *Constitutional Process: A Social Choice Analysis of the Supreme Court Decision Making.* Ann Arbor: University of Michigan Press.

Steen, Lynn Arthur (1985). Private communication (August).

Steen, Lynn Arthur (ed.) (2001). *Mathematics and Democracy: The Case for Quantitative Literacy.* Washington, DC: National Council on Education and the Disciplines.

Stein, Janice Gross (1993). "The Political Economy of Security Agreements: The Linked Costs of Failure at Camp David." In Peter B. Evans, Harold Jacobson, and Robert Putnam (eds.), *Double-Edged Diplomacy: International Bargaining and Domestic Politics.* Berkeley: University of California Press.

Steinhaus, Hugo (1948). "The Problem of Fair Division." *Econometrica* 16 (January): 101–104.

Straffin, Philip D., Jr., and Bernard Grofman (1984). "Parliamentary Coalitions: A Tour of Models." *Mathematics Magazine* 57, no. 5 (November): 259–274.

Straszak, Andrzej, Marek Libura, Jaroslaw Sikorski, and Dariusz Wagner (1993). "Computer Assisted Constrained Approval Voting." *Group Decision and Negotiation* 2, no. 4: 375–385.

Strøm, Kaare (1990). *Minority Government and Majority Rule.* Cambridge, UK: Cambridge University Press.

Stromquist, Walter (1980). "How to Cut a Cake Fairly." *American Mathematical Monthly* 87, no. 8 (October): 640–644.

Su, Francis Edward (1999). "Rental Harmony: Sperner's Lemma in Fair Division." *American Mathematical Monthly* 106, no. 2 (December): 922–934.

Sung, Shao Chin, and Milan Vlach (2004). "Competitive Envy-Free Fair Division." *Social Choice and Welfare* 23, no. 1 (August): 103–111.

Svensson, Lars-Gunnar (1999). "Strategy-Proof Allocation of Indivisible Goods." *Social Choice and Welfare* 16, no. 1 (August): 557–567.

"Symposium: Divided Government and the Politics of Constitutional Reform" (1991). *PS: Political Science and Politics* 24, no. 4 (December): 634–657.

Taagepera, Rein, and Matthew Soberg Shugart (1989). *Seats and Votes: The Effects and Determinants of Electoral Systems.* New Haven, CT: Yale University Press.

Taylor, Alan D. (2005). *Social Choice and the Mathematics of Manipulation.* New York: Cambridge University Press.

Telhami, Shibley (1990). *Power and Leadership in International Bargaining: The Path to the Camp David Accords.* New York: Columbia University Press.

Tsebelis, George (2002). *Veto Players: How Political Institutions Work.* Princeton, NJ: Princeton University Press.

Tullock, Gordon (1967). *Toward a Mathematics of Politics.* Ann Arbor: University of Michigan Press.

Tullock, Gordon (1979). "Comment on Brams and Fishburn and Balinski and Young." *American Political Science Review* 73, no. 2 (June): 552–553.

Ullman-Margalit, Edna (1977). *The Emergence of Norms.* Oxford, UK: Oxford University Press.

Van Damme, Eric (1991). *Stability and Perfection of Nash Equilibria,* 2nd ed. Heidelberg: Springer-Verlag.

Van Deemen, Adrian (1993). "Paradoxes of Voting in List Systems of Proportional Representation." *Electoral Studies* 12, no. 3 (September): 234–241.

Van Deemen, Ad. M. A. (1997). *Coalition Formation and Social Choice.* Boston: Kluwer Academic Publishers.

Vickrey, William (1961). "Counterspeculation, Auctions, and Competitive Sealed Tenders." *Journal of Finance* 26 (March): 8–37.

Wagner, Carl (1983). Anscombe's Paradox and the Rule of Three-Fourths." *Theory and Decision* 15, no. 3 (September): 303–308.

Wagner, Carl. (1984). "Avoiding Anscombe's Paradox." *Theory and Decision* 16, no. 3 (May): 223–235.

Wagner, Clifford H. (1982). "Simpson's Paradox in Real Life." *American Statistician* 36, no. 1 (February): 46–48.

Warwick, Paul V. (1994). *Government Survival in Parliament Democracies.* Cambridge, UK: Cambridge University Press.

Warwick, Paul V. (2001). "Coalition Policy in Parliamentary Democracies: Who Gets How Much and Why." *Comparative Political Studies* 34, no. 10 (December): 1212–1236.

Warwick, Paul V., and James N. Druckman (2001). "Portfolio Salience and the Proportionality of Payoffs in Coalition Governments." *British Journal of Political Science* 31, no. 4 (October): 627–649.

Weber, Robert J. (1995). "Approval Voting." *Journal of Economic Perspectives* 9, no. 1 (Winter): 39–49.

White, Stephen (1989). "Reforming the Electoral System [USSR]." *Journal of Communist Studies* 4 (December): 1–17.

Woodall, D. R. (1986). "How Proportional Is Proportional Representation?" *Mathematical Intelligencer* 8, no. 4: 36–46.

Woodward, Bob, and Scott Armstrong (1979). *The Brethren: Inside the Supreme Court.* New York: Simon and Schuster.

Wright, Jeff (1990). "School Funding Reform Options Unpopular." *Register-Guard* (Eugene, OR) (May 16): 1C, 3C.

Yilmaz, Mustafa R. (1999). "Can We Improve upon Approval Voting?" *European Journal of Political Economy* 15, no. 1 (March): 89–100.

Young, H. Peyton (1987). "On Dividing an Amount According to Individual Claims or Liabilities." *Mathematics of Operations Research* 12: 398–414.

Young, H. Peyton (1991). *Negotiation Analysis.* Ann Arbor: University of Michigan Press.

Young, H. Peyton (1994). *Equity in Theory and Practice.* Princeton, NJ: Princeton University Press.

Yunfeng, Luo, Yue Chaoyuan, and Chen Ting (1996). "Strategy Stability and Sincerity in Approval Voting." *Social Choice and Welfare* 13, no. 1: 17–23.

Zuppan, Mark A. (1991). "An Economic Explanation for the Existence and Nature of Political Ticket Splitting." *Journal of Law and Economics* 34, pt. 1 (October): 343–369.

Zwicker, William S. (1991). "The Voters' Paradox, Spin, and the Borda Count." *Mathematical Social Sciences* 22, no. 3: 187–227.

Index

Milton Keynes UK
Ingram Content Group UK Ltd.
UKHW011219150624
444218UK00003B/123